THE ECONOMETRICS OF ENERGY DEMAND

THE ECONOMETRICS OF ENERGY DEMAND

A Survey of Applications

William A. Donnelly

PRAEGER

New York
Westport, Connecticut
London

Library of Congress Cataloging-in-Publication Data

Donnelly, William A.
The econometrics of energy demand.

Bibliography: p.
1. Energy consumption—Econometric models.
I. Title.
HD9502.A2D67 1987 333.79'12 86-25220
ISBN 0-275-92610-9 (alk. paper)

Library of Congress Catalog Card Number: 86-25220
ISBN: 0-275-92610-9

First published in 1987

Praeger Publishers, 521 Fifth Avenue, New York, NY 10175
A division of Greenwood Press, Inc.

Printed in the United States of America

The paper used in this book complies with the Permanent Paper Standard issued by the National Information Standards Organization (Z39.48-1984).

10 9 8 7 6 5 4 3 2 1

Contents

List of Figures and Tables

FIGURES

TABLES

List of Equations

Acknowledgments

The development of this book came about, to a great extent, through my collaboration with several researchers, most of whom are co-authors and whose works are listed in the bibliography as well as in the text. But much credit is owed to others as well, others too numerous to list, who have offered extensive suggestions and corrections and other substantive assistance. Among these latter I thank especially P. G. Chandler, V. B. Hall, C. C. Harris, S. Harris, and D. James. I am grateful to M. Fisher, Economics Editor for Praeger. Thanks are due to Croom Helm, Publishers, for their permission to incorporate portions of my research published previously. The National Energy Research Development and Demonstration Council of the Commonwealth of Australia and the Centre for Resource and Environmental Studies of the Australian National University provided primary funding and support for this research. The Department of Economics at the University of Wisconsin–La Crosse funded a partial summer research fellowship that facilitated completion of this manuscript. Naturally, all errors of omission and commission are my sole responsibility.

THE ECONOMETRICS OF ENERGY DEMAND

One

Policy Modeling

OVERVIEW

Modeling is as much an art as it is a science. The models under discussion[1] are used to assist in developing policy alternatives, and thus they incorporate variables over which the decision maker may exercise some sort of control. The process of model development includes (1) problem identification, (2) model selection, (3) database construction, and (4) implementation and documentation. The "energy crisis" is the background for the discussion, with some general insights on modeling offered to prospective analysts and decision makers.

INTRODUCTION

The objectives, strategies, and modes of construction and evaluation of policy models that are to be used in energy analyses are the bases for the discussion. Questions relating to energy supply and demand became topical during 1973 because of the war in the Middle East that threatened the supplies of oil to Europe, Japan, and the United States [Energy Crisis 1 (EC1)]. In the United States the uncertainty of supply led to the declaration by President Richard Nixon of Project Independence, a program intended to provide the policies necessary to achieve energy self-sufficiency by eliminating the country's reliance upon imported oil. In pursuit of the objective of energy independence, the U.S. Federal Energy Office was established, then reconstituted as the U.S. Federal Energy Administration (FEA), finally becoming a permanent fixture, the U.S. Department of Energy (DOE).[2]

The basic objective of promoting energy independence was clearly defined, and it was incumbent upon DOE to formulate policy alternatives for consideration and for potential submission to Congress. (In order to accomplish this task, the energy problem and its ramifications required explication.) It was concluded that ad hoc approaches would not suffice, and the recognition that this was the case brought about the establishment of the Project Independence Evaluation System, or PIES project, and the development of an overall modeling strategy. The acronym PIES refers to a set of hybrid econometric/linear programming (L-P) models constructed and used by DOE.

The DOE strategy of employing quantitative models for policy analyses entailed the construction of a comprehensive database from which information concerning the interrelationships among the sectors of the economy, the substitution potential, and the supply alternatives could be derived, extracted, and used to model energy scenarios. Because prior to 1973 relative energy prices exhibited a continuing decline, and because the share of energy in the cost of production was small, limited data were then available on energy flows and the prices of alternatives. This paucity of information required new primary data collection through the use of periodic surveys and censuses; and whereas modelers in other areas of interest are often able to utilize secondary data sources, this was not then and generally continues now not to be the case in regard to energy analysis.

Whatever energy statistics become available are used in determining the parameters of the specific models. The parameters may be developed from knowledge of engineering relationships, from application of standard mathematical forms, from comprehensive statistical analyses, or from some combination of these. In fact, while PIES relies upon all of these approaches, most energy models utilize only a single approach; and while the formal techniques for constructing models are widely discussed, the contribution of a modeler's intuition and experience is infrequently recognized or admitted. Quite often model results are modified after the fact by analysts because the raw output appears to be wanting. Such adjustments might involve the intercept term of a linear model, to account for changes in levels, or proportions or elasticities that may have changed. Thus, intuition is involved in modeling, and its application is an art that makes model evaluation difficult. Formal models are verified and exercised to provide insights into their behavior, and should be fully documented so that interested persons can understand their assumptions, strengths, and weaknesses; but this process is the least glamorous aspect of modeling, and therefore it is the most neglected. Some of these facets of modeling are considered in this chapter, with the implications for the appropriate directions for future research outlined. The components of model construction (problem identification, selection of modeling strategy, and model implementation) are discussed, and general observations on the modeling process are made.

POLICY PROBLEMS

The use of modeling to address policy issues dates from antiquity when astrologers were queried about propitious times for conducting religious, social, and political ceremonies. Modeling, by definition, involves constructing a simplified representation of the real world from a set of factors believed to be the most essential elements influencing a particular set of events. Parsimony is a general objective. This is implicit from the basic rationale for the use of models, namely, the desire to identify the determinants of the system and their influences. This tenet was summarized by Goldberger (1968): "Other things being equal a simple functional form is to be preferred to a complex one. This dictum is in accord with a general principle of philosophy and scientific method known as *Occam's razor*: Do not compound hypotheses unnecessarily" (p. 133).

Thus, the simpler and more accurate the model, the more elegant: for example, the Copernican versus the Ptolemaic theory of astronomy, as Adam Smith (1982) observes: "Copernicus (1473–1543) began to mediate a new system, which should correct together with celestial appearances, in a more simple as well as a more accurate manner, than that of Ptolemy [circa 140, A.D.]" (p. 71). The Copernican theory provided better predictions and is distinctly less convoluted than its predecessor. But models go through cycles, coming in and going out of fashion. Although Aristarchus of Samos, "the mathematician" (circa 310–230 B.C.), is credited as the originator of the helicocentric hypothesis, (Heath 1981, p. 301), there occurred a two-millennium interregnum between the abandonment of the Ptolemaic geocentric model of cosmology in favor of the former view. This rediscovery of the heliocentric model may be attributed to the diminishing predictive capacity of the geocentric model as more astronomical observations became available and the requirements for more accurate predictions increased. Increasing needs combined with the improvement of data may facilitate consideration of alternative modeling strategies that can provide new insights into the behavior of the system under study, and this is equally true for today's policy models.

In economics, modeling has a relatively shorter history, dating from Sir William Petty's seventeenth-century population estimates for the city of London, which were derived by extrapolating from the number of chimney stacks (Hull 1964, pp. 535–36). Late in the next century, Thomas Malthus postulated the inevitable conflict between an arithmetic growth rate of food production and a geometric growth rate in population, leading to the widespread characterization of economics as the "dismal science" (Malthus 1966). The policy implications of such conflicts are obvious, but contemporary (as I am sure do every generation's) modelers like to believe that their art has progressed significantly since those earlier times and appreciate more fully that many factors influence population growth. However, these simplistic, purely deterministic approaches reappeared in the 1970s in the guise of "systems dynamics models." Fortunately, such forms of judgmental modeling represent a declining minority of extant policy models.

The resurgence in economic modeling in the 1940s coincided and represented a symbiotic relationship with the development of reliable national (and regional) economic data series. One example of this is Wassily Leontief's seminal work of that period (1951), the development of the first input-output (I-O) model for the United States. The I-O model, whose antecedents may be traced to François Quesnay's *Tableau Economique* of 1758 (Kuczynski and Meek 1972), explicitly recognizes sectoral interrelationships, and an outgrowth of the I-O technique comprises the mathematical programming models that began finding acceptance during the Second World War in addressing questions of military logistics. Ever since, government analysts have employed formal quantitative models to assist them in considering various policy questions.

Frequently these models are highly aggregate attempts at providing insights into the functioning of the overall economy, attempting to model domestic production, consumption, government revenues and expenditures, and exports and imports. Numerous research groups have expanded the scope of economic modeling. These include federal government agencies such as the Federal Reserve Board of Governors, the Treasury, the Bureau of Economic Analysis, and DOE; as well as commercial organizations such as Data Resources Inc. (DRI), Chase Econometrics, and Evans Econometrics; and academic researchers such as those affiliated with the National Bureau of Economic Research and Wharton Econometrics. In addition to the many aspects of macroeconomics, these groups are addressing issues in the field of microeconomics; and, in general, evaluating these latter questions involves the adoption of new modeling strategies. The work of Hudson and Jorgenson (1974) at DRI under the auspices of the Ford Foundation and the work of DOE (FEA 1976) are examples of the innovative hybrid modeling approaches recently developed to address questions relating to energy demand and supply.

The first "energy crisis" of the 1970s gave rise to these models designed to consider questions concerning the allocation of energy resources among the various sectors of the economy and among primary sources of energy. Questions relating to industry's ability to substitute among fuels as relative prices change, the accommodation of shortages of specific fuels, and the usage of domestic and transport fuels were not immediately amenable to existing data or models. Research activity progressed on several fronts simultaneously by way of considering both the aggregate and the sectoral implications of changing economic relationships on energy consumption patterns.

Decision makers want to know the potential impacts of alternative policies that might be adopted; and while such information is required to address energy shortages that occur in the short run, insight is needed also into those policies that will promote long-run adjustments with a minimal amount of disruption. The government policy analyst therefore seeks information on individuals' responses to variables such as prices that may be affected by government action and on the technical and managerial substitution possibilities existing within industry. To accomplish these objectives, models that provide demand elasticities and elasticities of substitution are often posited. Particular questions necessitate different modeling approaches; a general strategy applicable to all issues is not possible. The next section provides an overview of the modeling strategies that have been applied in the analysis of energy policy.

MODELING STRATEGIES

Introduction

The development of effective policy, whether in terms of the general economy or specific energy aspects of it, requires knowledge about the influential factors and the behavioral response to those control variables providing for that policy's implementation. Decision makers need an understanding of the interaction of variables in the system with factors that can be controlled through government action. Models can be categorized in various ways: ad hoc or formal; deterministic or stochastic; static or dynamic; single equation or system of equations; partial or general equilibrium. The policy models to be discussed here fall within several of these categories; thus, none suffices to define the strategies adequately. The selection of the appropriate modeling approach depends upon the objectives of the analysis and upon the resources available in terms of time, money, and data. As modeling analysts understand: "There is no completely flexible 'universal' model that can answer all policy questions on energy, economics and the environment. Models should be constructed that are specially suited to particular kinds of policy analysis" (James 1983, p. viii). Different questions may be addressed by different modeling strategies, as in the analysis of electricity demand. Questions relating to the distinction between the stock of electricity—or, as the electrical engineer would say, "power" (the instantaneous requirement for kilowatts)—and the flow of "energy" (the cumulative usage of kilowatts over some period of time, i.e., kilowatt-hours) require that alternative modeling strategies be adopted. Questions relating to the load characteristics are amenable to time-series analysis, while others concerning the demand for energy may be modeled by econometric methods.

Forecasts of energy consumption to address policy issues that do arise under differing circumstances are handled using the "scenario" approach. This approach allows analysis of the effects of differing assumptions about alternative future conditions. Projections based upon scenarios are interpreted as conditional forecasts and not as predictions, allowing sensitivity analyses to be performed that may identify important factors warranting further attention. Green (1985, p. 172) refers to these "hot spots" so identified when verifying and exercising models. Whenever forecasts are insensitive to a wide range of values for a variable, then it may be assumed that either that aspect of the observations is of little importance in the system or that the model has been misspecified. Theoretical considerations may suggest the latter and provide insights into possible reformulation. Thus, the use of scenarios and sensitivity analysis may help in ameliorating some of the uncertainties that inevitably occur in the process of developing forecasts.

The various modeling strategies depend upon the specific assumptions required of each methodology, and these necessarily affect the resultant forecasts. A more complete understanding of these assumptions is developed in this chapter and the next. Five general modeling strategies, briefly identified here and further discussed in Chapter 2, are

Judgmental forecasting
Mathematical programming
Time-series analysis
Econometric methods
Hybrid models

Judgmental Forecasting

This first category contains several techniques, none of which may be made explicit in the presentation of an analysis. One example is trend extrapolation, which utilizes the simplest assumptions in forecasting: namely, that the latest response in the system, such as a growth rate, will continue. Historically electric utilities rely upon this approach using exponential growth rates to forecast demand. These extrapolations are used to establish capital requirements and to develop construction schedules, but such a forecasting technique ignores the learning curve growth pattern in the development of a market for any product: Over time, the demand for a good changes.

For example, overseas in Australia the average annual growth rate in electricity consumption during the 1950s was 9.5 percent. This declined to 8.6 percent in the next decade, to 5.8 percent in the 1970s, and to 4.4 percent for the four-year period in the 1980s from 1979/80 to 1983/84. Information on the growth rates in electricity production is presented in Table 1.1.

Trend extrapolation is a deterministic forecasting technique, and its results will depend only upon the initial conditions of the system and the assumed rate of growth; adaptation and adjustment to changing conditions are ignored in forecasts. Shocks occurring to the system cannot be adequately handled by trend extrapolation, and this includes the changes in oil prices after both the 1973 embargo (EC1), and the 1979 supply disruptions caused by the 1978 Iranian revolution [Energy Crisis 2 (EC2)], where substitution among the inputs to the production process occurred. In addition, deterministic models ignore random aberrations that arise in the realization of a process.

Because of their ad hoc nature, judgmental forecasts do not emphasize the assumptions that are inherent but generally not clearly stated. The individual analyst is often unaware of some of the assumptions being used; in addition, the causal factors and the variable interrelationships are never defined. Forecasts developed in

Table 1.1. Selected Growth Rates in the Production of Electricity (annual percentages)

	1938-50	1950-60	1960-70	1970-80	1980-83
Australia	2.0	7.1	7.9	5.8	3.3
Canada	5.7	7.2	6.2	6.2	2.6
England	1.5	7.1	6.2	1.4	-1.0
France	1.8	5.5	6.7	5.6	4.7
New Zealand	2.2	5.6	6.7	4.7	5.0
Soviet Union	3.3	7.8	9.3	5.6	2.8
U.S.A.	8.5	7.7	5.9	4.3	0.2

Source: Compound growth rates were calculated from data provided in the following: United Nations, 1985, *1983 Energy Statistics Yearbook* (New York: UN), Table 34; W. L. Liscom, 1982, *The Energy Decade 1970–1980: A Statistical and Graphic Chronicle* (Cambridge, MA: Ballinger); and J. Darmstadter, with P. D. Teitelbaum and J. G. Polach, 1971, *Energy in the World Economy: A Statistical Review of Trends in Output, Trade, and Consumption Since 1925* (Baltimore: Johns Hopkins University Press), Table XI.

this way are subjective, and their quality reflects the insights of the particular forecaster rather than the soundness of the procedure. Thus, statistical testing cannot be employed. The fundamental assumption of extrapolation, that all relevant interrelationships remain invariant over the forecast period, is seldom made explicit. In a critique of the systems dynamics methodology as espoused by Forrester (1971) and Meadows et al. (1972), Berlinski (1976) states that "these are volumes of studied defects . . . valuable as an example of the unencumbered differential [equation] method" and argues that "both Forrester and Meadows are innocents of rigor (*naifs statistiques*): missing from their work is some sense that a substantial body of statistical technique must mediate between an original theory and its applications" (pp. 52, 71).

In a somewhat more constructive vein, Nordhaus (1973) offers alternative modeling strategies and reaches the conclusion that "Forrester is apparently content with the subjective plausibility . . . [of his own] intuitively plausible [relationships] . . . whereas most scientists would require empirical validation of either the assumptions or the predictions of the model before declaring its truth content" (p. 1183). Similarly, Simon (1981) finds the U.S. Government *Global 2000 Report*, which is based upon the systems dynamics methodology, "shoddy," and he adds to the "First Great Law of Models (GIGO, garbage in, garbage out) . . . another law: PIPO, prejudice in, prejudice out" (p. 6). He points out that the "thoroughly discredited" *Limits to Growth* document has now even been repudiated "by the sponsoring Club of Rome" (p. 6).

The multitudinous factors affecting the growth in energy consumption are never directly considered in extrapolative forecasting procedures; for example, growth is treated as an autonomous phenomenon. The demand for energy as a derived demand depends upon the social, economic, and physical conditions and the pattern of use of the energy-consuming machines; therefore, these factors must be taken into consideration. Yoke (1984) provides a "parable" of an energy planner who uses extrapolation, misses the potential for technical change and substitution, and makes some naive recommendations based upon these omissions. His critique of the 1983 International Energy Workshop held by the International Institute for Applied Systems Analysis illustrates the continuing reliance upon extrapolative techniques by many energy analysts. Judgmental forecasts are simple to prepare with only a minimal amount of investment required, but, the quality of such "educated guesses" being difficult to assess, the simplicity of the approach represents a well-honed dual-edged sword all the same.

Mathematical Programming

Mathematical programming in policy analysis is also deterministic in nature, and while there are stochastic forms of programming models, these are not currently in widespread use. Operations research, the applied field of mathematical programming, is an outgrowth of military planning. The I-O model of Leontief may be considered as well a form of mathematical programming, interpreted as the solution to a particular optimization problem. In I-O analysis the economy is divided into individual sectors. The interrelationships among the sectors are defined by the flow of goods among them measured in monetary units. The relationships defined by the I-O matrix then represent those processes in the programming model that have been selected, or entered, in the optimization, and these may be assumed to pertain over some range of values. Outside this range a new solution to the programming model would provide another I-O table. Thus, the I-O technique explicitly recognizes that economic activity involves sectoral interrelationships.

Hannon (1973) adapts the technique to construct an energy I-O table for the United States, the elements of which are defined in terms of energy units. Energy as the "numeraire" for the I-O table implies an "energy theory of value" and has the same shortcomings as does the Marxist assumption of a "labor theory of value" or indeed the assumption that any other factor input is the only scarce resource; but since scarcity is a relative concept, then the most appropriate numeraire should be the universally convertible commodity, money (see Georgescu-Rogen 1979). On the other hand, Marxists sometimes argue that money is indeed *too* convertible; that is, that its value is manipulated by agencies and means that do not result from market forces. I-O modeling has been used extensively for economic forecasting, but the approach is suitable for short- to medium-range forecasting where technologies are well understood and relatively constant. Conversely, mathematical programming models allow the economic evaluation of new technologies, and therefore these are better suited for medium- to long-range forecasting.

The mathematical programming matrix has additional rows over and above those included in the I-O model, an objective function, and specified constraints. The selection among the available production processes is based upon the optimization of an objective function subject to the system constraints. The reasonableness of any programming solution, or I-O model for that matter, depends upon how well technologies are depicted and on how flexible in actuality industries are, given existing institutional constraints. When utilized in a forecasting context, these models require information concerning future technologies. Such information is often difficult to obtain. Until the construction of the supply side of the PIES model at DOE, the L-P framework had not been used to model the entire energy supply system of a country (see FEA 1976).

Implementing a model requires adequate information. I-O analysis suffers particularly here, and this is a major problem in energy analysis when relationships are expected to change. Various indirect estimation procedures have been developed to update I-O matrices, and these procedures often employ an **rAs** iterative matrix-balancing routine, where *r* and *s* are row and column scalars used to adjust porportionately the original table to new row and column totals (see Bacharach 1970, chap. 4). There is little reason to believe that such proportional adjustments accurately reflect the technological changes and improved business practices introduced by industry in response to system shocks. Therefore, the I-O technique is unsuitable for the evaluation of major shocks to the system such as resulted from EC1 and EC2. Such shocks induce industries to adopt alternative production technologies and to improve energy management techniques, thereby changing the underlying interindustry relationships. As Bullard and Herendeen (1977) recognize:

> Input-output coefficients change with time, yet we hope to use the results to predict the consequences of hypothetical future consumption patterns. Can one quantify their loss of reliability with time? This is a major point, for which much work is needed. Our feeling is that our results are most sensitive to changes in direct energy use coefficients, which may change faster than others due to fuel substitutability and the potential for energy conservation. [The] use of price indices and an overall energy/GNP factor was a very poor updating method . . . very little better than doing nothing at all . . . [; however] the use of actual 1967 energy-use technology as well as price indices, was *quite successful* [emphasis added]. (pp. 76, 80).

Those authors never define what is meant by "quite successful"; unfortunately, even if such an adjustment procedure is successful, it cannot be used in a forecasting context.

While mathematical programming models can handle system shocks, these models may tend to overestimate the adjustment potential of the system, since although a strategy may be optimal in the modeling sense, constraints in the real world may preclude the adoption of the new technologies suggested. In addition, much of the information needed for long-range forecasting about new, untested technology may be quite hypothetical and therefore have a large variance. Still, the programming approach can serve to identify areas of potential policy concern that merit further attention.

Time-Series Analysis

The data used in time-series analysis are ordered by a time parameter with the concepts of theoretical process and realization introduced. Thus, classical statistical procedures that assume order to be unimportant are no longer strictly relevant. In its most rudimentary univariate form, the theoretical process, which includes a stochastic component, is defined by the data series, as Nelson (1973) explains: "The past history of the time series is called upon to do double duty: First, it must inform us about the particular mechanism which describes its evolution through time, and, second, it allows us to put that mechanism to use in forecasting the future" (p. 19).

The assumptions of stationarity and ergodicity are required for modeling (these concepts and assumptions will be explained more fully in Chapter 2). Alternative specifications include autoregressive (AR), moving-average (MA), and combined autoregressive moving-average (ARMA) processes. The standard process is univariate in that the current value of the dependent variable is defined on its historical values, and Goldberger's dictum of parsimony (1968) is important in selecting the most appropriate form. A more comprehensive multivariate time-series approach allows for the inclusion of behavioral relationships in the process being modeled. The variables used in the modeling may be separated into exogenous variables, those determined by factors outside of the model, and endogenous variables, those determined by the model structure. Whereas substantial work has been directed toward univariate time-series analysis, much less effort has been expended on the multivariate approach. This is because the multivariate generalization introduces many analytical and computational problems. Johnston (1984) states that the "transfer function approach [to achieving stationarity] has been extensively developed for the *single-input* case (that is, one explanatory variable with various lagged values), and there is no firm agreement yet on the appropriate extension to cope with two or more inputs, each with a set of lags" (p. 372).

Economic theory can provide the requisite model structure; however, identifying the most appropriate order of the process and the model structure requires much insight and experience. The multivariate process can include several orders of differencing—one for each variable and several orders of AR processes. An advantage of the multivariate over the univariate approach, as Granger and Newbold (1977, p. 218) point out, is that the former, when combined with the structure imposed by economics, allows the model to be decomposed into blocks for the behavioral relationships in the system and for the exogenous variables. When applied to well-defined stable systems, univariate time-series analysis has proved a successful tool for forecasting. As long as the assumption that relationships remain unchanged can be reasonably maintained, the univariate represents a powerful tool. However, many economic phenomena generally do not justify the univariate approach, since these procedures fail to cope with the effects of exogenous factors.

The questions of the availability of data and the amount of modeling effort required for successful analysis are also issues, and these problems tend to increase geometrically with the dimensions of the model. This may be seen by comparing the number of unknown parameters in the AR, MA, ARMA, and integrated ARMA (ARIMA) models as defined in Equations 2.14, 2.16, 2.18, and 2.19, respectively. Another problem is the selection of the appropriate model from the many possible alternative formulations. This involves determining the order of the model parameters, the causal variables, and the appropriate model structure. The number of combinations of differencing and lag structure imposes severe computational burdens, and whereas parsimony in model parameters is a desirable objective, alternative model structures will often provide similar results in terms of forecast values and accuracy.

Econometric Methods

Econometric techniques rely upon hypothesis testing of behavioral relationships as posited by economic theory using statistical methods. Berlinski (1976) refers to the "metrical sciences . . . econometrics for economics; polimetrics for political science; biometrics for biology; even cliometrics for history–whose task is to interpose themselves between the broad-ranging and frequently untestable assumptions of a given theory and the mass of data the theory is meant to confront" (p. 71).

The econometric model contains an endogenous variable and a set of exogenous explanatory variables. In specifying a model of demand, important factors are selected based upon utility theory in the case of individuals (or production theory in the case of firms), which underlies economics, and the results are evaluated statistically. Information on the dependent variable and the explanatory variables represents the observation set, and the parameters of the relationship are the unknowns. The theoretical basis of the model and the inclusion of a stochastic error term differentiate this approach from the deterministic techniques discussed earlier. Ordinary least squares (OLS) provides estimates of the parameters by minimizing the sum of squares of the difference between the observed value of the endogenous variable and the linear relationship specified on the exogenous variables. In the two-dimensional case, the procedure is to minimize the vertical distance between the observed values of the dependent variable and the regression line. Alternative objective functions might be specified, but such approaches may lack the theoretical foundations attributed to this application of OLS by economists. The exact mathematical form of the relationship is usually not defined by economic theory; however, assumptions of linearity are the norm. It is asserted that linearity holds within the relevant range of the data, but additional difficulties arise in specifying the most appropriate functional form for the relationship. Box and Cox (1964) recommend a transformation that can be used to test for functional form, as will be explicated in Chapter 2.

In theory, hypothesis testing is straightforward; however, in practice, "data mining" introduces statistical problems that should not be ignored (see Lovell 1983). Data mining involves the estimation of several alternative relationships and then the reporting of the single equation that the analyst considers to be most appropriate.

Such Bayesian estimates depend upon all the models estimated and rejected, and therefore the statistical tests must be adjusted to reflect those attempts (see Leamer 1978). Few analysts actually follow such practice but rather present just the results of the preferred relationship. The pitfalls of econometric methods are numerous to the unwary, which is not to say that these are absent from the other modeling strategies already discussed.

Hybrid Models

The hybrid modeling strategy involves applying various combinations of techniques, and it should be emphasized that these are simple combinations of techniques. An example is the dynamic I-O Interindustry Forecasting Model of the University of Maryland (INFORUM), developed by Almon et al. (1974), which relies upon econometric methods to derive the I-O coefficients. Hudson and Jorgenson (1974) used another mixed I-O and econometric modeling strategy by deriving endogenous I-O coefficients from the dual of an econometrically estimated production function. The previously cited PIES, with its Econometric Regional Demand Model (ERDM), is another hybrid strategy (see FEA 1976, Appendixes A and C). The PIES L-P model depicts different supply technologies that can be utilized to satisfy the energy demands by ERDM. The solution algorithm is iterative. Preliminary price information in the econometric model generates initial energy demands, driving the PIES model and consequently providing new estimates of the opportunity costs of energy (the shadow prices obtained from the L-P solution). The shadow prices are used in the demand module to construct another set of energy requirements. Iterations between the supply and demand modules continue until convergence in prices and quantities occurs, a condition for the system that is not ensured. The PIES/ERDM model has evolved into the Mid-range Energy Forecasting System (MEFS), which incorporates a macroeconomic model (see deSouza 1981, chap. 7).

Uri (1977, 1979) experiments with "mixed" and "combined" time-series/econometric models of electricity demand. In one paper the time-series model parameters are explained as functions of economic variables, while in the other first econometric methods are used to estimate a demand function and then time-series analysis is employed to explain the disturbance term. He concludes that the hybrid models outperform either individual technique used separately.

The hybrid modeling strategies have been developed to rectify some of the deficiencies in traditional modeling, such as the failure of univariate time-series analysis to provide insights into behavioral relationships. Johnston (1984) argues that

> there is as yet no clear cut consensus on the relative roles of time-series techniques and the more orthodox econometric methods. Some mistakenly view them as competitive rather than complementary. Each is still an 'art', as distinct from a 'science', in that time-series practitioners have to make various judgments in the course of their analyses just as econometricians 'choose' between different regressions and specifications. (p. 381)

Those "more orthodox econometric methods" can be readily adapted to handle AR and MA error models while exploiting the underpinnings of economic theory, but it is not evident that time-series procedures are quite so malleable, this being due to the complexities of multivariate estimation. Thus, one may have to resolve a trade-off between theory and predictive powers.

Likewise, although dynamic I-O coefficients developed econometrically improve the I-O model's forecasting ability when compared with the static, fixed coefficient descriptive approach, pioneered by Hudson and Jorgenson (1974) in energy analysis through explicit modeling of the production relationships (see also Hudson 1981), the costs of such modeling increase dramatically over those of either approach applied separately. The PIES/ERDM models consider successively the supply and demand sides of the energy market, applying different modeling techniques to each component, but all such hybrid approaches require large investments of time and resources.

The mathematical programming approach is popular because of its inherent flexibility, which takes into account a wide range of alternatives. These may be of interest to decision makers, and, when combined with econometric demand estimates, consumer behavior is incorporated in the forecasts. The programming model–generated opportunity costs are the prices used in the demand equations. Changing technology and demand are thereby reflected in the factor input and output prices and product mixes. However, currently the feedback between the models is an interactive one. Thus, Smith's "blades of the scissors" may not cut, and computational limitations mean that for a model of the size and complexity of PIES/ERDM a simultaneous solution is infeasible.

A final issue needing consideration is that the data requirements of a Hudson-Jorgenson hybrid model are enormous; those authors' work was facilitated by the availability of annual I-O tables for the United States covering the years 1947–71 constructed by Jack Fauchett and Associates (1973) and by their association with DRI and access to its database and macroeconomic models. Unfortunately, such I-O data are not always available. Also the procedures used to generate I-O tables, usually matrix-balancing routines, may not accurately reflect the actual changes in the coefficients as they occur in the real world, because these are quite arbitrary, mechanistic procedures.

The hybrid approaches have been developed in an attempt to offset the weaknesses of one technique against the strengths of another, but, as always, the objectives of the modeling exercise should determine the approach, with the most appropriate modeling strategy adopted for a particular policy issue. For example, if the objective is to determine short-run consumer response to a change in price, then a static partial-equilibrium single-equation model may well provide sufficient information into the operation of the market for insights to be drawn. In such an instance, that modeling strategy would then be preferable to a more complicated dynamic general-equilibrium multi-equation approach. Also, analysts must recognize that the complexity of a model increases as techniques are combined. Sophisticated hybrid models often require large teams of diverse modelers and the development of simultaneous solution methods; therefore, "appropriate" remains the byword to all model builders.

MODELING DATA

Quality models are impossible without quality data. Some problem areas that may be identified and that remain are (1) confidentiality; (2) consistency, and (3) applicability. These categories are not rank ordered, exhaustive, or necessarily mutually exclusive. The primary impediment to modeling policy issues is the refusal of access to publicly collected data series. Federal and state government agencies operate under stringent, and often very narrowly interpreted, legislative guidelines on the publication or release of data. Usually data cannot be released if individual persons or companies can be identified. While the philosophical bases for maintaining such confidentiality may not in particular instances be questioned, the application of this stricture often precludes many potentially valuable analytic exercises.

Several approaches might be considered to the problems posed by this issue. The first, and most costly in both private and social terms because of the economies of scale involved in data collection, is for each analyst to collect primary data rather than to rely upon secondary sources; this approach is the one used by commercial research groups. These groups maintain two databases: one available to the public for a fee and another one used internally for modeling and analysis purposes. The confidential database contains sensitive information that subscribing organizations provide especially for modeling purposes but do not want disclosed. Such arrangements work successfully when the organization is a private profit-making entity charging for regular forecasts, for special analyses, and for access to public data, but such entrepreneurial approaches are unavailable to government analysts and are of little interest to many academics.

A variant of this purely commercial approach might be for governments to collect and maintain the data, publishing those data that do not violate confidentiality and releasing any of the other prescribed data on a case-by-case basis under very strict usage guidelines. Such release should be conditional upon agreement that the third party obtaining the information not publish or otherwise disclose or use it for private commercial gains; failure to fulfill the terms of the agreement would entail loss of access to similar data in the future. This approach has been successful with government-employed consultants and between government agencies. During the development of the Regional, Energy, Activity, and Demographic (READ) model, intergovernmental agreements among the Bureau of Economic Analysis, the Bureau of Labor Statistics, and the Bureau of the Census were entered into for the release to DOE of data for modeling purposes that could not be published (see Donnelly et al. 1977), and DOE was held responsible for ensuring the security of those data. That agreement (entered into more than a decade ago) has operated successfully without any abuse of the privilege or compromise of the data; indeed, few abuses of such arrangements with government contractors occur, for the obvious reasons. It would therefore appear that some sort of similar arrangement with individuals or groups involved in scholarly research might be likewise successful.

Finally, data collection by government agencies is usually undertaken in order to satisfy statutory requirements and is not specifically developed with modeling in mind. This situation often results in the series so obtained being of limited usefulness and requiring transformations with inadequate information. The solutions to these problems require considerable effort in developing new series and in revising existing ones. The first step would be to include modeling practitioners on the committees responsible for overseeing data collection activities so that modeling needs might be understood and incorporated, as well as possible, in any changes or additions to census or survey definitions. As Moses (1981) believes: "A good model can inform rational useful data collection planning as nothing else can. And on the other hand, there can be no adequate substitute for good data (real data, accurate data) as a foundation for forecasts and other model applications" (p. 21).

The problem of the suitability of the data is a fundamental one. Let us consider briefly the question of developing appropriate policy models with which to analyze questions relating to the demand for automotive gasoline. It is well accepted that motor gasoline is a derived demand, reflecting its use in satisfying transportation needs. The analysis of the demand for transport services would appear to be the logical starting point for developing a model for the demand for motor gasoline, such end-use analysis being a popular modeling strategy in addressing questions relating to the derived demand for energy. But difficulty arises with respect to obtaining data on the stock of energy-using appliances (automobiles and trucks), their energy efficiency, and the intensity of their use. The Sweeney "vintage capital" model of gasoline demand (1978) is a good example of the approach and some of the data problems that may be expected from this genre of engineering/economic model [see Green (1985, chap. 4) for a good discussion of this model]. The model was developed and specified at DOE during 1974 in response to the EC1-induced price rise in petroleum products. It identifies the effects of mandated automobile efficiency standards on the demand for gasoline, and is characteristic of the engineering-based end-use type of model specification. Sweeney, however, extends that generic approach explicitly to include behavioral relationships missing from simple engineering models. In this model the gasoline demand identity is

$$\mathbf{G} \equiv \sum \frac{\mathbf{M}_i \cdot \mathbf{S}_i}{\mathbf{mpg}_i} \tag{1.1}$$

where **G** = gasoline demand
M = vehicle miles traveled
i = vehicle vintage
S = stock of vehicles
mpg = miles per gallon.

This formulation allows for different fuel efficiency and usage patterns among various vintages of motor vehicles. In order to determine the level of gasoline demand, one models the demand for vehicle miles traveled, changes in vehicle efficiency, and stock and usage patterns for different vintages of vehicles. Detailed information on all of these variables is required, but while total gasoline consumption is available, even that information may be of questionable value; for example, the motor gasoline data often represent wholesale distributions and not retail sales. The other data required by an end-use model become progressively more difficult to obtain. For example, "total vehicle miles estimates" are developed by the individual states with little uniformity in the procedures used; some methodologies are sophisticated, and others are simplistic. The vehicle mile data for Kentucky are derived by multiplying a stock of motor vehicle series by an arbitrary fixed scalar, which to say the least involves some circularity in reasoning when the gasoline demand identity is applied to the subsequent values. The vehicle efficiency information is generally derived from manufacturers' specifications or obtained from static tests of sample vehicles. The Environmental Protection Agency's dynamometer tests are to be interpreted only in a comparative context, since these laboratory data are likely to be biased and to exhibit large variance when extended to actual vehicles. Furthermore, vehicle usage patterns are sampled only infrequently, so assumptions are required concerning their temporal patterns and their stability.

The stock of vehicles variable poses even more severe problems, because no annual reporting of the registered vehicles is made.[3] The calculation of changes in the stock of vehicles between survey dates is usually effected by noting that the stock of a particular efficiency category of vehicle in time period t is defined by the stock in the previous time period, plus new vehicle registrations in that category during the interval, less scrappage, plus previously registered vehicle imports, less registered vehicle exports. Equation 1.2 provides this identity:

$$\Sigma_i \mathbf{S}_i^t \equiv \Sigma_i \mathbf{S}_i^{t-1} + \Sigma_i \mathbf{N}_i^t - \Sigma_i \mathbf{R}_i^t + \Sigma_i \mathbf{M}_i^t - \Sigma_i \mathbf{E}_i^t, \quad (1.2)$$

where **S** = vehicle stock
i = efficiency category
t = time
N = new vehicle registrations
R = vehicles scrapped
M = vehicle imports
E = registered vehicle exports.

If modeling is done on a national basis, then those imports minus exports of previously registered vehicles are negligible, but this is not so if regional modeling is attempted and local data are not generally reported for those components of the change in the stock. This problem of the interstate movements of registered vehicles notwithstanding, scrappage by efficiency category—or any other category, for that matter—is not usually available.

These deficiencies force assumptions about (1) the shape and characteristics of the scrappage function, one necessarily independent of any economic factors; (2) usage patterns; (3) interstate vehicle movements; and (4) vehicle efficiency. Insufficient data are available to provide good estimates of any of these. After carefully considering the heroic assumptions and maintained hypotheses required by the paucity of these data, it is reasonable to question whether true explanatory power is being derived from the increased complexity of the vintage capital, stock of vehicles approach. Drollas (1984) argues, quite correctly in my view, "that one need not resort to particularly elaborate equation systems based upon such [miles traveled and usage] data to explain a high proportion of the variation in gasoline consumption over time. Furthermore, one need not use vehicle stock data either" (p. 74).

An alternative less-data-demanding, reduced form approach has been adopted by many energy modelers in order to determine the behavior of the consumer of motor gasoline, namely, the Balestra and Nerlove dynamic demand model formulation (1966). The inclusion of the lagged dependent variable in the single-equation demand model formulation is interpreted by Houthakker and Taylor (1970) and Verleger and Sheehan (1976) as reflecting the efficiency and usage of the unobservable stock of vehicles. The dynamic demand model is of the form

$$\mathbf{G}_j^t = \alpha_j + \beta_{ij} \cdot \mathbf{P}_j^t + \beta_{2j} \cdot \mathbf{Y}_j^t + \beta_{3j} \cdot \mathbf{G}_j^{t-1} + \mathbf{u}_j, \qquad (1.3)$$

where **P** = price of motor gasoline
Y = income
j = regional subscript
u = stochastic error term.

This formulation can provide policymakers with some insight into the anticipated affects of alternative pricing and taxing schemes on motor vehicles or gasoline.

A second example that might be cited concerns the use of production theory in analyzing energy demand, an approach that has become more prevalent since the beginning of the "energy crisis." While the empirical history of production function dates from the work of Cobb and Douglas (CD) (1928), as Heady and Dillon (1961, p. 15) point out, the CD functional form can be traced to Wicksell's 1916 work where, in the analysis of the decline in agricultural productivity, he specifies a three-factor production function that is homogeneous of degree one.[4]

The two-factor value-added model that Cobb and Douglas specify has proved to be amazingly robust in nature and continues to provide valuable insights. Madan (1984) uses the CD form to derive estimates of capital utilization rates; however, the inability of the formulation to be generalized to include more than two inputs and the unitary value for the elasticity of substitution mean that it is unsuitable for analyzing questions relating to a changing energy market. The basic relationship is specified as

$$\mathbf{V} = A \cdot \mathbf{L}^{\alpha} \cdot \mathbf{K}^{\beta}, \qquad (1.4)$$

where **V** = value added
L = labor
K = capital.

Recall that the elasticity of substitution measures the relative change in the optimal input ratio to the relative change in the input prices, and is, in the instance of the CD, always equal to 1:

$$\sigma_{CD} = \frac{d(\mathbf{L}/\mathbf{K})}{d(\mathbf{r}/\mathbf{w})} \cdot \frac{\mathbf{r}/\mathbf{w}}{\mathbf{L}/\mathbf{K}} \equiv 1 \tag{1.5}$$

where $\mathbf{r}$ = capital rental rate
$\mathbf{w}$ = wage rate.

Arrow et al. (1961) specify an alternative functional form commonly referred to as the "constant elasticity of substitution" (CES) production function, which allows for alternative substitution values, constant ones across the production surface. In addition, the CES functional form is easily generalized to more than two factor inputs as illustrated in Equation 1.6 with the inclusion of materials in the model:

$$\mathbf{Q} = \gamma \cdot [\delta_1 \cdot \mathbf{K}^{\rho} + \delta_2 \cdot \mathbf{L}^{\rho} + (1 - \delta_1 - \delta_2) \cdot \mathbf{M}^{\rho}]^{1/\rho}, \tag{1.6}$$

where $\mathbf{Q}$ = output
$\mathbf{M}$ = materials input
γ = efficiency parameter
ρ = substitution parameter
δ = distribution parameter.

In the CES function, the elasticity of substitution between all pairs of inputs is calculated as

$$\sigma_{CES} = \frac{1}{\rho} \; . \tag{1.7}$$

Thus, while these elasticities are constant along the production surface and not necessarily equal to unity, they do not provide useful substitution insights when more than two factors are being analyzed, since the value of σ_{CES} is the same for every and all pairs of inputs. This shortcoming led Christensen, Jorgenson, and Lau (1973) to specify a function that was a second-degree polynomial in logarithms, the so-called "transcendental logarithmic," or simply "translog," function. This translog form was originally defined by Heady and Dillon (1961, p. 205) to explore agricultural production, but did not achieve prominence in general economic or energy-modeling circles until the work of Berndt and Wood (1975) demonstrated its usefulness in modeling energy substitution. This provides yet another example of changing fashions in empirical models as the questions become more sophisticated and the data improve. Duality theory is invoked in deriving empirical results in the translog to specify the cost function from the production function. The translog is a "flexible" functional form since a minimal number of a priori restrictions is required, these involving symmetry and homogeneity in order to satisfy desirable regularity conditions of the production surface. The translog represents a local second-order approximation to any arbitrary function, but globally the translog is not well behaved, which can pose problems (see Diewert and Wales 1984). Also, the production function specification and the dual-cost function form need not reflect the same technology, as the translog is not self-dual: The elasticities of substitution among factors may, and generally do, differ between the two formulations. Admittedly these are shortcomings, but the translog's virtues include its general robustness and its not being particularly data demanding. To estimate the cost model, one requires only cost-share and price information for the factor inputs, but this differentiation converts the form to a first-order approximation only (see Theil 1980). The translog cost function is of the form

$$\mathbf{C} = \prod_{i=1}^{n} \mathbf{P}_i^{(\alpha_i + \frac{1}{2}\sum_{j=1}^{n} \beta_{ij} \cdot \ln \mathbf{P}_j + \theta_i \cdot \ln \mathbf{A})} \cdot$$

$$\prod_{j=1}^{m} \mathbf{Y}_j^{(\gamma_j + \frac{1}{2}\sum_{j=1}^{m} \delta_{ij} \cdot \ln \mathbf{Y}_i + \sum_{i=1}^{n} \rho_{ij} \cdot \ln \mathbf{P}_i + \theta_{j+n} \cdot \ln \mathbf{A})} \cdot$$

$$\prod_{i=1}^{2} e^{[\theta_{1+m+n} \cdot (\ln \mathbf{A})^1]} \cdot e^{\alpha_o}, \tag{1.8}$$

where **C** = total cost
P = input prices
Y = output prices
A = technical change
n = number of inputs
m = number of outputs
e = base of the natural logarithms
ln = natural logarithms.

The total cost function that allows for multiple outputs and multiple inputs has $(n^2 + 2n + n \cdot m + 2m + m^2 + 2)$ unknown parameters. This number is reduced by the imposition of the symmetry assumptions that $\beta_{ij} = \beta_{ji}$ and $\delta_{ij} = \delta_{ji}$. The above cost equation thus has $[(n^2 + n)/2 + 2n + 2m + n \cdot m + (m^2 + m)/2 + 3]$ parameters to be estimated.

The marginal productivity relationships derived from the cost function are combined with Shephard's lemma (1981, pp. 28, 52) to yield a set of cost-share equations, one for each input, whose shares sum to unity.[5] These are the share equations modeled and are of the form

$$\frac{\partial \ln \mathbf{C}}{\partial \ln \mathbf{P}_i} = \mathbf{M}_i \equiv \mathbf{S}_i = \alpha_i + \sum_{j=1}^{n} \beta_{ij} \cdot \ln \mathbf{P}_j + \sum_{j=1}^{m} \rho_{ij} \cdot \ln \mathbf{Y}_j + \theta_i \cdot \ln \mathbf{A}, \quad \text{for } i = 1, \ldots, n. \tag{1.9}$$

This specification includes the *m* product output terms in each share equation. These disappear when a single homogeneous output is modeled under an assumption of linear homogeneity in output. Also, the model results are not sensitive to the point of expansion around which estimation proceeds; that is to say, neither the regressors nor the derivative measures, such as elasticities or marginal costs, are affected by the specific expansion point that is selected. The translog function is a quadratic approximation around the expansion point of "one" for the input data (see Denny and Fuss 1977, p. 407). As Berndt and Christensen (1973) explain: "The formula for the AES [Allen Elasticity of Substitution] is a function only of the M_i and γ_{ij} [in Equation 1.9 the β_{ij}s]. Since the regressors are logarithmic, estimates of γ_{ij} are independent of units of measurement. The fitted values M_i are also invariant to scaling of regressors. Therefore, our estimates of the σ_{ij} are independent of units of measurement" (p. 97).

Some of the derivative measures obtained from the translog model are defined below. First, there are the elasticities of substitution, which are couched in terms of AESs (Allen 1938, pp. 340–43). The AESs are defined as

$$\sigma_{ii} = \frac{\beta_{ii} - \mathbf{S}_i}{\mathbf{S}_i^2} + 1, \text{ for } i, \ldots, n \tag{1.10}$$

and

$$\sigma_{ij} = \frac{\beta_{ij}}{S_i \cdot S_j} + 1, \text{ for } i = 1, \ldots, n;\ i \neq j.$$

These values are derived from the cost-share equations and do not require any of the additional information contained only in the total cost function. The AES measures are the substitution among pairs of inputs while holding output constant, and they represent the curvature of the industry's isocost curves. An I-O relationship would exhibit an AES σ equal to 0 since factor inputs are used in fixed proportions with no substitution possible. A linear homogeneous function would have a σ equal to 1, as is the case with the CD form. As the value of σ approaches infinity, substitution between the factors becomes progressively easier. The translog AESs vary over the production surface as relative prices and factor shares change, and these will differ among pairs of inputs as well as providing increased flexibility over the CD and CES functional forms. This added flexibility is purchased at the price of a small increase in the amount of modeling data required. In addition, as was shown in Equation 1.8, multiple outputs are readily accommodated by the translog form. However, this generalization greatly increases the amount of data required because some of the other derivative measures defined are based upon parameters contained solely in the total cost function and not available in cost-share equations. The factor demand elasticities are calculated from the AES values and information on factor shares of the inputs. Equation 1.11 provides the formula for the Marshallian income-compensated demand elasticities:

$$\eta_{ij} = S_j \cdot \sigma_{ij}, \text{ for } i, j = 1, \ldots, n. \qquad (1.11)$$

Although the multiple-output model can provide additional information, the modeling strategy becomes very data demanding and the ever-present problem of collinearity is exacerbated, which problem persists even to the case of the single-output formulation when the total cost function itself is to be estimated. Recall that Equation 1.8 can be expanded into

$$\ln \mathbf{C} = \alpha_1 \cdot \ln \mathbf{P}_1 + \ldots + \beta_{11} \cdot (\ln \mathbf{P}_1)^2 + \ldots$$
$$+ \gamma_1 \cdot \ln \mathbf{Y}_1 + \delta_{11} \cdot (\ln \mathbf{Y}_1)^2 + \ldots$$
$$+ \theta_{m+n+2} \cdot \ln \mathbf{A} + \theta_{m+n+2} \cdot (\ln \mathbf{A})^2 \,. \qquad (1.12)$$

In the explanatory variables ln $\mathbf{P}_i$ and $(\ln \mathbf{P}_i)^2$, ln $\mathbf{Y}_i$ and $(\ln \mathbf{Y}_i)^2$, and ln **A** and $(\ln \mathbf{A})^2$, there is high collinearity, obviously. Some of the price parameters, however, appear in both the total cost equation and one of the share equations; the additional information provided in that relationship tends to mitigate the problem of precise identification for the α_i and β_{ii} parameters, but will not assist in identifying the γ_i and δ_{ii} parameters, which appear only in the total cost function. Thus, the increased information being sought from inclusion of the total cost relationship imposes data requirements on the modeler that may not be achievable. The paucity of data and, in fact, their time-series nature may render the results open to question. As the time span increases, technological change can be expected to extract a progressively greater toll. In a perceptive paper, Klein (1947) argues for the increased reliance upon cross-sectional data in production analyses, and Brown, Caves, and Christensen (1979) implement that strategy for a multiproduct translog model, but frequently such data are unavailable.

Finally, even if the requisite modeling data are available to implement the adopted modeling strategy, the results may prove quite variable. This point is effectively demonstrated by Wibe (1984) and by Lutton and LeBlanc (1984) and should be well understood by modelers. It remains the exception to the rule when it occurs that the data unequivocally suggest specific functional forms without the modeler's formally testing their appropriateness.

MODEL IMPLEMENTATION

The development of model parameters will be touched upon only lightly here, as the specific procedures required differ among the various modeling strategies and some of these will be covered in later sections. What is of importance in the immediate discussion is some rudimentary understanding of the relative costs of the alternative methodologies. Rank order in terms of the total cost of implementing and operating specific models from the least costly to the most expensive would be (1) judgmental, (2) time series, (3) econometric, (4) mathematical programming, and (5) hybrid. This is a general ranking that naturally depends upon the complexity of the specific model being considered; thus, a particular hybrid model could well be less costly than some econometric model. However, for roughly similar models (if such actually exist), the ordering would be as stated, since complexity tends to follow, again, that particular pattern; here, the byword "appropriate" arises once more.

Unfortunately, in practice, the modeling strategy sometimes dictates the questions to be addressed, which is a perverse approach to policy analysis. Much modeling might be categorized as a technique searching for a purpose. The national nuclear research laboratories in the United States during the 1970s were confronted with the need to diversify research activities from traditional areas of applied physics and engineering, and some of the modeling expertise was redirected to address questions relating to socioeconomic policy. The Reference Energy System (RES), Brookhaven's initial foray into the field of energy analysis, is a descriptive model of the energy flows in the United States. The colorful flow diagrams that characterize that work are continually updated and still grace the covers of government reports; they may be seen prominently displayed on bureaucrats' walls, and the model is expanded into the mathematical programming one referred to as the "Market Allocation Model" (MARKAL). MARKAL has been converted for several International Energy Agency member countries, not necessarily at the behest of decision makers, but rather owing to the persistent diligence of particular modelers, and, at least in the United States, neither MARKAL nor any of its predecessors were ever used in government policy development. Presumably the lesson is that decision makers should be wary of modelers when clear-cut policy objectives have not been previously identified and defined, as the modeling strategy should evolve from the policy questions being raised by governments and not the converse. As has been explained, Project Independence provided the impetus for the PIES/ERDM and the MEFS approach. Now, while some analysts disagree with the approach adopted by DOE, it cannot be argued that it was a model searching for an application.

A significant shortcoming of PIES/ERDM was that no explicit interdependence existed between the hybrid L-P/econometric models and the macroeconomic models. This deficiency was cavalierly dismissed by some analysts at the time with the somewhat amazing assertion that the feedbacks from the energy system to the macroeconomy were in reality minimal, thereby, in effect, negating the raison d'être of the entire effort. Fortunately, the weight of opinion was otherwise and the READ model project was initiated to attempt to fill some of the gaps in PIES/ERDM, and MEFS subsequently integrated the DRI macroeconomic forecasts into the overall system. Still it remains that the entire PIES enterprise is a multimillion-dollar modeling effort employing many analysts for a number of years. It is unlikely that in the contemporary political climate such a project would even be seriously suggested.

Another lesson to be learned is that after any preliminary decision on modeling strategy is made, it must be determined that the requisite data are readily available or can be developed within the scope of the project; unfortunately, those funding such modeling projects may by necessity have to rely upon the knowledge and judgment of those who are the supplicants in order to assess the feasibility of the undertaking. And as is sometimes the case, the incipient modelers may in all good conscience err in their evaluation. Thus, funding review panels that consider modeling proposals should include as members individuals who are active modelers.

Once models are constructed, they need to be validated, exercised, documented, and maintained. None of these aspects of modeling are particularly attractive to modelers, and they are therefore frequently neglected. Validation involves establishment of the robustness of the adopted modeling strategy and demonstration that it succeeds in replicating the observed system behavior (see Labys 1982). Exercising a model is done to identify hot spots and potential weaknesses in the formulation. Most models used in policy analysis are exercised, but quite often this is not done in any systematic manner, and so the general behavioral characteristics of the model may not be identified. Whenever models are used in a "crisis management" or "fire-fighting" mode, some facets of modeling, including validation, exercising, and documentation, are relegated to the future. The adage that modelers may be "good at modeling but not at documenting" is lamentable but all too true. Green, who does a good job of documenting the Sweeney model, does not do the same for his own natural gas model (see Green 1985, chap. 2).

When such models as PIES are used in formulating policy, the results must be presented in a condensed and interpreted form. Model results in the form of stacks of computer printouts are unintelligible even to the most experienced analyst; thus, the essence of the results must be captured and presented for consideration. PIES did this with reports of decreasing detail referred to as "wonderbread," "wondercookie," and "wondercrumb," and all declared to be "better seven ways." Ordinary analysts were supposed to understand the complete wonderbread results, with the most senior-level decision maker able to comprehend the implications of the condensed wondercrumb reports.

CONCLUSIONS

It should be reiterated that the costs of modeling are frequently underestimated because all aspects of the process are not fully considered when selection is made among the alternative approaches. The actual specification of the model and its estimation represent only a small portion of the expense of building and operating a policy-oriented model. The components involving problem identification, data development, validation, exercising, documentation, and maintenance will represent by far the most costly aspects of modeling. And while none of these activities is particularly glamorous, either to the funding organization or to the modeler, each must be done if the potential of modeling in policy analysis is to be fully achieved.

Finally, decision makers need to have some insights into the strengths and weaknesses of particular models and recognize that as their questions become more sophisticated, those models should necessarily evolve. It is sometimes difficult to convince some senior-level officials that model maintenance is as important an aspect of the modeling process as the original construction and development phases were. Many persons view models as analogous to pieces of durable capital equipment and unrealistically assume that once constructed they have a long life expectancy with virtually no maintenance required. This is far from the truth. Therefore, the expectations of those commissioning model development must be conditioned to the fact that models are short-lived entities and need constant care and feeding if consistent high-quality results are to be attained. Those associated with the development and use of policy models understand their continuing requirements, but decision makers do not always fully appreciate this. This is particularly important if a model is developed under a grant or contract and the sponsoring entity does not have the "in-house" capability to run or maintain it. In such an instance, the model, no matter how good, may not prove useful to the organization.

Formal modeling is a valuable technique in policy analysis wherein relationships are explicitly defined. The appropriate model must be constructed to address specific questions about the behavior of the system under analysis. General all-purpose models are to be eschewed, since no model will be able to answer all questions equally well. If such an entity existed, by definition it would not be a model but the actual system itself, and there are philosophical limitations to defining a system completely from within.

NOTES

1. Versions of this chapter have benefited from seminars at Ball State University, the University of Wisconsin–La Crosse, and the University of New South Wales. The last is incorporated in *Energy Modeling in Australia* (1985), sections of which appear in this chapter and the next with permission.

2. For convenience throughout the book, the various agencies that were the precursors of DOE will be referred to merely as the latter agency.

3. Quite similar problems are inherent in demographic models of population change, and those issues have never been adequately addressed either.

4. Wicksell's notation was

$$a^{\alpha} \cdot b^{\beta} \cdot c^{\gamma} = p,$$

$$\text{where } |\alpha| < 1, |\beta| < 1, |\gamma| < 1,$$

$$\alpha + \beta + \gamma = 1.$$

He analyzed the change in output to be expected if each input were changed by a factor m to be $m \cdot p$ (Wicksell 1916, p. 287).

5. Shephard's lemma relates an indirect cost function to conditional factor demand equations utilizing the envelope theorem. Essentially, the lemma states that the partial derivative of the indirect cost function with respect to the price of a particular factor input yields that factor's demand equation *conditional* upon a given level of output being maintained. Thus, the marginal productivity of the input is equated to the factor's share of output, and the duality between the direct and indirect cost functions depends upon evaluating the functions only at optimal cost, minimizing the set of values for the inputs (see Beattie and Taylor 1985, pp. 226–27, 232–34).

REFERENCES

Allen, R. G. D. 1938. *Mathematical Analysis for Economists.* London: Macmillan.

Almon, C., M. R. Buckler, L. M. Horwitz, and T. C. Reimbold. 1974. *1985: Industry Forecasts of the American Economy.* Lexington, MA: Lexington Books.

Arrow, K. J., H. B. Chenery, B. S. Minhas, and R. M. Solow. 1961. "Capital-Labor Substitution and Economic Efficiency." *Review of Economics and Statistics,* 43:225-50.

Bacharach, M. 1970. *Biproportional Matrices and Input-Output Change.* Cambridge: Cambridge University Press.

Balestra, P., and M. Nerlove. 1966. "Pooling Cross Section and Time Series Data in the Estimation of a Dynamic Model: The Demand for Natural Gas." *Econometrica,* 34:585-612.

Beattie, B. R., and C. R. Taylor. 1985. *The Economics of Production.* New York: John Wiley & Sons.

Berlinski, D. 1976. *On Systems Analysis: An Essay Concerning the Limitations of Some Mathematical Methods in the Social, Political and Biological Sciences.* Cambridge, MA: MIT Press.

Berndt, E. R., and L. R. Christensen. 1973. "The Translog Function and the Substitution of Equipment, Structures, and Labor in U.S. Manufacturing 1929-68." *Journal of Econometrics,* 1:81-114.

Berndt, E. R., and D. O. Wood. 1975. "Technology, Prices and the Derived Demand for Energy." *Review of Economics and Statistics,* 57:259-68.

Box, G. E. P., and D. R. Cox. 1964. "An Analysis of Transformations." *Journal of the Royal Statistical Society,* B26:211-43.

Brown, R. S., D. W. Caves, and L. R. Christensen. 1979. "Modeling the Structure of Cost and Production for Multiproduct Firms." *Southern Economic Journal,* 46:256-73.

Bullard, C. W., and R. A. Herendeen. 1977. "The Energy Cost of Goods and Services: An Input-Output Analysis for the U.S.A., 1963 and 1967." In *Energy Analysis,* edited by J. A. G. Thomas, pp. 71-81. Surrey: IPC Science and Technology Press.

Christensen, L. R., D. W. Jorgenson, and L. J. Lau, 1973. "Transcendental Logarithmic Production Functions." *Review of Economics and Statistics,* 55:28-45.

Cobb, C. W., and P. H. Douglas. 1928. "A Theory of Production." *American Economic Review,* 18(suppl):139-65.

Darmstadter, J. with P. D. Teitelbaum and J. G. Polach, 1971, *Energy in the World Economy: A Statistical Review of Trends in Output, Trade, and Consumption since 1925,* Baltimore: Johns Hopkins University Press.

Denny, M., and M. A. Fuss. 1977. "The Use of Approximation Analysis to Test for Separability and the Existence of Consistent Aggregates." *American Economic Review,* 67:404-18.

DeSouza, G. R. 1981. *Energy Policy and Forecasting: Economic, Financial and Technological Dimensions.* Lexington, MA: Lexington Books.

Diewert, W. E., and T. J. Wales. 1984. "Flexible Functional Forms and Global Curvature Conditions." Technical Working Paper No. 40. Cambridge, MA: National Bureau of Economic Research.

Donnelly, W. A. 1985. "Selecting Appropriate Policy Modeling Strategies." In *Energy Modeling in Australia,* Centre for Applied Research. Kensington: University of New South Wales.

Donnelly, W. A., F. Hopkins, A. Havenner, B. Hong, and T. Morlan. 1977. "Estimating a Comprehensive County-Level Forecasting Model for the U.S." In *Socio-economic Impact of Electrical Energy Construction,* pp. 193–255 Washington, D.C.: Association of University Bureaus of Economic Research.

Drollas, L. P. 1984. "The Demand for Gasoline: Further Evidence." *Energy Economics,* 6:71–82.

Fauchett, J., and Associates. 1973. "Data Development for the I-O Energy Model." Final Report to the Energy Policy Project, Washington, D.C.

Forrester, J. W. 1971. *World Dynamics.* Cambridge, MA: Wright-Allen Press.

Georgescu-Roegen, N. 1979. "Energy Analysis and Economic Valuation." *Southern Economic Journal,* 45:1023–58.

Goldberger, A. S. 1968. *Topics in Regression Analysis.* New York: Macmillan.

Granger, C. W. J., and P. Newbold. 1977. *Forecasting Economic Time Series.* New York: Academic Press.

Green, R. D. 1985. *Forecasting with Computer Models: Econometric, Population and Energy Forecasting.* New York: Praeger.

Hannon, B. 1973. "An Energy Standard of Value." *Annals of the American Academy of Political and Social Science,* 410:139–53.

Heady, E. O., and J. L. Dillon. 1961. *Agricultural Production Functions.* Iowa State University Press.

Heath, T. 1981. Reprint of 1913 edition. *Aristarchus of Samos: The Ancient Copernicus.* New York: Dover.

Houthakker, H. S., and L. D. Taylor. 1970. *Consumer Demand in the United States.* 2nd ed. Cambridge, MA: Harvard University Press.

Hudson, E. A. 1981. "Modeling Production and Pricing Within an Interindustry Framework." In *Energy Policy Planning,* edited by B. A. Bayraktar, E. A. Cherniavsky, M. A. Laughton, and L. E. Ruff, pp. 201–14. NATO Advanced Research Institute on the Application of Systems Science to Energy Policy Planning, New York, 1979. New York: Plenum Press.

Hudson, E. A., and D. W. Jorgenson. 1974. "U.S. Energy Policy and Economic Growth, 1975–2000." *Bell Journal of Economics and Management Science,* 5:461–514.

Hull, C. H., ed. 1964. *The Economic Writings of Sir William Petty, Vol. II.* Reprint of *Five Essays in Political Arithmetick.* 1687. New York: August M. Kelley.

James, D. E. 1983. *Integrated Energy-Economic-Environmental Modeling with Reference to Australia.* Department of Home Affairs and Environment. Canberra: Australian Government Publishing Service.

Johnston, J. 1984. *Econometric Methods.* 3rd ed. New York: McGraw-Hill.

Klein, L. R. 1947. "The Use of Cross-Section Data in Econometrics with Application to a Study of Production of Railroad Services in the United States." New York: National Bureau of Economic Research. Mimeo.

Kuczynski, M. and P. L. Meek, editors, 1972, *Quesney's Tableau Economique,* London: Macmillan.

Labys, W. C. 1982. "Measuring the Validity and Performance of Energy Models." *Energy Economics,* 159–68.

Leamer, E. E. 1978. *Specification Searches: Ad Hoc Inference with Nonexperimental Data.* New York: John Wiley & Sons.

Leontief, W. W., 1951, *The Structure of the American Economy 1919–1939,* New York: Oxford University Press.

Liscom, W. L., 1982, *The Energy Decade 1970–1980: A Statistical and Graphic Chronicle,* Cambridge, MA: Ballinger.

Lovell, M. C. 1983. "Data Mining." *Review of Economics and Statistics,* 65: 1–12.

Lutton, T. J., and M. R. LeBlanc. 1984. "A Comparison of Multivariate Logit and Translog Models for Energy and Nonenergy Input Cost Share Analysis." *Energy Journal,* 5:35–44.

Madan, D. B. 1984. "The Measurement of Capital Utilization Rates." Department of Econometrics. Sydney: University of Sydney. Mimeo.

Malthus, T. R. 1966. Reprint of 1798 edition. *First Essay on Population.* New York: St. Martin's Press.

Meadows, D. H., D. L. Meadows, J. Randers, and W. W. Behrens. 1972. *The Limits to Growth.* New York: Universe Books.

Moses, L. E. 1981. "Keynote Address: One Statistician's Observation Concerning Energy Modeling." In *Energy Policy Planning,* edited by B. A. Bayraktar, E. A. Cherniavsky, M. A. Laughton, and L. E. Ruff, pp. 17–33. NATO Advanced Research Institute on the Application of Systems Science to Energy Policy Planning, New York, 1979. New York: Plenum Press.

Nelson, C. R. 1973. *Applied Time Series Analysis for Managerial Forecasting.* San Francisco: Holden-Day.

Nordhaus, W. D. 1973. "World Dynamics: Measurement Without Data." *Economic Journal,* 83:1156–83.

Shephard, R. W., 1981, *Cost and Production Functions,* Reprint of the First Edition, Berlin: Springer-Verlag.

Simon, J. L. 1981. "Global Confusion, 1980: A Hard Look at *The Global 2000 Report.*" *The Public Interest,* 62:3–20.

Smith, A. 1982. Reprint of 1795 edition. *Essays on Philosophical Subjects.* Indianapolis: Liberty Press/Liberty Classics.

Sweeney, J. L. 1978. "The Demand for Gasoline in the United States: A Vintage Capital Model." In *Workshops on Energy Supply and Demand,* pp. 240–77. International Energy Agency. Paris: Organization for Economic Cooperation and Development.

Theil, H. 1980. *The System-wide Approach to Micro-economics.* Chicago: University of Chicago Press.

United Nations, 1985, *1983 Energy Statistics Yearbook,* New York: UN.

Uri, N. D. 1979. "A Mixed Time-Series/Econometric Approach to Forecasting Peak System Load." *Journal of Econometrics,* 9:155–71.

———. 1977. "An Integrated Box-Jenkins/Econometric Model for Forecasting Time Series." *Proceedings of the American Statistical Association,* 1:404–7.

U.S. Federal Energy Administration. 1976. *National Energy Outlook.* FEA-N75/713. Washington, D.C.: U.S. Government Printing Office.

Verleger, P. K., and D. P. Sheehan. 1976. "A Study of the Demand for Gasoline." In *Econometric Studies of U.S. Energy Policy,* edited by D. W. Jorgenson, pp. 179–241. Amsterdam: North-Holland.

Wibe, S. 1984. "Engineering Production Functions: A Survey." *Economica,* 51:401–12.

Wicksell, K. 1916. "Den 'kritiska pukten' i lagen fur jordbrukets aftagande produktivitet." *Ekonomisk Tidskrift,* 10:285–92.

Yoke, G. W. 1984. "Interpreting the International Workshop Survey Results—Uncertainty and the Need for Consistent Modeling." *Energy Journal,* 5:73–77.

Two

Modeling Theory

OVERVIEW

This chapter discusses several methodologies used in modeling. The techniques include judgmental forecasting, mathematical programming, time-series analysis, econometric methods, and hybrid procedures. They are analyzed by way of considering the assumptions and identifying the strengths and weaknesses of each.

INTRODUCTION

Modeling is more sophisticated now than in the past; as a result, close scrutiny of the forecasting procedure that is chosen is an increasingly important aspect of research, as well as due note of the process' tendency toward demanding more and improved data. In this chapter there will be considered in more detail the various policy modeling strategies contingent upon specific assumptions required by each methodology, for these necessarily influence the results. A complete understanding of these assumptions, many of which are usually implicit, will assist in the development of a more rational decision-making process; the strengths and weaknesses of several modeling techniques are discussed in the context of energy analysis. These attributes affect the quality of forecasts that are developed and thus subsequent decisions relating to those forecasts. It is therefore important for analysts to understand the foundations of the methodology adopted when they are using forecasting techniques to help formulate policy decisions, and additionally to remember that parsimony is an important aspect of modeling. The value of giving simplicity a place in energy modeling augments clarity of thought. In that which follows, the five modeling approaches discussed are those previously mentioned.

JUDGMENTAL FORECASTING

Judgmental forecasting involves diverse techniques. The simplest forecasting model is one that assumes that the last value of the variable under study will recur. Such an approach assumes that no learning occurs in the system or process whenever previous forecasts are not realized. Often military decision makers prepare for the next war by analyzing the last one, and thus the blitzkrieg of World War II easily overran the Maginot Line. Agencies with experience in constructing hydroelectric dams sometimes restrict their thinking to building more such facilities oblivious to changing circumstances; for example, in the United States the U.S. Corps of Engineers and overseas the Tasmanian Hydro-electric Commission (HEC). Mathematically, this type of forecast is stated as

$$\hat{y}_t = y_{t-1}, \tag{2.1}$$

where $\hat{y}$ = forecast of y
t = time period.

Such an estimator is useful whenever no other information or insight into the particular process is available, and then it may be the only reasonable modeling approach. An obvious extension would be to utilize additional historical information in developing the forecast. A simple average, such as Equation 2.2, defines the expected value of y based upon the experience as defined by the arithmetic mean of the last $(t - 1)$ time periods:

$$\hat{y}_t = \sum_{i=1}^{t-1} \frac{y_i}{t-1}. \tag{2.2}$$

While this scheme utilizes additional information, it fails to allow for any changes in the system; it is a descriptive statistic. If changes occur over time, a moving-average that considers only the latest, and perhaps the most relevant, observations might inprove the estimator. Such a moving-average estimator would be

$$\hat{y}_t = \sum_{i=t-n+1}^{t-1} \frac{y_i}{n-1} . \tag{2.3}$$

The moving-average technique (simply a linear filter) estimator provides for growth in the series, and the implicit rate of growth is continually updated as recent observations are added and early observations dropped, but this approach still remains a descriptive statistic rather than a forecasting tool.

Conversely, an analyst might wish to include some anticipated growth in developing the forecast. As indicated in Chapter 1, long-run electricity demand forecasts have traditionally relied upon exponential trend extrapolation models that assume a continuing rate of growth. The assumption in regard to the rate of growth once proved a reasonably reliable procedure for forecasting electicity consumption and determining capital requirements during the growth phase of the commodity. Still, trend extrapolation remains a deterministic procedure, depending upon the initial condition of the system and a constant growth rate, and this assertion is unrealistic, although not uncommon in many models.[1] The product life cycle demand for a good evolves with rapid initial growth followed by eventually declining rates that reflect replacement demand and new household formations rather than experiencing continually increasing rates of growth. This product life cycle phenomenon is applicable to electricity demand as well. Socioeconomic factors will affect demand, and these include income changes, price fluctuations, and other market disruptions. (None of the deterministic methods is able to explicitly accommodate these types of effects.) Mathematically, trend extrapolation is defined as in Equation 2.4. The endogenous variable $\hat{y}$ for a particular year t is forecast as a function of the initial conditions of the system y_0 and the selected rate of growth r:

$$\hat{y}_t = y_0 \cdot e^{rt} , \tag{2.4}$$

where y_0 = initial conditions
e = base of the natural logarithms.

The problems with exponential growth rate models are similar to those of the simplest naive, no change model, and such models have been widely abused by electricity authorities in developing self-serving forecasts of consumer demand. The approach may be compared to a "cargo cult" as in a few primitive societies. The analogy lies in the belief that if a particular single-input resource is provided, such as electricity, and then the presumed subsequent availability of that resource cheaply will prove sufficient enticement to manufacturers to one's region and thereby stimulate lagging local economies. Unfortunately, all too often, the analysis that serves to justify additional expenditures is based upon unrealistically low discount rates and inflated estimates of secondary and tertiary benefits. When these are combined with the fixed growth rate assumption, projects show appropriate benefit-cost results.

Exponential smoothing is a mechanism that continually updates forecasts, putting the greatest weight on the most recent observation and geometrically declining weights on the older historical data (see Granger and Newbold 1977, pp. 163–65). This is a logical extension of the moving-average approach. Equations 2.5 and 2.6 represent still further extensions wherein the smoothing applies also to an explicit trend term, thus a combination of the moving-average and exponential growth models:

$$\hat{y}_t = \alpha \cdot y_t + (1 - \alpha)(\hat{y}_t + \hat{s}_{t-1}), \qquad (2.5)$$

where s = trend term
$0 < \alpha < 1$;

$$\hat{s}_t = \beta \cdot (\hat{y}_t - \hat{y}_{t-1}) - (1 - \beta) \cdot \hat{s}_{t-1}, \qquad (2.6)$$

where $0 < \beta < 1$.

The above procedure fails to address any of the shortcomings identified with the simpler approaches above. Other deterministic approaches rely upon a particular mathematical form, such as the logistics curve depicted in Equation 2.7, or a particular distribution, such as Poisson distribution shown in equation 2.8:

$$\hat{y}_t = \frac{k}{(1 + e^{(a+b \cdot t)})}, \tag{2.7}$$

where a, b, and k = constants
$b < 0$;

$$\hat{y}_t = (e^{-m} \cdot m^t) \cdot t!, \tag{2.8}$$

where m = constant
! = factorial.

These techniques also ignore, as do all deterministic approaches, random aberrations that occur in the realization of a process. Smith and Hill (1985) discuss how sharing techniques might be evaluated, since they cannot be checked in applications for allocating totals in regard to the synthetic data constructed. This is important, since reliance on such data is commonplace in empirical studies of energy demand. The authors specify a particular production function exhibiting constant returns to scale of "Cobb-Douglass form with a constant elasticity of substitution subfunction"[2] (p. 36). Forty simulations providing for various elasticity of substitution parameterizations run by the authors using the first 30 observations estimate transcendantal logarithmic (translog), "log-odds" (logistic), and probabilistic models with the remaining observations left available for predictive evaluation.[3] Smith and Hill conclude that the logistic and translog models are superior to the probabilistic model formulation with "only a slight margin of superiority for the log-odds model" (p. 42). They thus provide an experimental approach that may be used to validate models based upon synthetic data. The generality of their results in correctly identifying production function forms other than the one they specified, however, would still have to be demonstrated.

Because of the nature of judgmental forecasts, it will always remain impossible to determine what the assumptions are unless they are clearly stated, which, in general, they are not; and since such forecasts are subjective, their quality reflects more nearly the insights of the forecaster than the robustness of the procedure. Such limitations are too infrequently discussed by analysts, and ignoring the factors that affect a system can led to erroneous policy prescriptions. For example, when energy consumption growth is treated as an independent phenomenon, substantial overexpansion and development will occur. For any such derived demand as energy, the social, economic, and physical conditions are all important determinants; therefore, these factors must be taken into consideration when developing forecasts designed to assist in making policy decisions. The continuing interest in the Kondratieff Wave Theory, as reported in the *Wall Street Journal* (Malabre 1986, p. 1), is an example of the difficulties posed by such judgmental models; it propounds a 56-year cycle of economic growth and depression–one lacking statistical verification while yet gaining some degree of popular acceptance.

Admittedly, judgmental forecasts are simple to prepare, and only a minimal amount of investment is required to produce them. While everyone uses judgmental forecasting to some extent, the results provided by the diverse techniques are extremely variable. The deterministic nature of the models (once the model is specified) allows for the replication of results but not of the specification. Thus, the very simplicity of judgmental techniques represents both the strength and the weakness of the approach.

MATHEMATICAL PROGRAMMING

The next methodology considered, mathematical programming, is also deterministic and represents one area of the field of systems analysis.[4,5] A widely used modeling technique, I-O analysis, is a subset of the general category of mathematical programming.[6,7] In I-O models the system under study is divided into a set of mutually exclusive, exhaustive sectors. The interrelationships of the economy described by the I-O model are measured in terms of dollar flows, although recent work by engineers engaged in energy studies has advocated, and quite incorrectly so, using joules or therms instead [see Hannon (1973) who constructed an energy I-O table for the United States].[8] The individual elements of Hannon's flow matrix are defined in terms of energy units rather than monetary units. As explained in Chapter 1, such a methodology raises problems. To recapitulate briefly, energy units do not serve as a convenient medium of exchange; therefore, their use as a numeraire is inappropriate, as this view implies that energy is the only scarce resource. Recognition that scarcity is only relative, and that such relationships will inevitably change over time, requires that the numeraire be some commodity that is readily convertible into other commodities, which implies the use of monetary units and *not* units of a particular factor input (see Griffin and Steele 1980, pp. 224–26); Webb and Ricketts 1980, pp. 243–47).

The I-O model divides the gross output for a particular industry $\mathbf{Y}_i$ between final (consumption) demand $\mathbf{X}_i$ and intermediate (production) demands. The intermediate demands $\mathbf{y}_{ij}$ represent the sales (flows) of product i as an input to the sector producing j as an output. Since the I-O model covers all sectors, changes in final demand will necessarily mediate among all the intermediate demand sectors on which final demand sectors depend. These relationships may be written as

$$\mathbf{Y}_i = \mathbf{X}_i + \sum_{j=1}^{n} y_{ij}, \text{ for } i = 1, \ldots, n, \tag{2.9}$$

where $\mathbf{Y}_i$ = gross output
$\mathbf{X}_i$ = final demand
y_{ij} = intermediate demand.

General solutions to the I-O model can be obtained by collecting terms and calculating the so-called I-O coefficients $\mathbf{a}_{ij}$, and the equation may be then rewritten as

$$\mathbf{Y}_i - \sum_{j=1}^{n} \mathbf{a}_{ij} \cdot y_j = \mathbf{X}_i, \text{ for } i = 1, \ldots, n, \tag{2.10}$$

where $\mathbf{a}_{ij} = \dfrac{x_{ij}}{\mathbf{X}_j}$.

This relationship, when solved for gross output and written in matrix notation, is

$$\mathbf{Y} = (\mathbf{I} - \mathbf{a})^{-1} \cdot \mathbf{X}, \tag{2.11}$$

where $\mathbf{Y}$ = vector of total demand
$\mathbf{I}$ = identity matrix
$\mathbf{a}$ = I-O coefficient matrix
-1 = matrix inverse operator
$\mathbf{X}$ = vector of final demand.

Equation 2.11 states how total demand in the economy is determined, particularly highlighting the interrelationships among industry sectors. The level of the final demand for all commodities plus the I-O relations is reflected in the fixed I-O coefficients, which are assumed to be fixed for all levels of output. The $(\mathbf{I} - \mathbf{a})$ matrix is constructed with the rows representing the producing industries and the columns being the purchasing sectors of the economy. The fixed I-O coefficients reflect the product flow information provided in the raw I-O table and represent the solution of some unspecified process. This matrix contains the technical process information, the I-O coefficients, which, when they are assumed to be constants, allows the determination of the effects that changes in final demand will have on both the intermediate and the total demands of the economy. Any adjustment to changes in final demand is achieved by proportional changes in total demand, with the inverse of the $(\mathbf{I} - \mathbf{a})$ matrix defining the magnitude of these impacts and referred to as the matrix of "direct and indirect multipliers." Thus, the effect of an increase in demand for oil can be traced through the changes necessitated in the various intermediate industries that are suppliers to the oil sector. These would include exploration and drilling equipment manufacturers, heavy truck makers, paper mills, steel producers, and the oil industry itself, among numerous other sectors. The I-O technique is an explicit representation of the interrelationships in the production of intermediate goods and the production of final demand commodities. Every interindustry relationship ties the various sectors together, and a change in the level of production of one industry affects the level of production of the other industries. Once the interindustry flows have been measured, that information is converted into the fixed I-O coefficients. The static multipliers that appear in the inverse of the $(\mathbf{I} - \mathbf{a})$ matrix reflect the direct and indirect effects of changes in final demand, and these are constants for all levels of demand.

As an example, let us assume that there exist four mutually exclusive and exhaustive sectors in the economy: agriculture, manufacturing, services, and energy. The final demands in each of these sectors are 10, 15, 20, and 5, respectively. Thus, in this economy, agriculture accounts for 20 percent of final demand, manufacturing 30 percent, services 40 percent, and energy the remaining 10 percent. The intermediate demands or flows are shown in Table 2.1. The gross

Table 2.1. Input-Output Intermediate Demand Flow Matrix

	Buyer			
Seller	Agriculture	Manufacturing	Services	Energy
Agriculture	2	6	3	4
Manufacturing	1	10	5	6
Services	0	5	12	2
Energy	1	4	1	3

Source: Compiled by the author.

output for each of the four sectors of this economy is determined by summing across the respective rows and adding that sector's final demand, namely, 25, 37, 39, and 14, respectively. Then using Equation 2.10 the I-O coefficients may be calculated.[9]

The matrix of direct and indirect multipliers is determined from Equation 2.11. The $(\mathbf{I} - \mathbf{a})^{-1}$ matrix is shown in Table 2.2. The elements on the diagonal of this matrix provide the direct impact on gross output to be expected from a change of 1 in the final demand of the sector for that row, while the off-diagonal elements in the row provide the indirect increases that will occur to that sector when the final demand for another sector increases by 1. Thus, if the final demand for energy increases by 20 percent from 5 to 6, total output in energy will increase by 1.457 (the direct impact), and, in addition, the gross output for the agricultural sector will increase by 0.667; for the manufacturing sector, the increase will be 0.979, and for the service sector, the increase will be 0.492 (the indirect impacts). The new levels of total demand that result from the change in the final demand for energy for the various sectors are thus 25.7, 38.0, 39.5, and 15.5, respectively. The percentage changes in total

Table 2.2. Input-Output Matrix of Direct and Indirect Multipliers

	Agriculture	Manufacturing	Services	Energy
Agriculture	1.133	0.392	0.223	0.667
Manufacturing	0.112	1.604	0.346	0.979
Services	0.037	0.365	1.534	0.492
Energy	0.074	0.253	0.109	1.457

Source: Compiled by the author.

demand by sector are 2.7, 2.6, 1.3, and 10.4, respectively. As can be seen, the multipliers are static and the absolute change in the total demands will be the same even if the increase in the final demand for energy represents only an increase of 6.7 percent from 15 to 16 or even just an increase of 0.9 percent from 115 to 116.[10]

The constancy of the I-O multipliers has been addressed by some researchers by specifying a dynamic model wherein the multipliers adjust as the level of production changes, as technology changes, as management practice changes, and as other factors such as relative cost change (see Hudson and Jorgenson 1974; Almon et al. 1974). Hudson (1981) describes how production theory and I-O analysis are combined in the Hudson-Jorgenson model in order to provide variable I-O coefficients that respond to changing economic conditions. Other forms of extensions to a dynamic I-O model (L-P formulations) as well as this hybrid approach greatly increase the complexity of models, requiring substantially more data than are necessary to implement a static I-O model.

As indicated, mathematical programming models are a superset of the I-O model described above. The programming matrix includes additional rows that specify alternative technologies, an objective function, and a set of constraint equations.[11] In the mathematical programming approach, the selection among the available production processes is based upon optimization of an objective function through either the minimization of a cost function or the maximization of a profit function, subject to the system constraints that describe production technologies and resource availability. The basic forms of mathematical programming models are L-P, nonlinear programming, and integer programming.

Using L-P as the example, the model is described by

$$\text{Minimize:} \quad z = \mathbf{c} \cdot \mathbf{Y} \tag{2.12}$$

$$\text{Subject to:} \quad \mathbf{A} \cdot \mathbf{Y} \leqslant \mathbf{X}, \; x_i \geqslant 0, \tag{2.13}$$

where z = objective function
$\mathbf{A}$ = matrix of processes
$\mathbf{X}$ = vector of final demands
$\mathbf{Y}$ = vector of total demands
$\mathbf{c}$ = vector of costs.

In the above modeling strategy, the choice among the various processes is made so as to minimize the cost function z.[12] While optimization techniques provide the "best" solution for the specified objective function and the technological relationships defined by Equations 2.13, the result will not necessarily reflect the operation of the market. That "real-world" solution will reveal the effects of institutional constraints, market rigidities, and any and all other market imperfections, and such factors as these are not always modeled or, when they are, are not always modeled correctly.

The rows of a mathematical programming matrix are constructed to define a particular production process.[13] The reasonableness of a solution depends upon how well the technologies are depicted in the model equations and on the effects of the omitted factors mentioned above. When forecasting, programming models require information concerning future technologies.[14] New solutions occur whenever an alternative technology is introduced into (or is said to enter) the model solution based upon the benefit of adopting that process to its opportunity cost (or shadow price). Successful solution of the mathematical programming model requires the assumption that all production surfaces are regular and of the appropriate shape and exhibit a global extremum—either a maximum or a minimum, depending upon how the objective function is couched—that clearly dominates any local extrema. Numerous computational algorithms are available to the modeler, and the calculation of a solution is not a problem. In the field of energy analysis, L-P models have been used since the early 1960s in determining the optimum operations of petroleum refineries in the light of changing factor prices and qualities of crude oil inputs. It was not until the mid-1970s, however, with the construction of PIES, that a large-scale L-P model was used to model the entire energy supply system (see FEA 1976).[15] MARKAL is a mathematical programming model of the supply side of the energy system currently being promulgated by the International Energy Agency.[16] Analysts debate whether such optimization models as MARKAL, which are largely descriptive, should be considered really to be forecasting models. Although MARKAL is not predictive, it can, and is, utilized to provide conditional forecasts for different scenarios, but MARKAL has never been used in the formulation of government energy policy in the United States, to the author's knowledge.

To review, mathematical programming models can theoretically handle system shocks through the selection among alternative production approaches; however, the optimal model solution may not be a feasible real-world solution. Since this modeling strategy allows for the economic evaluation of new technologies, as far as they can be defined, it is therefore suited for medium- to long-range forecasting. Unfortunately, much of the information on new technologies that is required for long-range forecasting is hypothetical and would be expected to have a large variance. Conversely, the I-O approach is most suitable for short- to medium-range forecasting where technologies are well known and relatively unchanging.

Finally, it should be noted here that all modeling strategies require data for implementation and some are more data demanding than others. Selection of an appropriate modeling approach depends upon the existence of adequate data; I-O analysis usually suffers from out-of-date interindustry data. Because it is costly to collect the requisite interindustry flow data, governments have given a low priority to the acquisition of up-to-date information. Whenever primary current data are not available, estimating procedures are employed to revise existing I-O matrices and to construct regional I-O tables.[17] These procedures are often mechanical, as per the iterative matrix-balancing routine, the **rAs** technique mentioned in Chapter 1. But even with a "perfect" updating procedure, the fixed coefficient nature of the static I-O model remains; therefore, the I-O modeling strategy is rendered unsuitable for the evaluation of major shocks to the economic system such as resulted from the Organization of Petroleum Exporting Countries (OPEC) increase in crude oil prices of EC1, the curtailment of production of EC2, and the major reductions in price experienced in the first quarter of 1986 [Energy Crisis 3 (EC3)]. Such shocks induce substitution, altering the interindustry coefficients. As Bullard and Herendeen (1977) readily admit, the I-O approach is severely problematic in regard to modeling such changes accurately (see the discussion in Chapter 1). Modeling strategies all fall prey to this problem of handling shocks; however, some are better adapted to address this problem than are others, as we will see later.

TIME-SERIES ANALYSIS

The time-series approach utilizes the information contained in the data series.[18] Such data are ordered by time, and the traditional statistical concepts of population and sample are replaced with analogous concepts of theoretical process and realization. This theoretical process, which includes a stochastic component, is defined by the data series, but modeling the process requires both the assumption of stationarity and that of ergodicity. A stochastic process $\mathbf{x}_t$ is said to be stationary if its distribution function $F(\mathbf{x}_t)$ does not depend upon time; and the process is ergodic if the observed values are separated by sufficient time to be "almost uncorrelated" (Granger and Newbold 1977, p. 4). Ergodicity is required for separating the meaning from system noise.[19] These alternative concepts and assumptions of time-series analysis allow the use of a single realization of a process, such as a set of observations, in developing the model rather than requiring sampling, as is necessitated by traditional statistics. This is a major advantage of time-series analysis because sampling is an approach that is generally impossible when a large range of phenomena, including many economic issues, is being considered.

The time-series may be specified as **AR**, **MA**, or, more generally, **ARMA**. The fundamental univariate process in the case of the **AR** time-series model relates the current value of the dependent variable as a polynomial-distributed lag of its own historical values; recall Nelson's "first duty" of the data (1973) is to inform about its evolution (see Chapter 1). Thus, $\mathbf{y}_t \sim \text{AR (p)}$ is generated by the equation:

$$y_t = \sum_{j=1}^{p} a_j \cdot y_{t-j} + \epsilon_t , \qquad (2.14)$$

where
- y = observed series
- p = order of the process
- **a** = roots defined outside the unit circle
- ϵ = zero mean, white noise.

In matrix notation this AR time-series process may be written as

$$\mathbf{A} \cdot \mathbf{Y} = \epsilon , \qquad (2.15)$$

where **A** = diagonal matrix with identical elements being a polynomial in the backward shift operator and coefficients **a**.

The alternative MA process, $y_t \sim MA(q)$, is defined in terms of successive values from a zero-mean, white noise process, and may be written as

$$y_t = \sum_{j=1}^{q} b_j \cdot \epsilon_{t-j} + \epsilon_t , \tag{2.16}$$

where q = order of the process
$\mathbf{b}$ = roots defined outside the unit circle,

which in matrix notation would be written as

$$\mathbf{Y} = \mathbf{B} \cdot \epsilon , \tag{2.17}$$

where $\mathbf{B}$ = diagonal matrix with identical elements being a polynomial in the background shift operator and coefficients **b**.

It can be demonstrated that a stationary AR(p) process may be written as an MA (∞) form, and conversely that an MA(q) process is equivalent to an AR (∞) one, although one form is normally of much lower order than the other (see Granger and Newbold 1977, pp. 20–24; Box and Jenkins 1970, p. 79, Table 3.2). Parsimony in model structure is one important criterion of selection. The ARMA process, $y_t \sim ARMA(p, q)$, which assumes stationarity in the series, achieved through some appropriate differencing of the data, generalizes the AR(p) and the MA(q) processes. The ARMA process is denoted as

$$\mathbf{A} \cdot \mathbf{Y} = \mathbf{B} \cdot \epsilon . \tag{2.18}$$

A useful extension of the ARMA process, particularly when dealing with economic phenomena, is the ARIMA model, $y_t \sim ARIMA(p,d,q)$. Integrated processes exhibit nonstationarity; one example is a random walk.[20] When using a random walk model for forecasting, as the forecast time horizon increases, the variance increases arithmetically and the confidence intervals, which are based upon the standard deviation, increase with the square root of the time horizon. Thus, the precision of the forecast decreases with the root of the time horizon (see Nelson 1973, pp. 8–11).

In general, economic data include trend components and periodic cycles, and the ARIMA process may provide the nonstationarity with which to depict the observed realization. The trend component of a time series may be thought of as a random walk with adjustment for drift, while the cyclical portion may be understood as relating to the season of the year or to the time of day.[21] In general, then, economic time-series data include trend, cyclic, and random components. The order of the ARIMA process includes the parameter d, where d relates to the number of times that differencing is required to achieve stationarity. The generating function for the ARIMA model is written as

$$\mathbf{A} \cdot \nabla^{d} \cdot \mathrm{Y} = \mathbf{B} \cdot \epsilon\,, \tag{2.19}$$

where ∇^{d} = order of differencing.

In other words, the operator ∇^{d} is used to achieve stationarity in the ARIMA(p,d,q) process, converting it to an ARMA(p,q) one. A more comprehensive multivariate time-series approach would allow for the inclusion of behavioral relationships in the process being modeled. The variables used in such models are separated into exogenous variables, those determined by factors other than behavior, and endogenous variables, those determined by the model structure. Whereas substantial work has been applied to developing the univariate techniques of time-series analysis, much less effort has been expended on the multivariate method. This is due to the generalization of the univariate model, which introduces many analytical and computational problems. Granger and Newbold (1977) discuss the analytical problems in terms of "time-series (TS)-identification" and "economic (E)-identification" (pp. 219–24). Economic identification relates to having the system sufficiently well structured that a unique model is described (the matrix is not singular); such definition, as was noted in Chapter 1, can be provided by economic theory.

Identifying the most appropriate order of the process and the model structure requires insight and experience. Economic theory will assist in defining the structure; intuition is required for this component as well and is paramount in the selection of the model order. The advantage of the multivariate formulation, as mentioned in Chapter 1, is that the model can be decomposed into a block for the endogenous variable (expressing the behavioral relationships of the model) and a block for the exogenous variables. This latter block is strictly of the univariate ARIMA form. Thus, the multivariate time-series modeling strategy provides a mechanism for generating future values of the totally exogenous variables in the system, as opposed to the exogenous control variables, an advantage lacking in the econometric approach, which will be reviewed in the next section. The multivariate ARIMA (MARIMA) process is

$$\mathbf{A}_1 \cdot \nabla^{d_1} \cdot Y + \sum_{j=2}^{s} \mathbf{A}_j \cdot \nabla^{d_j} \cdot X_j = \mathbf{B} \cdot \epsilon , \qquad (2.20)$$

where **A** and **B** = nondiagonal matrices of polynomial backward shift operators
∇^d = differencing operator
d_1 and d_j = order of differencing
Y = endogenous variable
$\mathbf{X}_j$ = exogenous variables.

The MARIMA(p_i,d_j,q), as represented above, can include several orders of differencing d_j, one for each variable, and several orders of AR processes p_i. The i subscript can take on the values from 1 to s, where s is the total number of endogenous plus exogenous variables in the model. The MARIMA is consequently considerably more complex than the univariate ARIMA process, and for this reason the techniques do not enjoy widespread use.

To summarize, the traditional univariate time-series techniques imply either that other factors in the system have no influence upon the series under study or that the influence they exert is expected to remain unchanged throughout the forecast period. When studying purely physical phenomena, as is the basic origin of time-series analysis, this assumption may be sensible; however, for socioeconomic activity, this is not a tenable hypothesis. Recent research in the field of time-series analysis has been directed toward developing the multivariate approach into a useful tool for analysis. Advances in this research could provide insights into policy issues that are sensitive to behavioral influence, something that is not to be found in the univariate approach. Since many problems are still to be resolved in this largely experimental work, the development of the MARIMA approach is not very far advanced, and its effectiveness in regard to econometric methods is yet to be demonstrated. A major problem is in selecting the time-series and economic structure of the multivariate model, which involves determining the order of the model parameters (p, d, and q) and the causal variables of the model. Parsimony notwithstanding, alternative model structures will too often provide similar results, and this remains the "art" in the time-series approach.

ECONOMETRIC METHODS

Econometric techniques test hypotheses by applying statistical methods in accordance with the economic theories that specify the model itself.[22] An econometric model contains a dependent variable, a set of independent variables, *and* a stochastic error term. The explanatory variables incorporated in the model depend upon the phenomena—economic, social, political, or whatever—being analyzed. Consumer utility theory specifies final demand equations, and production theory models the demand behavior of companies. The appropriate theory determines the set of potentially important variables to be tested in the model, and the influence of these factors is evaluated statistically. As we know, economic theory postulates that the final demand for a commodity is related to the price of the good, the price of other goods, consumer income levels, and other factors such as consumer tastes, socioeconomic status, demographics, and physical conditions. Alternative theories may suggest a relationship between labor value and capital expenditure. While theory indicates which variables are to be considered, it does not establish the mathematical form of the relationship, and this latter consideration is one source of the diversity of analysis that may be seen in the literature. The econometric approach is couched in matrix notation as

$$\mathbf{y} = \mathbf{X} \cdot \mathbf{b} + \mathbf{u}, \tag{2.21}$$

where $\mathbf{y}$ = vector of dependent variable observations
$\mathbf{X}$ = matrix of observations on the independent variables

$\mathbf{b}$ = vector of parameters
$\mathbf{u}$ = stochastic error term vector.

Information on the dependent and independent variables represents the observation set, and the observed values are taken to be a sample drawn from the population under analysis; but, as was mentioned in regard to one advantage of time-series techniques in the substitution of the concepts of theoretical process and realization for those of population and sample, experiments generally are not possible when the modeler analyzes economic issues, since economics is not a laboratory science. Consequently, additional sampling is not possible, and econometric methods often conveniently ignore this shortcoming. The model parameters **b** are the "unknowns" to be determined statistically. The assumption of a stochastic component and the provision for more than a single explanatory variable preclude a simple algebraic solution of the relationships. Multiple regression or OLS provides a method for deriving estimates of these in regard to the parameters of the model.[23] The objective function can be specified as

$$\text{Minimize } \mathbf{U} = \sum_{i=1}^{n} (\mathbf{y}_i - \sum_{j=1}^{m} \beta_j \cdot \mathbf{x}_{ij})^2 , \qquad (2.22)$$

where $\mathbf{U}$ = sum of squares of the error terms, u_i
n = number of observations
m = number of exogenous variables
β = unknown parameters.

It can be seen from Equation 2.22 that the error term $\mathbf{u}_i$ is the square of the difference between the observed value of the endogenous variable $\mathbf{y}_i$ and the relationship, as depicted by the β_j terms, on the exogenous variables x_{ij}. In the two-dimensional case, the OLS objective function minimizes the vertical distance between the observed values of the dependent variable; by tradition, economists graph this on the vertical axis, and the independent variable on the horizontal axis.

Each observation describes a realization of the stochastic process, and the error term of each is assumed to be normally distributed and to have a mean of 0 and a variance of σ^2. The calculus is used to find the solution to this minimization problem, by taking the first derivative of the equation with respect to each of the independent variables, setting these derivatives equal to 0, and solving the resulting set of simultaneous equations for the unknown β parameters. The derivatives of the quadratic function presented in Equation 2.22 are linear relationships, and these are referred to as the set of normal equations of multiple regression analysis. In matrix notation the normal equations are

$$(\mathbf{X}^{\mathbf{T}} \cdot \mathbf{X}) \cdot \mathbf{b} = \mathbf{X}^{\mathbf{T}} \cdot \mathbf{y}, \qquad (2.23)$$

where $\mathbf{T}$ = matrix transpose operator.

The solution of Equation 2.23 in terms of the unknown parameters is then

$$\mathbf{b} = (\mathbf{X}^{\mathbf{T}} \cdot \mathbf{X})^{-1} \cdot \mathbf{X}^{\mathbf{T}} \cdot \mathbf{y}. \qquad (2.24)$$

The various assumptions made about the stochastic error (or disturbance) term result in best linear unbiased estimates (BLUE) for the parameters; that is to say, no other linear procedure will provide consistent estimates with smaller variances for the coefficients. This attractive property of the OLS model is premised on the following five assumptions relating to the distribution of the disturbance term and the nonstochastic nature and independence of the regressors:[24]

$$E(\mathbf{u}) = 0 \qquad (2.25)$$

$$E(\mathbf{u}_t^2) = \sigma^2, \text{ for all t} \qquad (2.26)$$

$$E(\mathbf{u}_t \cdot \mathbf{u}_s) = 0, \text{ for all } s \neq t \qquad (2.27)$$

$\mathbf{X}$ is a T-by-K matrix that is fixed in repeated samples (2.28)

The rank of $\mathbf{X} = K \leqslant T$. (2.29)

Condition 2.25 is the assumption that the estimated relationship is unbiased; the mean of the stochastic error term is 0. The next two requirements are that the error term be homoscedastic; that is, the variance of the error term is constant across observations and there exists no pairwise autocorrelation among observations. In the instance of time-series observations, this latter condition is referred to as the "absence of serial correlation." The assumption that a stable relationship (nonstochastic) between the endogenous variable and exogenous variables exists over time and across observations is expressed in condition 2.28. This postulate can represent a shortcoming whenever econometric methods are applied to the analysis of the effects of major shocks to the system under study. The final assumption is that no exact linear relationship exists among the regressors; that is, the level of collinearity is 0 [see Goldberger (1964, chap. 4, sect. 3) for a complete presentation of these conditions]. Briefly, the major criteria, outlined above, with respect to the error term are (1) its mean is 0 and variance a constant; (2) no correlation exists between paired combinations of error terms; (3) the observation set is stable from sample to sample; and (4) the matrix is not singular and therefore can be inverted. In addition to these assumptions required for the parameter estimates to be considered BLUE, the variables selected in the modeling, which represent their theoretical counterparts, are assumed to be free of observation errors [see Morgenstern (1963) for a most informative discussion of the accuracy of economic data].

Numerous testing procedures have been developed in order to identify the sources of "single" violations of the BLUE requirements. An example is the Durbin-Watson (DW) test statistic (1950) [or Durbin's **h** statistic (1970) when a lagged dependent variable is included in the explanatory variable set] for first-order serial correlation, which assumes the presence of normality, homoscedasticity, and a correct specification of the functional form. Durbin's **h** statistic can be derived as a Lagrangian multiplier (LM) statistic having a chi-square distribution (see Breusch and Pagan 1980, p. 244). Bera and Jarque (1981) propose an additive statistic based upon the LM principle as an alternative to the single type of test, such as Durbin's, to test simultaneously for normality, homoscedasticity, serial independence, and functional form. They argue that "one-directional" tests that assume these conditions lack robustness, and therefore the results are suspect in the presence of more than one type of model misspecification. In addition, experiments by Granger and Newbold (1974) demonstrate that the presence of serial correlation can lead to artificially high values for the coefficient of multiple determination $\mathbf{R}^2$, and that erroneous conclusions might consequently ensue. Thus, additional problems arise even when only a single violation of the BLUE conditions occurs. The general consequence of an autocorrelated error term is that the regression parameter estimates are inefficient, albeit unbiased and consistent: Coefficient estimates are unduly large, variances are biased, and suboptimal forecasts result.[25] The existence of heteroscedasticity likewise leads to unbiased and consistent but inefficient parameter estimates (see Kmenta 1971, chap. 8). The presence of these problems in regard to the error term renders the usual statistical tests invalid, but adjustment procedures are available to the analyst to correct for individual problems.

In the instance of autocorrelation, Cochrane and Orcutt (1949) propose a correction for first-order serial correlation. When $p = 1$, Equation 2.30 depicts the Cochrane-Orcutt form of AR relationship:[26]

$$u_t = \sum_{i=1}^{p} \rho_i \cdot u_{t-i} + \epsilon_t , \tag{2.30}$$

where ϵ = white noise.

The condition for a nonexplosive solution of the Cochrane-Orcutt AR model is that the roots of the polynomial equation,

$$X^p - \sum_{i=1}^{p} \varphi_i \cdot X^{p-i} = 0,$$

all lie in the unit circle (Maddala 1977, p. 275). An alternative form for the autocorrelation structure is that of an MA scheme:[27]

$$u_t = \epsilon_t + \sum_{i=1}^{q} \theta_i \cdot \epsilon_{t-1} . \tag{2.31}$$

The condition for a stable solution for the MA model is subject to

$$\sum_{i=1}^{q} \theta_1^2 < \infty .$$

Pagan (1974) suggests a generalized approach to the treatment of autocorrelation to account for both the AR and the MA structures, and he points out that any AR process can be modeled by a higher-order MA process and vice versa, as was noted in the discussion of the time-series models above. Thus, the handling of correlation problems associated with disturbance terms is similar for econometricians and for time-series analysts, although the basic theoretical foundations do differ. Time-series analysts view the problem of autocorrelation merely as requiring correction through statistical techniques, while econometricians argue that it is the misspecification that needs to be identified.[28]

Another inherent problem in economic data is that of collinearity (a condition also referred to as "multicollinearity" or "ill conditioning"). While the existence of an exact linear relationship among the regressors is admittedly rare, and is assumed away by Equation 2.29, its occurrence means that the $\mathbf{X}^T\mathbf{X}$ matrix is singular and therefore cannot be mathematically inverted. A common situation is that some degree of interdependence occurs among the explanatory variables. By their very nature many economic data series move together temporally, either in the same or in the opposite direction. Depending upon the severity of the collinearity, the analyst may not be able to identify accurately the individual explanatory variable relationships with respect to the endogenous variable. Often in a forecasting context it is assumed that this is not a major problem if the collinear relationship is expected to remain invariant over time. However, if the nature of the collinearity is expected to vary, then it will pose a problem even in a forecasting setting, and in such an instance collinearity can lead to incorrect conclusions. A gross measure of the degree of collinearity, suggested by Farrar and Glauber (FG) (1967), is to calculate the coefficient of multiple determination, the $\mathbf{R}^2$, for each of the exogenous variables regressed against all of the other exogenous variables simultaneously. These $\mathbf{R}^2$ values are then compared with the one obtained from the Equation 2.24 estimate.[29] A statistic proposed by Haitovsky (1969) considers the degree of deviation from perfect singularity; however, as Kumar (1975) points out, the assumption inherent in the FG procedure is that the data matrix $\mathbf{X}$ is stochastic. This latter condition would necessarily violate the assumption made in Equation 2.28 (see Belsley, Kuh, and Welsh 1980). In other words, the results are no longer invariant among repeated samples when the matrix is stochastic. Therefore, reliance on the FG or Haitovsky measures of collinearity implies the violation of one of the necessary conditions for parameter estimates to be BLUE, and thus the OLS approach is strictly interpreted as inappropriate.

Economic theory says little about the specific mathematical form of demand or other relationships, but it has become convention to assume linearity either in the actual levels of the data or in the logarithms of the observations. The implication of such a hypothesis is that the linear relationship holds within the range of values relevant to the observed data and to the potential forecast values, if forecasting is an objective. This simplifying assumption reflects to some extent the convergence problems associated with solving nonlinear equations. Box and Cox (BC) (1964) recommend a transformation of the data that can be used to test for alternative functional forms. The transformation they suggest is

$$y_i^{(\lambda)} = \frac{y_i^{\lambda} - 1}{\lambda}, \qquad \text{for } \lambda \neq 0 \tag{2.32}$$

$$y_i^{(\lambda)} = \ln(y_i), \qquad \text{for } \lambda = 0 . \tag{2.33}$$

The introduction of the additional parameter λ, which is estimated simultaneously with the unknown β values, allows for identification of some specific functional forms, but inevitably increases the problem of collinearity (see Box and Cox 1982; Bickel and Doksum 1981). For example, whenever the estimated value of λ is not statistically different from 1, a linear model specification seems appropriate, and if the estimate is not statistically different from 0, the logarithmic form is suggested by the data. Other values for λ imply alternative functional forms (see the discussion in Chapter 1). Zellner (1971, sect. 6.1) discusses the BC transformation and shows how the values of the log-likelihood function may be used in identifying the most appropriate form of the equation.[30] He suggests a statistic in the form of a multivariate Student's t distribution with which to study the sensitivity of the inferences about the parameters in reference to different assumptions about the value of λ. It is only recently that this transformation has been used in empirical analyses to identify functional form in light of some of the other potential problems associated with the disturbance term (see Seaks and Layson 1983; Ericsson 1982; Savin and White 1978). Since many economists still rely upon linear formulations in demand analyses, this maintained hypothesis of linearity fails under scrutity for the same reasons as does the "constant-coefficient" assumption of the classic I-O model: Namely, whenever major shocks occur to the system or the direction of movement in a variable changes, then questions of linearity and symmetry of behavior may be relevant. It may be reasonable to accept the idea that the elasticity of demand is one value with respect to a price decrease and another value for a price increase (recall that this type of behavior is the basic concept underlying the "kinked demand curve"), and some attempts have been made to develop reasonable model formulations that provide for variable elasticities of demand linked to the level or direction of change in explanatory variables (see Betancourt 1981). As will be demonstrated in Chapter 3 on electricity demand modeling, careful attention must, however, be paid to the proper specification of such variable elasticity of demand models so that the interpretation of the elasticity measure is not invalidated (see Donnelly and Diesendorf 1985).

The burdens of econometric modeling should not pose a problem to the analyst, because numerous computational packages are readily available.[31] Econometric models are somewhat less data intensive than I-O or mathematical programming models, but they necessitate employing a theoretical context, an aspect missing in those other approaches. In energy analysis, however, the availability of data will still pose a problem, as will be seen in the models presented in subsequent chapters. Also, the quality of the data is a concern, since proxy variables have to be defined carefully when information cannot be obtained for the particular influential variable that theory specifies that the model require. Specification of the appropriate functional form and its estimation, if it is nonlinear, may also be a problem. Finally, if an econometric model is to be utilized in a forecasting context, then some mechanism must be provided for forecasting the model's explanatory variables. Time-series techniques might be employed to serve this end, as well as the approach of scenario analysis, which is discussed in one of the models presented in Chapter 3 on electricity modeling.

Most of the modeling techniques discussed suffer from the problem brought about by changes that occur outside the range of values in the observation set. Mathematical programming is the possible exception to this rule, in that it attempts to incorporate explicit information on new, unused technologies. However, in long-range forecasting, even this approach is faced with uncertainty, and analysts are forced to rely upon Delphi techniques.

HYBRID MODELS

The final modeling strategy discussed is the hybrid method, which involves applying various combinations of the preceding techniques, such as

I-O tables and econometric methods
Econometric methods and L-P models
Econometric methods and time-series analysis

It should be emphasized that these hybrid approaches are combinations of techniques and do not represent syntheses. An example of the use of econometrics to develop the I-O coefficients for a dynamic I-O model is INFORUM, developed by Almon et al. (1974). Dynamic I-O coefficients derived econometrically improve the I-O model's forecasting ability when compared with the static fixed-coefficient (purely descriptive) approach. Hudson and Jorgenson (1974) extend this strategy in their mixed I-O and econometric model of national energy demand. They derive the I-O coefficients as endogenous variables obtained from the dual of a set of production functions defined as a transcendental logarithmic, or translog, "price possibilities" frontier. The demand functions obtained from the translog estimation are used to calculate the input elasticities of substitution incorporated in the I-O table.

Another hybrid energy model is PIES/ERDM developed by DOE (see FEA 1976, Appendixes A and C; Borg et al. 1975). The PIES/ERDM model considers separately the supply and demand sides of the energy market using different modeling strategies for each component. PIES is a large-scale L-P model of the U.S. energy supply system, depicting the different technologies that can be utilized to satisfy the economy's demand for energy. The energy demands are provided by ERDM. An iterative, rather than simultaneous, solution is derived, using preliminary energy price information in the econometric demand model to generate initial energy demands. These "first-round" demand estimates are fed to the PIES model, which furnishes fresh estimates of energy prices–these are the opportunity costs or shadow prices–obtained from the L-P solution. The shadow prices are then used in ERDM to construct a revised set of energy demands. The procedure is continued until the prices and quantities provided by the two models converge, although convergence is merely assumed and not ensured.

A "mixed" time-series/econometric model of monthly electricity demand is presented by Uri (1977). In this paper the peak demand is modeled using a Box-Jenkins time-series approach with the ARIMA parameters updated at quarterly intervals; for forecasting purposes, these ARIMA parameters are "explained" econometrically as functions of the price of electricity, consumer incomes, and climatic variables. In a subsequent paper, Uri (1979) compares an alternative "combined" modeling approach with his former results. This time, econometric techniques are utilized first to estimate the demand relationship, with the time-series analysis employed to explain the econometric model's disturbance term.[32] Uri concludes that both of these hybrid modeling strategies outperform either the time-series or the econometric methods used singly. Although his mixed approach performs better than the combined approach, Uri argues that the latter may still be preferred because of its ease of calculation, given the existing state of computer modeling packages.

In summary, dynamic I-O models using econometrically derived coefficients are an advance over the purely static I-O model, but represent a significant increase in the modeling effort that must be expended. The mathematical programming approach, because of its flexibility, can adjust for a wide range of alternative scenarios and, when combined with econometric demand estimates, may include information on the behavioral response of consumers. This strategy allows for changing technology and changing demands, these being reflected in the factor input mix and the output price and product mix. Again, the modeling costs are greatly increased in this form of hybrid model, and the feedbacks between the models are iterative and convergence is not ensured.[33] Computational limitations currently mean that models as large and complex as PIES/ERDM do not allow for simultaneous solutions.

A final important issue in considering alternative model strategies is that of data. For a hybrid model such as the Hudson-Jorgenson one, the data requirements are severe. The research of the aforementioned authors was facilitated by the availability of a temporal series of I-O tables for the United States (see Chapter 1). Unfortunately, up-to-date interindustry data are not always so readily available. Similarly, the requisite modeling data needed for estimating production functions generally have to be developed from scratch, but this approach, while technically feasible, is costly. Many authors recognize that high-quality models are an impossible objective without good-quality data: See Moses (1981) and, overseas, Harris (1982),

Hall (1984), and James (1980), who all deplore the existing state of energy data and strongly advocate that renewed emphasis be placed on extending the price and quantity data used in energy models.

The hybrid approaches seen to date have been developed in an attempt to offset the weaknesses of one of the modeling strategies against the strengths of another. As always, the objectives of the modeling exercise will determine the appropriate approach. However, the degree of art involved in modeling increases exponentially as the techniques are combined.

SUMMARY

Traditional modeling approaches need to be reviewed for their appropriateness, and it should be recognized that modelers have personal biases in the selection among the alternative strategies. This is not necessarily bad; however, the failure to consider other modeling approaches can often hinder the decision-making process rather than contribute to it. Unfortunately, modeling is not a single discipline, but one with practitioners from many disparate fields—a fact that is evident in several of the hybrid models mentioned above. Maturity in modeling for policy analysis purposes will necessitate a synergistic evolution of the techniques being currently employed by the various independent disciplines involved when decision makers address policy issues. Subsequent chapters consider several econometric models of energy demand that have been developed for use to assist in policy analysis. These include models of

Electricity demand
Gasoline demand
Industrial energy demand

It should be remembered in considering the models presented that the appropriate modeling strategy differs for each particular policy question. For example, a static partial-equilibrium single-equation approach can provide sufficient insight into the response of the market when the interest is consumer reaction to changes in the price of the good. When this is the case, that simpler approach should be adopted rather than a more elegant dynamic general-equilibrium multiequation model. Thus, the following econometric models will be discussed in the context of their applicability in addressing specific questions of energy policy import.

NOTES

1. As mentioned earlier, the systems dynamics models utilize this assumption of constant rates of growth, achieving a seeming complexity of interactions among several deterministically forecasted series from which insights can supposedly be derived

2. The production function is of the form

$$Q_i = X_i^{a_i} (b_i \cdot E_1^{c_i} + (1 - b_i E_2^{c_i})^{(1 - a_i)/c_i},$$

where Q_i = output of the ith firm
X_i = nonenergy input
E_1, E_2 = energy inputs
a_i, b_i, c_i = parameters.

3. The model forms tested are

$$S_i = \alpha_i + \sum_{j=1}^{n} \beta_{ij} \cdot \ln P_j + \epsilon_i, \qquad \text{for the translog}$$

$$\log \frac{S_i}{S_e} = \alpha_0 + \alpha_1 \cdot \log \frac{P_i}{P_e} + u, \qquad \text{for the log-odds}$$

$$S_i = \frac{P_i^{-\gamma}}{\sum_{j=1} P_j^{-\gamma}}, \qquad \text{for the probabilistic}$$

where S_i = share of the ith input
P_i = price of the ith input
P_e = price of the energy input
ϵ_i, u = stochastic error terms.

4. There are stochastic forms of mathematical programming models. However, these are not in widespread use by policy analysts and therefore will not be discussed here.

5. Systems analysis includes mathematical programming, cybernetics, queuing theory, graph theory, and decision theory among other separate disciplines.

6. The development of the I-O technique was the basis for Wassily Leontief's 1973 Nobel Prize in Economic Science. See Isard (1960, chap. 8) and Almon (1967, chap. 1) for good discussions of I-O modeling procedures.

7. As Dorfman, Samuelson, and Solow (1958) indicate, the "theory of input-output can also be regarded as a peculiarly simple form of linear programming" (p. 204).

8. James (1980) outlined a similar energy I-O matrix for Australia, and Karunaratne (1981) constructed one based upon a 1974/75 I-O table. The effects of technical change and interfuel substitution are ignored, on the assumption that the long lead times imply stable technical coefficients. However, as will be shown in Chapter 5, by the elasticity of substitution figures in the iron and steel industry that contention of stable coefficients in effectively refuted.

9. The I-O coefficients are

	Agriculture	Manufacturing	Services	Energy
Agriculture	0.080	0.162	0.077	0.286
Manufacturing	0.040	0.270	0.128	0.429
Services	0.000	0.135	0.308	0.143
Energy	0.040	0.108	0.026	0.214

10. It can be seen from the I-O coefficients in Table 2.2 that a final energy demand of 15 implies total demand for all sectors of 31.7, 46.8, 43.9, and 28.6, and a unit increase in that final demand would result in total demands of 32.3, 47.8, 44.4, and 30.0, respectively. In this case, the percentage changes are 2.1, 2.1, 1.1, and 5.1, respectively. Likewise, for the initial level of 115, the initial solution would be 98.3, 144.7, 93.1, and 174.3, while the aftershock levels would be 99.9, 145.7, 93.6, and 175.8 (or 0.7, 0.7, 0.5, and 0.8 percent), respectively.

11. References include Almon (1967); Baumol (1977); Dantzig (1963); Dorfman, Samuelson, and Solow (1958); and Hadley (1962, 1964).

12. These equations will collapse to the I-O form when the system has been solved.

13. The I-O table relationships, which include only a single technological relationship for each sector, may be thought of as the solution to some unspecified optimization problem. The I-O coefficients are assumed to be fixed, and this constancy is a major shortcoming. The mathematical programming approach allows for alternative technologies to be introduced whenever required by changes in the economic relationships.

14. Information about new technologies is sometimes obtained using the Delphi approach, wherein a consensus of experts is sought in an attempt to define the most probable technological developments expected to become available in the future.

15. Conversely, I-O modeling has been used extensively for economic forecasting as well as for energy analysis (see Dixon et al. 1982; Miernyk, Giarratani, and Socher 1978; Almon et al. 1974; Hannon 1973).

16. MARKAL is the latest of a series of optimization models developed by the Brookhaven National Laboratory. These models began with the "snapshot," a purely descriptive RES, and are referred to variously as the Brookhaven Energy System Optimization Model (BESOM) (Hoffman 1973), the Dynamic

Energy System Optimization Model (DESOM), and the Time-Stepped Energy System Optimization Model (TESOM) (Kydes and Rabinowitz 1979).

17. An updating procedure that has intuitive appeal is the one recently instituted by the Australian Bureau of Statistics, which utilizes survey flow data on one-third of the I-O table sectors to adjust the existing coefficients. This process operates on a three-year cycle, *but it ensures that the interindustry coefficients are wrong in two-thirds of the industry sectors at every point in time.*

18. Many texts discuss time-series analysis in greater detail than will be presented here, and the reader is referred to them: See Johnston (1984), Pindyck and Rubinfeld (1981), Granger and Newbold (1977), Nelson (1973), and Box and Jenkins (1970).

19. As Hannan (1960) explains heuristically, the import of this assumption to time-series analysis is that "one is inclined to believe that, shall we say, the behavior of the waves that beat upon the shores some age ago is totally unrelated to that of the waves that beat today and that this is true of all phenomena which those waves determined" (pp. 202–3).

20. A random walk is depicted by $\mathbf{x}_t = \mathbf{x}_{t-1} + \mathbf{u}_t$, where $\mathbf{u}_t$ is normally distributed with a mean of 0 and a variance of σ^2, denoted as $\mathbf{u}_t = N(O, \sigma^2)$. This means that the expected value of x_t is $E(\mathbf{x}_t) = E(\mathbf{x}_{t-1}) + E(\mathbf{u}_t)$; and since $E(\mathbf{u}_t) = 0$, then $E(\mathbf{x}_t) = E(\mathbf{x}_{t-1})$. The variance of x is then merely var $(\mathbf{x}_t) = \sigma^2_u$.

21. The drift, or trend, portion is merely added to the random walk equation, yielding $\mathbf{x}_t = \mathbf{x}_{t-1} + \mathbf{u}_t + \delta$ (see Nelson 1973, p. 11).

22. Any of these texts will provide a comprehensive discussion of econometric methods: Johnston (1984), Theil (1978), Maddala (1977), Kmenta (1971), and Goldberger (1964).

23. Classic regression analysis or OLS provides estimates of the unknown parameters by minimizing the sum of squared deviations of the stochastic error term about the computed regression line, that is, the vertical distance. Alternative objective functions might be to specify to minimize the horizontal distance or minimize the perpendicular distance. This latter is the instance of factor analysis, which is, among other things, an orthogonal regression minimizing

$$\mathbf{U} = \sum_{i=1}^{n} (\mathbf{A} \cdot \mathbf{x}_t - b)^{\mathbf{T}} (\mathbf{A} \cdot x_t - b), \text{ subject to } \mathbf{A} \cdot \mathbf{A}^{\mathbf{T}} = I.$$

(see Malinvaud (1980, pp. 36–39). (Psychological testing utillzes orthogonal regression to calculate intelligence quotients.)

24. *E* ( ) is statistical notation standing for the expected value of a variable.

25. Efficiency refers to minimizing the variance of the estimates of the parameters.

26. Note that this is similar to the time-series AR process described in Equation 2.14.

27. This is analogous to the time-series process in Equation 2.15.

28. Recall that Kmenta (1971) enumerates five types of potential misspecification (see the discussion in Chapter 1).

29. Farrar and Glauber (1967) argue that the use of the multiple correlation coefficient provides more information about the severity of collinearity than does merely analyzing simple and partial correlation coefficients.

30. The name "likelihood function" is given to the joint probability distribution of a sample. A class of estimators referred to as maximum likelihood estimators (MLEs) have as their objective function the selection of the parameter estimates that will maximize the joint probability of the sample; that is to say, the MLE provides those parameters that would generate the observed sample most often. If the conditions specified by Equations 2.25 through 2.29 are satisfied, then the OLS estimates are both BLUE and MLE as well (Kmenta 1971, pp. 174–82, 202, 247–48). The logarithmic transformation of a likelihood function is monotonic and facilitates the calculus required for optimization.

31; Among these are TSP, SHAZAM, and Fedeasy/Speakeasy, and TROLL for mainframe computers; and Micro-TSP, ESP, GAUSS, RATS, and xSTAT for microcomputers, to cite but a few of the readily available econometric packages.

32. This ARIMA model with a linear transfer function causes the econometric model's parameters to be inconsistent because conditions 2.25 to 2.27 are violated.

33. A description of the iterative solution method employed in the PIES/ERDM modeling is found in Askin (1976). DeSouza (1981, chap. 7) presents a discussion of the more recent MEFS model of DOE, which incorporates a macroeconomic model into the iterative process.

REFERENCES

Almon, C. 1967. *Matrix Methods in Economics.* Reading, MA: Addison-Wesley.

Almon, C., M. R. Buckler, L. M. Horowitz, and T. C. Reimbold. 1974. *1985: Industry Forecasts of the American Economy.* Lexington, MA: Lexington Books.

Askin, A. B. 1976. "The Macroeconomic Implications of Alternative Energy Scenarios." In *Econometric Dimensions of Energy Supply and Demand,* edited by A. B. Askin and J. Kraft, pp. 99–109. Lexington, MA: Lexington Books.

Baumol, W. J. 1977. *Economic Theory and Operations Analysis.* 4th ed. Englewood Cliffs, NJ: Prentice-Hall.

Belsley, D. A., E. Kuh, and R. E. Welsh. 1980. *Regression Diagnostics: Identifying Differential Data and Sources of Collinearity.* New York: John Wiley & Sons.

Bera, A. K., and C. M. Jarque. 1981. "Tests for Specification Errors: A Simultaneous Approach." Paper presented at the Tenth Conference of Economists of the Economic Society of Australia and New Zealand, Canberra, August 24–28. Mimeo.

Berndt, E. R., and D. O. Wood. 1975. "Technology, Prices and the Derived Demand for Energy." *Review of Economics and Statistics,* 57:258–68.

Betancourt, R. R. 1981. "An econometric Analysis of Peak Electricity Demand in the Short Run." *Energy Economics,* 3:14–29.

Bickel, P. J., and K. A. Doksum. 1981. "An Analysis of Transformations Revisited." *Journal of the American Statistical Association,* 76:296–311.

Borg, S., W. A. Donnelly, T. Eagan, D. Knapp, and D. H. Nissen. 1975. "The Econometric Regional Demand Model–ERDM." Energy Information and Analysis Technical Report No. EIATR 75-18. Washington, D.C.: U.S. FEA.

Box, G. E. P., and D. R. Cox. 1982. "An Analysis of Transformations Revisited, Rebutted." *Journal of the American Statistical Society,* 77:209–10.

____. 1964. "An Analysis of Transformations." *Journal of the Royal Statistical Society,* B26:211–43.

Box, G. E. P., and B. M. Jenkins. 1970. *Time Series Analysis, Forecasting and Control.* San Francisco: Holden-Day.

Breusch, T. S., and A. R. Pagan. 1980. "The Lagrangian Multiplier Test and Its Application to Model Specifications in Econometrics." *Review of Economic Studies,* 47:239–53.

Bullard, C. W., and R. A. Herendeen. 1977. "The Energy Cost of Goods and Services: An Input-Output Analysis for the U.S.A., 1963 and 1967." In *Energy Analysis,* edited by J. A. G. Thomas, pp. 71–81. Surrey: IPC Science and Technology Press.

Cochrane, D., and G. H. Orcutt. 1949. "Application of Least Squares Regressions to Relationships Containing Autocorrelated Error Terms." *Journal of the American Statistical Association,* 44:32–61.

Dantzig, G. B. 1963. *Linear Programming and Extensions.* Princeton: Princeton University Press.

DeSouza, G. R. 1981. *Energy Policy and Forecasting: Economic, Financial and Technological Dimensions.* Lexington, MA: Lexington Books.

Dixon, P. B., B. R. Parmenter, J. Sutton, and D. Vincent. 1982. *ORANI, A General Equilibrium Model of the Australian Economy.* Amsterdam: North-Holland.

Donnelly, W. A., and M. Diesendorf. 1985. "Variable Elasticity of Demand Models for Electricity." *Energy Economics,* 7:159–62.

Dorfman, R., P. A. Samuelson, and R. M. Solow. 1958. *Linear Programming and Economic Analysis.* New York: McGraw-Hill.

Durbin, J. 1970. "Testing for Serial Correlation in Least Squares Regression When Some of the Regressors Are Lagged Dependent Variables." *Econometrica,* 38:410–21.

Durbin, J., and G. S. Watson. 1950. "Testing for Serial Correlation in Least Squares Regressions." *Biometrika,* 37:409–28.

Ericsson, N. R. 1982. "Testing Linear Versus Logarithmic Regression Models: A Comment." *Review of Economics and Statistics,* 69:477–81.

Farrar, D. E., and R. R. Glauber. 1967. "Multicollinearity in Regression Analysis: The Problem Revisited." *Review of Economics and Statistics,* 49:92–107.

Goldberger, A. S. 1964. *Econometric Theory.* New York: John Wiley & Sons.

Granger, C. W. J., and P. Newbold. 1977. *Forecasting Economic Time Series.*

New York: Academic Press.

——. 1974. "Spuriuous Regressions in Econometrics." *Journal of Econometrics,* 2:111–20.

Griffin, J. M., and H. B. Steele. 1980. *Energy Economics and Policy.* New York: Academic Press.

Hadley, G. 1964. *Nonlinear and Dynamic Programming.* Reading, MA: Addison-Wesley.

——. 1962. *Linear Programming.* Reading, MA: Addison-Wesley.

Haitovsky, Y. 1969. "Multicollinearity in Regression Analysis: Comment." *Review of Economics and Statistics,* 51:486–89.

Hall, V. B. 1984. "Some Thoughts on Energy Modeling and Policy in Australia." *Australian Economic Papers,* 3:21–36.

Hannan, E. J. 1960. *Multiple Time Series.* New York: John Wiley & Sons.

Hannon, B. 1973. "An Energy Standard of Value." *Annals of the American Academy of Political and Social Science,* 410:139–53.

Harris, S. F. 1982. "Social Aspects of Energy in Australia: A Social Science Literature and Research Review." In *Liquid Fuels in Australia: A Social Science Research Perspective,* edited by J. Black, pp. 7–81. Sydney: Pergamon Press.

Hoffman, K. C. 1973. "A Linear Programming Model of the Nation's Energy System." Report No. BNL ESAG-4. Upton, NY: Brookhaven National Laboratory.

Hudson, E. A. 1981. "Modeling Production and Pricing Within an Interindustry Framework." In *Energy Policy Planning,* edited by B. A. Bayraktar, E. A. Cherniavsky, M. A. Laughton, and L. E. Ruff, pp. 201–14. NATO Advanced Research Institute on the Application of Systems Science to Energy Policy Planning, New York, 1979. New York: Plenum Press.

Hudson, E. A., and D. W. Jorgenson. 1974. "U.S. Energy Policy and Economic Growth, 1975–2000." *Bell Journal of Economics and Management Science,* 5:461–514.

Isard, W. 1960. *Methods of Regional Analysis: An Introduction to Regional Science.* Cambridge, MA: MIT Press.

James, D. E. 1980. "A System of Energy Accounts for Australia." *Economic Record,* 56:171–81.

Johnston, J. 1984. *Econometric Methods.* 3rd ed. New York: McGraw-Hill.

Karunaratne, N. D. 1981. "An Input-Output Analysis of Australian Energy Planning Issues." *Energy Economics,* 3:159–68.

Kmenta, J. 1971. *Elements of Econometrics.* New York: Macmillan.

Kumar, T. K. 1975. "Multicollinearity in Regression Analysis." *Review of Economics and Statistics,* 57:365–66.

Kydes, A. S., and J. Rabinowitz. 1979. "The Time-Stepped Energy System Optimization Model (TESOM): Overview and Special Features." Upton, NY: Brookhaven National Laboratory.

Maddala, G. S. 1977. *Econometrics.* New York: McGraw-Hill.

Malabre, A. L. 1986. "Kondratieff Rolls on, As Does the Economy." *Wall Street Journal,* 66(January 20):1.

Malinvaud, E. 1980. *Statistical Methods of Econometrics.* 3rd rev. ed. Amsterdam: North-Holland.

Miernyk, W. H., F. Giarratani, and C. F. Socher. 1978. *Regional Impacts of Rising Energy Prices.* Cambridge, MA: Ballinger.

Morgenstern, O. 1963. *On the Accuracy of Economic Observations.* 2nd ed. Princeton: Princeton University Press.

Moses, L. E. 1981. "Keynote Address: One Statistician's Observation Concerning Energy Modeling." In *Energy Policy Planning,* edited by B. A. Bayraktar, E. A. Cherniavsky, M. A. Laughton, and L. E. Ruff, pp. 17–33. NATO Advanced Research Institute on the Application of Systems Science to Energy Policy Planning, New York, 1979. New York: Plenum Press.

Nelson, C. R. 1973. *Applied Time Series Analysis for Managerial Forecasting.* San Francisco: Holden-Day.

Pagan, A. R. 1974. "A Generalized Approach to the Treatment of Autocorrelation." *Australian Economic Papers,* 13:267–80.

Pindyck, R. S., and D. L. Rubinfeld. 1981. *Econometric Models and Economic Forecasts.* 2nd ed. New York: McGraw-Hill.

Savin, N. E., and K. J. White. 1978. "Estimation and Testing for Functional Form and Autocorrelation." *Journal of Econometrics,* 8:1–12.

Seaks, T. G., and S. K. Layson. 1983. "Box-Cox Estimation with Standard Econometric Problems." *Review of Economics and Statistics,* 65:160–64.

Smith, V. K., and L. J. Hill. 1985. "Validating Allocation Functions in Energy Models: An Experimental Methodology." *Energy Journal,* 6:29–47.

Theil, H. 1978. *Introduction to Econometrics.* Englewood Cliffs, NJ: Prentice-Hall.

Turnovsky, M. H. L., and W. A. Donnelly. 1984. "Energy Substitution, Separability and Technical Progress in the Australian Iron and Steel Industry." *Journal of Business and Economic Statistics,* 2:54–63.

Turnovsky, M. H. L., M. Folie, and A. Ulph. 1982. "Factor Substitutability in Australian Manufacturing with Emphasis on Energy Inputs." *Economic Record,* 58:61–72.

Uri, N. D. 1979. "A Mixed Time-Series/Econometric Approach to Forecasting Peak System Load." *Journal of Econometrics,* 9:155–71.

——. 1977. "An Integrated Box-Jenkins/Econometric Model for Forecasting Time Series." *Proceedings of the American Statistical Association,* 1: 404–7.

U.S. Federal Energy Administration. 1976. *National Energy Outlook.* FEA-N75/ 713. Washington, D.C.: U.S. Government Printing Office.

Webb, M. G., and M. J. Ricketts. 1980. *The Economics of Energy.* New York: John Wiley & Sons.

Zellner, A. 1971. *An Introduction to Bayesian Inference in Econometrics.* New York: John Wiley & Sons.

Three

Electricity Demand Modeling

OVERVIEW

The previous chapter looked closely at the several types of modeling approaches used in policy analysis. This chapter considers econometric techniques as applied to analyzing electricity consumption behavior. Alternative static and dynamic model specifications are considered, and tests of functional forms are presented. The use of scenarios in energy forecasting is illustrated, with both point and interval elasticity estimates calculated. The results of several electricity demand models are discussed, and areas deserving of future research are suggested.

INTRODUCTION

The factors affecting electricity demand are still poorly understood within the electricity generation and supply industry.[1] This, to a large extent, is the natural consequence of a group of professional engineers quite correctly concentrating their attention on the physical determinants of electricity demand, an area in which they enjoy a comparative advantage, and disregarding the socioeconomic factors in areas outside their field of expertise. Presumably, this is an area in which economists might be able to assist, since they accept as a working hypothesis that factors such as price and income will affect the demand for a commodity; and econometric models include these latter as explanatory variables, along with purely physical factors such as climatic influences, with such hypotheses being put to the statistical test. Electricity demand analyses can be divided into short- and long-run modeling. The former category is concerned with "explaining" diurnal and seasonal variations, or load factors, whereas the latter emphasizes the longer-term behavioral responses. The general functional form specified to identify the long-run behavioral response is

$$\mathbf{Q}_E = f(\mathbf{P}_E, \mathbf{P}_S, \mathbf{P}_C, \mathbf{Y}, \mathbf{X}), \tag{3.1}$$

where $\mathbf{Q}_E$ = quantity of electricity
$\mathbf{P}_E$ = electricity price
$\mathbf{P}_S$ = price of substitute good
$\mathbf{P}_C$ = price of complementary good
$\mathbf{Y}$ = consumer income
$\mathbf{X}$ = other factors.

As was discussed in Econometric Methods in Chapter 2, this hypothesized relationship can be defined to have various specific functional forms for estimation. Most econometric work on electricity demand to date has relied upon either the linear or the log-linear model with some attention directed toward incorporating time-series procedures in the analyses. A nonlinear model tested by Betancourt (1981) will be redefined so that the measures of elasticity retain their conventional interpretation and the results compared with a general nonlinear model specification. But first some other models will be reviewed, and for this some background information is provided herein.

In Australia, for example, electricity generation and supply are the responsibility of the individual states, and currently all of the generating authorities there are government-owned and -operated entities. The Electricity Commission of New South Wales (ElCom) at the close of the 1984/85 financial year had a total generating capacity of 11,639 megawatts.[2] ElCom wholesales to 27 retail supply authorities in that state: the city of Sydney, numerous county councils, the Australian Capital Territory (ACT),[3] the government railways, and certain large industrial customers, for example, Broken Hill Proprietary Company (BHP). In Victoria the State Electricity Commission (SECV) had installed capacity of 5,494 megawatts at the end of 1984/85 and sells directly to retail customers in all areas of the state outside of the 11 Melbourne metropolitan municipalities. The Queensland Electricity Commission (QEC) supplies bulk power to seven regional distribution boards, which in turn sell to retail consumers throughout the state. Total generating capacity of the QEC was 4,816 megawatts on June 30, 1985. Plant capacity of 2,443 megawatts was operated by the Electricity Trust of South Australia. The trust sells directly to retail consumers and supplies some bulk power to regional distributors. In Western Australia the State Energy Commission has capacity of 2,132 megawatts. The majority of this capacity is in the Western Australia–interconnected grid. Tasmania, the smallest state in Australia in terms of population and geographic size, is the only state that derives the majority (approximately 97 percent) of its 1,943-megawatt capacity from hydro-generating facilities. The HEC of Tasmania has total responsibility for that state's electricity needs. In the Northern Territory the installed capacity of oil-fired, diesel, and gas turbine generating facilities was 243 mW's in 1984/85. There are plans to replace the Darwin oil-fired facility with a natural gas-fired combustion turbine and combined-cycle plant in 1987. The remaining publicly owned generating capacity in Australia, constructed and operated by the Commonwealth (federal) government, is the Snowy Mountains Scheme (Snowy Scheme) developed to supply irrigation and hydroelectric power to satisfy peak demands.[4] The installed capacity of the Snowy Scheme of 3,740 megawatts is shared among the states of New South Wales, Victoria, and the ACT. Table 3.1 presents the capacity, primary energy source, and the 1984/85 percentage of the actual electricity generated for each of the separate regions. These data are from the Electricity Supply Association of Australia (ESSA) 1986 annual publication of statistical information.

In this chapter we will consider several Australian electricity demand studies including those of Hawkins (1975), Donnelly (1984, 1985), Donnelly and Diesendorf (1985), Donnelly and Leung (1983), Saddler et al. (1980), and Donnelly and Saddler (DS) (1982, 1984).[5] The Hawkins work is a cross-sectional study of the residential, commercial, and industrial demand in New South Wales, including the ACT, in 1971. Donnelly, Donnelly and Diesendorf, and Donnelly and Leung analyze the residential demand for electricity in the ACT. Saddler et al. and Donnelly and Saddler investigate retail electricity demand in Tasmania. This analysis is extended here to study the residential, commercial, and industrial demand in that state as well.

Australian electricity demand modeling has to date been directed toward understanding the importance of factors that affect long-

Table 3.1. Electricity Generating Capacity in Australia (1984/85)

State	Installed Capacity (megawatts)	Percent of Australian Capacity	Percent of Australian Generation	Percent Capacity by Primary Energy Source	
New South Wales	11,639	35.9	34.7	93.7	black coal
Victoria	5,494	16.9	23.4	83.0	brown coal
Queensland	4,816	14.8	16.8	81.9	black coal
South Australia	2,443	7.5	7.2	77.5	gas
Western Australia	2,132	6.6	6.2	89.8	black coal
Tasmania	1,943	6.0	7.4	87.5	hydro
Northern Territory	243	0.7	0.7	87.7	oil
Snowy Mountains Scheme	3,740	11.5	4.4	100.0	hydro
Total Australia	32,450	55.6 black coal 21.7 brown coal 12.3 hydro 8.6 gas 1.8 oil			

Source: ESSA, 1986, *The Electricity Supply Industry in Australia: 1984-85,* (Melbourne) p. 19, pp. 48–51, April.

term consumer behavior, and has been confined largely to academics, although research groups with SECV, the New South Wales Energy Authority, and HEC have recently been applying econometric methods. Unfortunately, none of the details of these studies have been released to the public. The studies discussed here are all econometric models using either cross-sectional data or annual time-series data. The model structures differ both conceptually and empirically, and the results derived are similarly diverse.

NEW SOUTH WALES

The Hawkins cross-sectional work (1975) represents the first model of electricity demand for Australia to be published, and the data are those that reflect the electricity sold by the New South Wales retail authorities. (In 1971 there were 44 such authorities throughout the state.) Sales to these authorities represent over 80 percent of New South Wales electricity consumption, with the balance of sales representing primarily bulk users not being modeled.

The ElCom data cover the broad categories of residential, commercial, and industrial consumption.[6] Linear models are specified and estimated. In the case of the residential model, the dependent variable is defined as the average annual consumption of electricity $\mathbf{Q_E}$, and the explanatory variables included are household average real income $\mathbf{Y}$, real electricity price $\mathbf{P_E}$, real prices of substitute goods (oil and natural gas) $\mathbf{P_O}$ and $\mathbf{P_G}$, and various demographic and geographic factors, for example, the proportions of occupied households $\mathbf{N_G}$ and unoccupied holiday houses $\mathbf{N_H}$, the ratio of occupied houses connected to both natural gas and electricity to those connected only to electricity $\mathbf{NO_{EG}/NO_E}$, and some climatic variables. The residential electricity demand equation estimated is of the form

$$\mathbf{Q_E} = \alpha + \beta_1 \cdot \mathrm{Y} + \beta_2 \cdot \mathbf{P_E} + \beta_3 \cdot \mathbf{P_O} + \beta_4 \cdot \mathbf{P_G} + \beta_5 \cdot \mathbf{N_G} + \beta_6 \cdot \mathbf{N_H} + \beta_7 \cdot (\mathbf{NO_{EG}/NO_E}). \quad (3.2)$$

Regional cross-sectional data for the year 1971 are used in the estimation. The price of the electricity variable selected for use in the model is the average revenue per customer.[7] The own-price elasticity—calculated at the sample means is −0.55, and the income (or expenditure) elasticity is 0.93. Hawkins finds that the price of elasticity becomes more inelastic as the number of gas connections and holiday homes increases, as would be expected as substitutes to electricity become more readily available. This finding has implications as to the ACT electricity demand, which could be expected to "be reduced by 800 kWh [kilowatt-hours] a year per household if the proportion of households with a gas connection was as high as in Sydney, and by 1,000 kWh a year if the price rose to the sample average" (Hawkins 1975, p. 10). This conclusion has important policy ramifications, as natural gas connections were first effected in the ACT in the latter part of 1984. Climatic factors and prices of substitute fuels do not appear, in the Hawkins analysis, to influence domestic electricity demand.

The results in the commercial and industrial sectors are disappointing, because of the price of electricity showing no statistical effect in the author's preferred models. This consideration leads to his conclusion that the "elasticity with respect to changes in the wage rate was suspiciously high at 1.4, and . . . retail sales have a low, insignificant coefficient" (p. 13). Also, neither the climatic variable nor the price of substitute fuel variable affects the demand for electricity in either of these sectors. These findings support Hawkins's argument that demand in the commercial and industrial sectors is related only to the "level of activity" and is therefore perfectly inelastic with respect to price and the other factors tested. Such conclusion is surprising and, if accurate, would serve to argue against the "traditional wisdom" of electricity-generating authorities that lower negotiated tariffs for large industrial consumers act as an incentive to those manufacturers, thereby helping to stimulate regional economic activity. As will be discussed in the modeling of the demand for electricity in Tasmania, this premise of encouraging regional growth is explicit in the forecasts developed by HEC there.

AUSTRALIAN CAPITAL TERRITORY

It is convenient, as we have observed already, to divide the economy into residential, commercial, and industrial sectors in electricity demand analysis, and so many modelers have done this.[8] The pattern of electricity usage varies significantly among these generic sectors, and therefore it makes sense to study them separately. The ACT has the highest proportion of residential electricity consumers of all of the capital cities in Australia, representing approximately 54 percent of the kilowatt-hours consumed. Most residential housing is of the single-family type, and none of the multifamily housing is bulk metered; thus, the generic definitional problems sometimes associated with residential electricity demand–modeling work are absent. In addition, the residents in the ACT are more homogeneous in socioeconomic terms than those of other Australian cities, because the major employer there is the Commonwealth government; the area therefore presents a microcosm in which to study residential electricity demand. A disadvantage is that the large seasonal and diurnal variations in electricity consumption cannot be analyzed, because data are published in insufficient temporal detail. From information provided recently, it is possible, however, to construct Figures 3.1 and 3.2, which depict the monthly electricity load during calendar year 1983 and the typical winter-day diurnal load curves for the ACT and New South Wales, respectively. The contrasts are interesting and reflect the differences underlying the sectoral demands in the two regions, as well as their differing climates. These load curves are indexed with the peak consumption set to 100. It can

Figure 3.1. Monthly Electricity Consumption (1983)

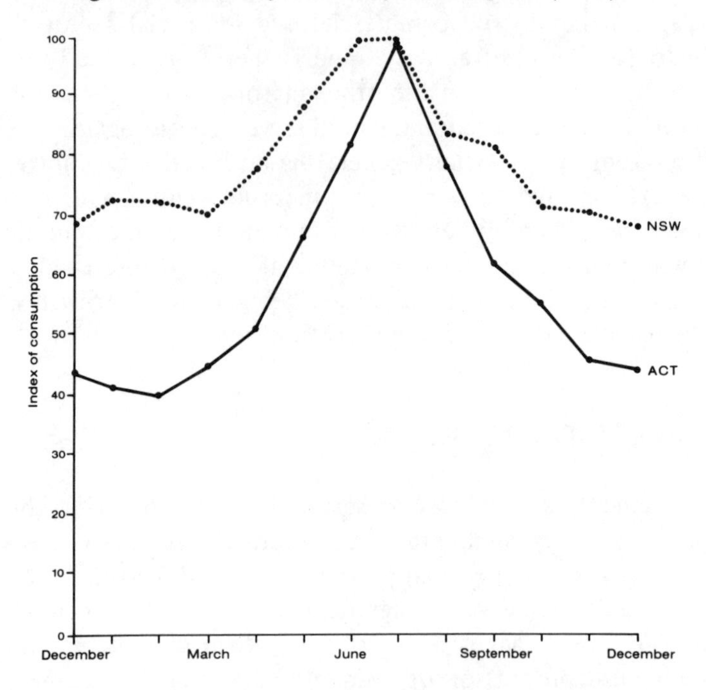

Figure 3.2. Diurnal Electricity Consumption (1983)

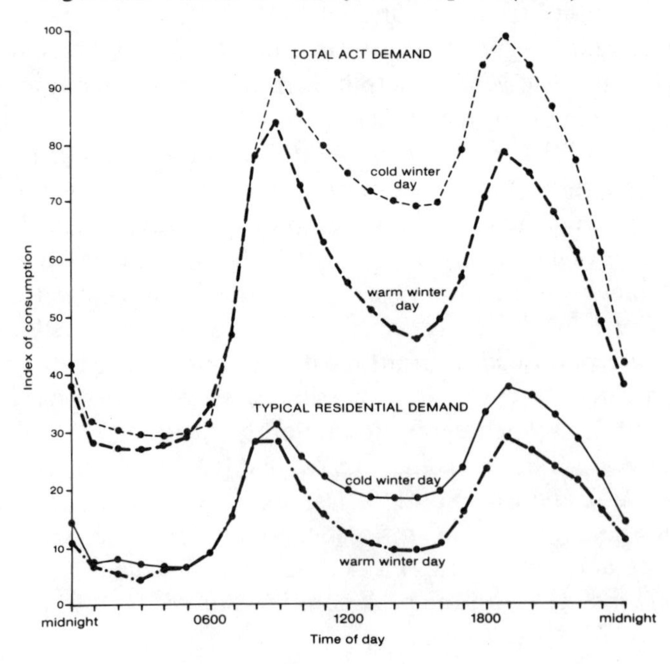

be seen that the annual peak-to-trough variation in the ACT demand of about two and one-half to one is much greater than the seasonal variation in New South Wales. Undoubtedly, this difference reflects to a large extent the harsher ACT climate, the general lack of central heating, and the reliance on resistance space heating, which is a relatively inefficient process. Similarly, the diurnal load curve for New South Wales shows less variation than the same curve for the ACT, a difference that may be attributed to the lack of any baseload industrial consumers in the ACT. The existence of an industrial baseload demand could tend to flatten out the diurnal load curve, depending upon the intensity of the manufacturing process and the structure of the industrial electricity tariff. The maximum diurnal variation in the ACT is about three and one-third to one overall and almost seven to one in a typical ACT residential neighborhood.

Data for the ACT residential sector were obtained annually for the 19 financial years from 1963/64 to 1981/82, and included electricity consumption, number of customers, electricity revenues, rate schedules, household disposable income, alternative fuel price index, consumer price index (CPI), and climatic factors. The electricity data for the ACT come from ESAA and the ACT Electricity Authority annual reports. Because of the simultaneity problem that results from the declining block rate price structure, causing the "quantity purchased" decision directly to affect the average unit price charged, the marginal price of electricity is derived by interpreting the actual tariff schedules, and this price series is used in the estimation. The marginal price is calculated as the charge for the next kilowatt-hour of electricity that would be consumed above the average monthly consumption figure that is actually observed, and when tariffs change during the year, the appropriate weighted average price is computed.

Figure 3.3 shows a typical declining block rate electricity tariff—this particular residential tariff was in effect in the ACT during the 1977/78 fiscal year—wherein three blocks of consumption are defined during the billing period (three months), namely, 50 kilowatt-hours or less, from 50 through 600 kilowatt-hours, and greater than 600 kilowatt-hours. The rate stated in 1977/78 current dollars within each of the blocks is 7.70, 3.03, and 1.95 cents per kilowatt-hour. The question of simultaneity with respect to quantity and price raised by Houthakker (1951b) can be demonstrated with the hypothetical demand curve (*d-d*) superimposed on the declining block rate tariff diagram. The consumer of electricity is not faced with a single

Figure 3.3. Declining Block Rate Electricity Tariff (Australian Capital Territory 1977/78)

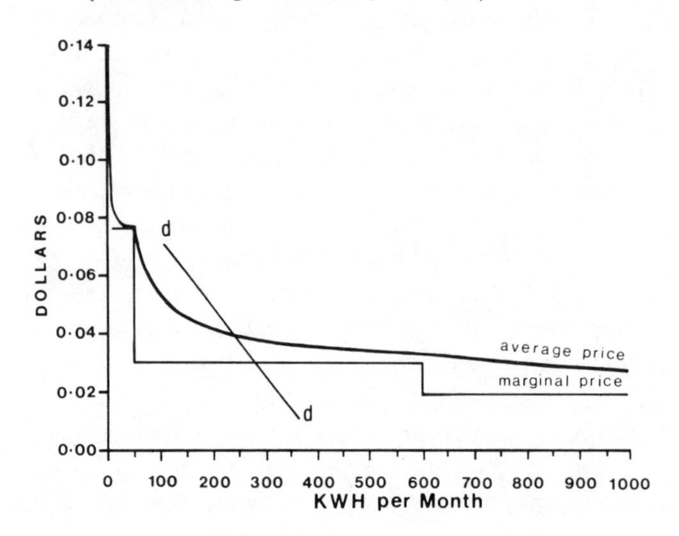

price for the commodity, but rather a schedule of prices based upon the level of consumption selected.[9] The *d-d* demand curve represents a household with a relatively low level of quarterly electricity demand.[10] A numerical example should illustrate the potential problem. Consider a consumption of 241 kilowatt-hours of electricity per quarter, which under that 1977/78 tariff would cost the householder 9 dollars and 64 cents, an average price of about 4.0 cents per kilowatt-hour. The price of the marginal kilowatt-hour, as defined above, would be 3.03 cents. A small shift in the demand function, either to the right (an increase in demand) or to the left (a decrease in demand), would not change the marginal price charged to the household. Thus, even a drop in consumption of roughly 10 percent to 218 kilowatt-hours would, while attracting a 4.1-cent average price, still maintain the 3.03-cent marginal price. Likewise, a similar percentage increase in consumption to 267 kilowatt-hours per quarter would yield an average price of 3.9 cents but no change in the marginal price. This interrelationship between the level of consumption and the average price paid is the basis of the simultaneity problem raised by Houthakker (1951a). All three levels of consumption cited (218, 240, and 267 kilowatt-hours) would pay the same marginal price (3.03 cents) for an additional kilowatt-hour, and this constancy of the marginal price argues on behalf of its use over the average price figure in any demand analysis.

Taylor (1975) suggests the use of two price variables constructed directly from the electricity tariff and not derived from ex post revenue figures. The first variable is the average price of the electricity purchased up to, but not including, the block in which the demand occurs,[11] and the second the marginal price of that particular sales block. Nordin (1976) argues that a better measure for the average price would be the excess amount paid for the initial blocks of electricity over the charge that would have occurred if all kilowatt-hours attracted the marginal price. This issue will be explored in the following empirical analysis.

A potentially more serious problem than the simultaneity one, at least for estimation purposes, occurs when consumption levels are in the neighborhood of the changing block rate points. In the above example, a discontinuity occurs at a level of demand of 600 kilowatt-hours per quarter. On average, quarterly consumption levels in the ACT are well above this level, so the fact that the tariff schedule is not continuous should pose no problem in practice. In the subsequent analysis, the assumption is one of a flat seasonal demand–for 1977/78 that level was 2,364 kilowatt-hours. Because of the historical usage patterns, the information also on the known seasonal variation would not be expected to affect the marginal price.[12] For different years, for different consumers, or for different tariffs, this situation might not hold; thus, the desirability of increased temporal and customer detail remains strong.[13]

In considering the other model variables, this work turns to the unpublished "other fuels" component of the CPI for Canberra used to reflect the price of substitute fuels, and obtains this material from the Australian Bureau of Statistics (ABS). This index comprises the price of home heating oil and kerosene, both being fuels that are used for residential space heating and in the limited number of homes that have central heating systems. The household disposable income series is developed from *Taxation Statistics* and the Australian National Accounts (ABS, quarterly). Since no stock of, or utilization rate for, electrical appliances is available, those elements of variability could not be included in the model specification, but two climatic variables, heating (HDD) and cooling (CDD) degree days, are included from "pseudodata" prepared by the Australian Bureau of Meteorology. These degree day data are calculated from monthly minimum and

Table 3.2. Australian Capital Territory Residential Electricity Dataset

Fiscal year	Annual residential electricity consumption (thousands kilowatt-hours)	Number of residential electricity customers	Average revenue (cents per kilowatt-hour)	Marginal price (cents per kilowatt-hour)	Lumpsum payment (dollars)	Household disposable income (dollars)	Fuels price index	Consumer price index	Degree days: Heating (12-degree base)	Degree days: Cooling (23-degree base)
1963/64	106,847	19,998	1.614	1.500	2.00	5,620	19.9	92.4	777.2	19.3
1964/65	125,223	21,577	1.597	1.475	2.15	5,825	19.9	95.2	851.7	14.8
1965/66	143,893	24,380	1.577	1.450	2.28	5,888	19.9	98.1	781.9	19.7
1966/67	165,722	26,551	1.574	1.450	2.26	6,201	20.6	100.0	745.1	19.0
1967/68	180,563	29,008	1.576	1.450	2.26	6,574	20.6	102.6	760.1	73.8
1968/69	213,668	31,969	1.568	1.450	2.26	6,926	20.3	104.4	901.8	45.3
1969/70	235,841	35,437	1.568	1.450	2.26	7,067	20.6	107.4	803.9	10.0
1970/71	268,643	39,386	1.541	1.400	2.51	7,655	20.9	113.0	868.3	20.9
1971/72	307,255	43,010	1.531	1.400	2.56	8,022	20.9	119.4	877.4	10.6
1972/73	333,226	47,088	1.567	1.412	3.05	8,855	20.9	126.3	647.3	75.5
1973/74	367,587	50,604	1.683	1.450	4.50	10,679	21.2	142.7	599.8	21.4
1974/75	456,089	55,463	1.824	1.550	6.27	13,149	24.0	164.9	798.2	12.0
1975/76	501,518	59,634	1.974	1.643	7.37	14,329	30.3	187.3	671.8	23.9
1976/77	580,752	63,732	2.123	1.785	8.08	14,770	34.1	212.9	802.4	33.1
1977/78	625,397	66,152	2.286	1.950	8.82	16,572	40.9	232.2	727.9	33.4
1978/79	695,980	69,954	2.399	2.067	9.35	17,081	52.7	251.0	798.5	66.5
1979/80	734,588	70,529	2.538	2.266	8.51	19,249	79.3	278.0	687.0	41.6
1980/81	793,972	72,518	2.792	2.615	6.37	22,149	100.0	305.0	624.4	108.4
1981/82	840,795	75,027	3.322	3.382	4.18	24,566	111.8	337.6	714.3	67.9

Source: Compiled by the author.

maximum temperatures, rather than from the actual daily values as referred to in the formal definition of a degree day.[14] The ACT modeling data appear in Table 3.2.

During the period over which data are available for the ACT, average household electricity consumption increased from about 5,340 to just over 11,200 kilowatt-hours per year, a compound annual growth rate of roughly 3.9 percent. Figure 3.4 illustrates the increasing household usage of electricity over those years from 1963/64 to 1981/82. The real price of electricity declined steadily from around 1.6 cents per kilowatt-hour in 1963/64, measured in 1966/67 constant dollars, to a low of approximately 0.8 cent per kilowatt-hour in 1979/80. The price then began to increase (reflecting the increased opportunity cost of coal), finishing the period at roughly 1 cent per kilowatt-hour. Figure 3.5 depicts the changing marginal price of electricity series. Figures 3.6 to 3.8 show how the relative price of other fuels, real household disposable income, and HDDs and CDDs varied over the 19-year period being modeled.

Figure 3.4. Average Household Electricity Consumption (Australian Capital Territory 1963/64 to 1981/82)

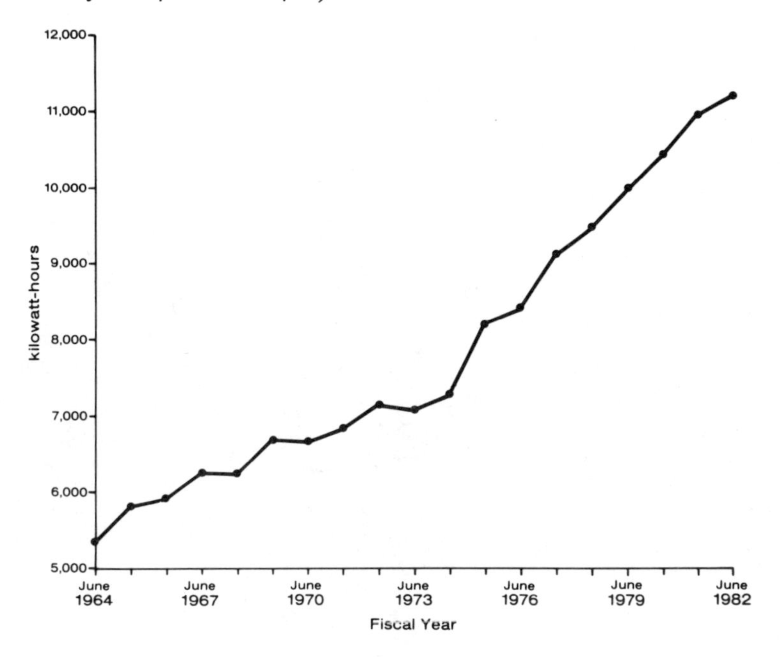

Figure 3.5. Relative Marginal Price of Electricity (Australian Capital Territory 1963/64 to 1981/82)

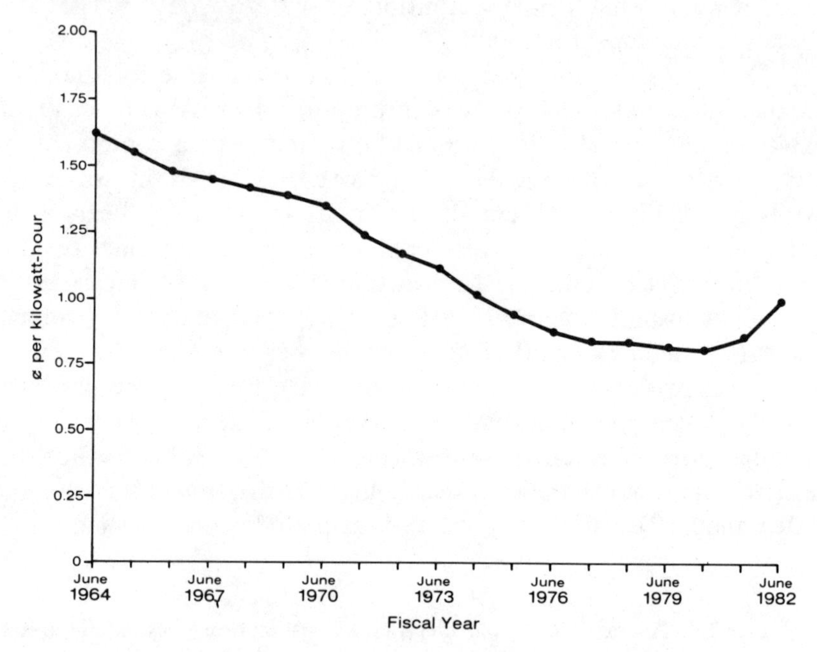

Figure 3.6. Relative Price Index for Other Fuels (Australian Capital Territory 1963/64 to 1981/82)

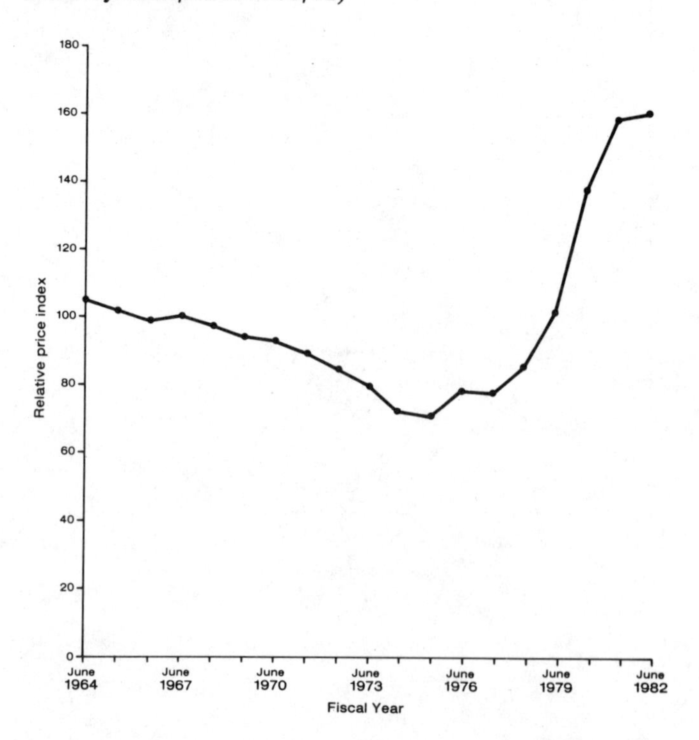

Figure 3.7. Real Household Disposable Income (Australian Capital Territory 1963/64 to 1981/82)

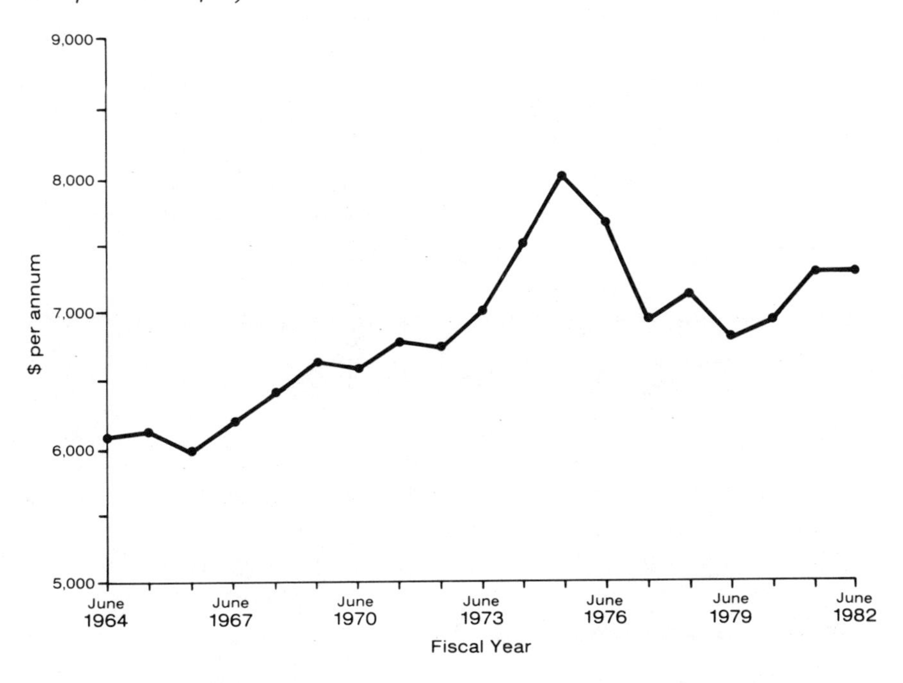

Figure 3.8. Heating and Cooling Degree Days (Australian Capital Territory 1963/64 to 1981/82)

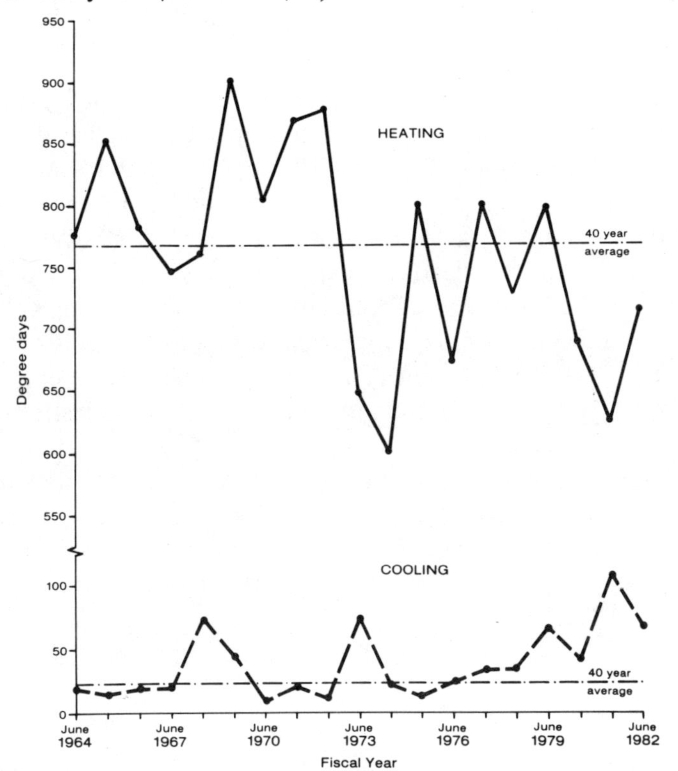

The basic demand model "reduced form" formulation relates the average consumption of electricity per household to the price of electricity, the price of substitute goods $\mathbf{P}_S$, income, and other factors. In the absence of appliance stock data, a dynamic-demand function, according to the form specified by Houthakker and Taylor (1970), is specified incorporating the lagged dependent variable in order to represent the effect of changes in the stock of electricity using appliances, their efficiency and their usage. Equation 3.3 shows the general functional form for the dynamic demand model:

$$\mathbf{Q}_{E,t} = f(\mathbf{P}_{E,t}, \mathbf{P}_{S,t}, \mathbf{Y}_t, \mathbf{X}_t, \mathbf{Q}_{E,t-1}). \tag{3.3}$$

In an attempt to determine which is the most appropriate functional form of this model to estimate, a BC analysis of the model (1964) is made using the Savin and White "likelihood ratio" statistic (1978) to test various conditional general and joint hypotheses. The data are transformed using the general nonlinear specification of Box and Cox, namely,

$$y_t^{(\lambda)} = \frac{y_t^{\lambda} - 1}{\lambda}, \qquad \text{for } \lambda \neq 0 \tag{3.4}$$

$$y_t^{(\lambda)} = \ln(y_t), \qquad \text{for } \lambda = 0. \tag{3.5}$$

Thus, as we saw in Chapter 2, if the estimated value of λ is not statistically different from 0 (1), the appropriate functional form is a logarithmic (linear) one. Savin and White (1978) and Seaks and Layson (1983) show how one adjusts the BC model for the presence of first-order serial correlation and for heteroscedasticity. Since the existence of heteroscedasticity can bias the BC solution toward acceptance of the logarithmic formulation, "because of the well-known result that the logarithmic transformation tends to stabilize the error variance" (Seaks and Layson 1983, p. 163), then wherever a logarithmic model is suggested, the presence of heteroscedasticity in the linear formulation should be tested.

In testing functional form, the maintained hypothesis Ω is tested against the restricted hypothesis ω and the null hypothesis H_0 yields a "limiting chi-square distribution with q degrees of freedom where q is the number of additional restrictions imposed" (Savin and White 1978, p. 6). The test statistic is

$$\theta = 2 \cdot [L(\Omega) - L(\omega)] > \chi^2_\alpha(q) . \quad (3.6)$$

Various combinations of the following restrictions are tested using the ACT data:

1. $\lambda = 1$—a linear model
2. $\lambda = 0$—a logarithmic model
3. $\rho = 0$—no first-order serial correlation

The λ variable is used to transform all of the regression variables in Equation 3.3 in what Savin and White refer to as an "extended BC" (BCE) model. Other forms of the BC transformation include transforming only the dependent variable, only the independent variables, or all of the model variables by different λ values. The BCE procedure is applied here because the lagged endogenous variable is included in the explanatory dataset, and logic dictates that it should be transformed by the same value of λ as is used on the dependent variable. The estimation results of the alternative BCE forms of the electricity demand relationship along with the relevant elasticities η_i are presented in Table 3.3.

The hypotheses tested and the chi-square results are presented in Table 3.4. Given an assumption of no serial correlation, that is, $\rho = 0$, only the linear model formulation of the conditional tests, model number 1, is rejected at the 0.01 significance level. The unrestricted and joint tests of the linear model, that is, $\lambda = 1$, are rejected at the 0.025 and 0.05 levels of significance, respectively. The unrestricted serial correlation test is rejected at the 0.10 level with $\theta = 3.14$. Thus, the BC procedure effectively rejects the linear

Table 3.3. Dynamic Demand Model (1963/64 to 1981/82)

Number	Model	BCE λ	AR ρ	$L(\lambda, p)$	Elasticity P_E	P_S	Y	W	Q_{t-1}
1	Linear	1	0	8.00238	-0.33	0.17	0.12[a]	0.19	0.60
2	AR, linear	1	0.33	8.56993	-0.34	0.18	-0.17[a]	0.23	0.58
3	BCE, log	0	0	9.57310	-0.35	0.19	0.31	0.20	0.55
4	BCE AR, log	0	-0.35	10.02221	-0.34	0.17	0.26	0.15	0.56
5	BCE	-0.04	0	9.57546	-0.35	0.19	0.32	0.20	0.55
6	BCE AR	-1.10	-0.65	11.24629	-0.29	0.15	0.43	0.11	0.58

[a] Not statistically different from 0 at the 0.05 level.
Source: Compiled by the author.

Table 3.4. Extended Box-Cox Test Statistics (1963/64 to 1981/82)

Null hypothesis, H_0	X^2
Conditional tests	
$\lambda = 1 \mid \rho = 0$	3.1462[a]
$\lambda = 0 \mid \rho = 0$	0.0047
$\rho = 0 \mid \lambda = 1$	1.1351
$\rho = 0 \mid \lambda = 0$	0.8982
Unrestricted tests	
$\lambda = 1$	5.3527[b]
$\lambda = 0$	2.4482
$\rho = 0$	3.3417[a]
Joint tests	
$\lambda = 1, \rho = 0$	6.4878[c]
$\lambda = 0, \rho = 0$	3.3464

[a]Reject at the 0.10 level of significance.
[b]Reject at the 0.025 level of significance.
[c]Reject at the 0.05 level of significance.
Source: Compiled by the author.

formulation. The unrestricted and joint tests do not reject the logarithmic formulation, and the absence of first-order serial correlation in that form is not rejected either. The LM chi-square test statistic for the presence of heteroscedasticity in the linear model is 1,249, with 1 degree of freedom (see Pagan and Hall 1983, p. 178).[15] This means that the null hypothesis of homoscedasticity cannot be rejected, and the extended BC results therefore should not be biased in favor of the log-linear formulation. The chi-square results, for the hypotheses tested and the elasticities measured at the means of the data appear in Table 3.3. Because these results do not reject the logarithmic formulation, only the results of the log-linear model will be reported, as shown in Equation 3.7:

$$\ln \mathbf{Q}_{E,t} = \alpha + \beta_1 \cdot \ln \mathbf{P}_{E,t} + \beta_2 \cdot \ln \mathbf{P}_{S,t} + \beta_3 \cdot \ln \mathbf{Y}_t + \beta_4 \cdot \ln \mathbf{W}_t + \beta_5 \cdot \ln \mathbf{Q}_{E,t-1} + \mathbf{u}_t , \qquad (3.7)$$

where
$\mathbf{Q}_E$ = average household electricity consumption
$\mathbf{P}_E$ = real marginal price of electricity
$\mathbf{P}_S$ = real price of substitute fuels
$\mathbf{Y}$ = real household income
$\mathbf{W}$ = HDDs
t = time period
u = stochastic error term
ln = natural logarithm.

The parameter estimates in this form of log-linear model are the elasticities of the respective exogenous variables, and these are constant over the entire range of the demand function.[16] The imposition of the constant elasticity assumption as a maintained hypothesis is acceptable if, as is the case here, only an indication of the relative magnitudes of consumer response is to be expected from the available data; and owing to the annual nature of the present database and the concomitant simplifying assumptions made in calculating the model's prices, it seems reasonable in this instance. The model includes lagged consumption as an explanatory variable and proxy for the missing information on the stock of electricity using appliances and their usage rates; its inclusion enables derivation of both short- and long-run elasticities of demand.[17] While the definition of the duration of the "long run" is imprecise, its merely being the time that elapses toward achieving a stable demand from the model after a system shock has occurred, it is possible to make a crude estimate of the number of time periods needed to achieve stability from the value of the coefficient of the lagged endogenous variable. Under a ceteris paribus assumption of all other factors remaining constant, a 0.55 coefficient for the lagged term would suggest that a 90 percent adjustment to an equilibrium level of consumption will be achieved after four time periods—in this model, "years"—and a 95 percent adjustment by the fifth year. The OLS estimation results are given below:

$$
\begin{array}{llllll}
\ln \mathbf{Q}_{E,t} = & -2.94 & - \; 0.35 \ln \mathbf{P}_{E,t} & + \; 0.19 \ln \mathbf{P}_{S,t} & & \\
 & (-4.273) & (-3.873) & (3.487) & t & \\
 & & 0.93 & 0.83 & \text{FG } R^2 & \\
 & & + \; 0.31 \ln \mathbf{Y}_t & + \; 0.20 \ln \mathbf{W}_t & & \\
 & & (2.413) & (3.607) & & \\
 & & 0.70 & 0.31 & & \\
 & & & + \; 0.55 \ln \mathbf{Q}_{E,t-1} & & (3.8) \\
 & & & (4.625) & & \\
 & & & 0.96 & &
\end{array}
$$

$$
\begin{array}{rcl}
\bar{R}^2 & = & 0.990 \\
\text{DW} & = & 2.149 \\
\text{Durbin's } h & = & -0.655 \\
\text{RMSE} & = & 1.8\% \\
\text{Theil's } U & = & 0.366
\end{array}
$$

The variables included in the model are as follow: dependent variable, annual electricity consumption per household; explanatory variables, the marginal price of electricity; the price of other fuels; household disposable income; and HDDs. The monetary variables are couched in constant dollar terms, that is, deflated to 1966/67 constant dollars, and the regression is run on the natural logarithms of the variables. In selecting the weather variable, a model including both the HDD and CDD data is tested, and the parameter estimate on the CDD variable is found to be insignificant. This is not a surprising result as few residences in the ACT have central, or even room, air conditioning, given that the arid climate of the region allows cooling through the use of relatively energy-efficient evaporative coolers. Because of these results, the CDD variable is omitted from Equation 3.8.

Interpreting the results indicates that this dynamic demand model formulation "explains" 99 percent of the observed variation in the dependent variable. $\bar{R}^2$, the coefficient of multiple determination adjusted for degrees of freedom, is the ratio of the model's explained to unexplained variation. Collinearity, as measured by the FG R^2, may pose an estimation problem, its being 0.96 for the lagged dependent variable and 0.93 for the electricity price variable; but a high level of collinearity is not unexpected in such a formulation. If the purpose of the model is forecasting and the collinearity relationships are expected to remain stable ones, then there should be no problem in using these results. However, if exact identification of the elasticities is desired, then some additional assessment of the severity of the problem is warranted. One procedure for this testing is to estimate a series of ridge regressions, and in this approach the diagonal elements of the $\mathbf{X}^T\mathbf{X}$ matrix, from Equation 2.23, are augmented by some small value **k**, the values for **k** being selected as occurring between 0 and 1. The increments selected for testing this model are 0.1, and the results are presented in Table 3.5. It can be seen that the variation in the own-price elasticity, the coefficient of the price of electricity variable, is over 45 percent, ranging from -0.24 to -0.35. A 95 percent confidence interval on the short-run price elasticity in Equation 3.8 is from -0.17 to -0.53, suggesting that while it is statistically different from 0, its precision may be questionable. The cross-price elasticity and the income elasticity variations in the ridge regressions are of similar magnitudes, being roughly 40 and 35 percent, respectively. The lagged dependent variable's variation is over 80 percent, and the HDD elasticities are generally insignificant. If identification of the elasticity values themselves is desired, then a more comprehensive analysis of the nature and the impact of the collinearity, as Belsley, Kuh, and Welsh (1980) recommend, would seem warranted.

Table 3.5. Ridge Regression Results (1963/64 to 1981/82)

	Elasticity				
k	P_E	P_S	Y	W	Q_{Et-1}
0.0	-0.35	0.19	0.31	0.20	0.55
0.1	-0.35	0.21	0.39	0.14	0.45
0.2	-0.33	0.20	0.41	0.09[a]	0.42
0.3	-0.32	0.19	0.42	0.06[a]	0.40
0.4	-0.30	0.18	0.42	0.03[a]	0.38
0.5	-0.29	0.18	0.42	0.01[a]	0.36
0.6	-0.27	0.17	0.42	-0.01[a]	0.34
0.7	-0.26	0.16	0.41	-0.03[a]	0.33
0.8	-0.26	0.16	0.41	-0.04[a]	0.32
0.9	-0.25	0.15	0.40	-0.05[a]	0.31
1.0	-0.24	0.15	0.40	-0.05[a]	0.30

[a]Not statistically different from 0 at the 0.05 level.
Source: Compiled by the author.

The test for first-order serial correlation as measured by Durbin's *h* statistic indicates that this is not a problem with these data, even though the DW statistic fails within the range of uncertainty. The former measure is considered by some to be the more powerful one to be used for testing models that include a lagged dependent variable, but this contention is open to debate. The root-mean-square error (RMSE) of the model is 1.8 percent of the household's electricity consumption. All of the elasticities, the coefficients in the model, are statistically significant and are different from 0 in regard to all of the Student's *t* statistics, which are in excess of 2.

Residential electricity consumption in the ACT is thus found to be inelastic in the short run, with an own-price elasticity of -0.35 (see Table 3.6). The long-run price elasticity is estimated to be -0.77, which indicates that a 1 percent increase in the price of electricity—as measured by the marginal price in constant dollar terms—to the household will induce a 0.77 percent decline in average consumption; necessarily, such a price increase thereby increases the revenues of the ACT Electricity Authority. The cross-price elasticity of demand for "other fuels" used in household space heating has the appropriate positive algebraic sign required for a substitute good, and its value, 0.19, confirms the hypothesis that the commodities, namely, electricity and other fuels, are not close substitutes. In the long run a 1 percent increase in the price of these other fuels would increase the electricity consumption of the average ACT household by 0.42 percent. In addition, residential electricity consumption is found to be income inelastic. In the long run, an equiproportionate increase in the price of electricity and in household disposable income would result

Table 3.6. Residential Electricity Demand Elasticities (1963/64 to 1981/82)

Variable	Short-run elasticity	t	Long-run elasticity	Long-run 95% confidence interval
P_E	-0.35	(-3.873)	-0.77	-1.04 to -0.58
P_S	0.19	(3.487)	0.42	0.27 to 0.57
Y	0.31	(2.413)	0.69	0.08 to 1.59
W	0.20	(3.607)	0.43	0.16 to 1.08

Source: Compiled by the author.

in a reduction in electricity consumption of roughly 10 percent of that change. The effect of the weather has only a slight influence on an ACT electricity consumer, in that a year that was 10 percent colder than normal would be expected to induce only a 2 percent increase in the demand for electricity. Table 3.6 summarizes these results and presents confidence interval estimates for the long-run elasticities, wherein a 95 percent confidence interval is derived for each of these from the formula provided by Pindyck and Rubinfeld (1981, pp. 269–70). These authors define the confidence interval around the expected value of the long-run elasticity with the quadratic equation that incorporates the variances and covariance of the short-run elasticity and the coefficient of the lagged dependent variable. As Pindyck and Rubinfeld point out, the interval so defined may not be symmetric, because the expected value of the long-run elasticity is not necessarily equal to the ratio of the short-run elasticity to 1 minus the coefficient of the lagged dependent variable (the point estimate is derived from the equation in note 17), the latter being the conventional measure of a long-run elasticity; this divergence is evidenced here in the case of the income and weather elasticities, which both appear to be asymmetric. However, the significance of the long-run point estimates must be interpreted in light of the ridge regression results, which suggest a stability problem in the equation parameters.

Two further static demand model analyses using these ACT data are of interest. First, the data are used to test Nordin's assertion (1976) that the marginal price and a variable reflecting the lump sum payment for electricity should be the only own-price variables included in a demand model for electricity. He reasons that the lump sum parameter estimate should be of the same magnitude as the income parameter but opposite in sign, and this appears to be the case in the AR BC estimate of a static demand function such as appears in Table 3.7. It is evident in these data that the Nordin hypothesis including the lump sum variable that reflects the income effect cannot be rejected; in addition, it is interesting to note that the own-price elasticity of demand in this formulation does not differ significantly from unity. The explanatory power of the static demand equation is similar to that of the dynamic demand model, but the policy implications of the two models are somewhat different. The static result suggests that an electricity price increase will result in a concomitant, exactly proportional reduction in consumption,

Table 3.7. Extended Box-Cox Estimates (1963/64 to 1981/82)

	P_E	P_S	Y	LS	W		
Elasticity	-0.79	0.42	0.27	NI	0.23		
t	(-15.658)	(13.596)	(1.679)		(3.149)		
FG R^2	0.68	0.17	0.70		0.30	$\bar{R}^2$	= 0.977
						DW	= 1.5979
						BCE λ	= 0.59
						LFF	= 0.787
Elasticity	-0.98	0.16	0.28	-0.22	0.20		
t	(-14.537)	(2.863)	(1.734)	(-5.356)	(4.034)		
FG R^2	0.85	0.68	0.73	0.77	0.31	$\bar{R}^2$	= 0.986
						DW	= 1.6948
						AR ρ	= 0.40 (1.902)
						BCE λ	= -0.34
						LFF	= 6.348

Notes: NI, not included. Elasticities measured at the means of the variables.
Source: Compiled by the author.

leaving revenues of the electricity authority unchanged. The consequences of a price increase will in all likelihood not reduce the revenues of the electric utility, but may or may not encourage conservation on the part of consumers. The static demand model does not suffer from the collinearity problems of the dynamic model, which might serve to explain some of the apparent differences in the elasticity estimates provided.

The BC results raise additional questions about the appropriateness of either a simple linear or a log-linear model formulation. This issue deserves further comment in view of the fact that little nonlinear model analysis is done. Recently, Betancourt (1981) proposed a nonlinear model with which to test the existence of variable price and income elasticities. The model he specifies in Equation 3.9 unfortunately does not yield measures of elasticities that are invariant to the units in which the modeling data are couched:

$$Q_E = \alpha_0 \cdot Y^{\alpha_1} \cdot P^{\alpha_2 \, (P_{-1}, Z)} \cdot X^{\alpha_3} \cdot S, \qquad (3.9)$$

where Q_E = electricity consumption
Y = income
P = average electricity price
P_{-1} = price lagged one period

X, Z = vectors of weather and other variables that might influence the price response

S = maximum kilowatt-hour consumption of existing stock of appliances.

One problem in this formulation to be addressed is developing nonlinear model specifications that provide derivative measures such as elasticities and elasticities that are invariant to the scaling of the modeling data. This issue arises in the Betancourt specification, as Donnelly and Diesendorf (1985) demonstrate in the elasticities that they derive. A specific model (Equation 3.10) suggested by the general formulation may be used to illustrate the problem that this particular nonlinear model introduces:

$$\mathbf{Q}_E = \alpha_0 \cdot \mathbf{P}^{\beta_1 \mathbf{P}_{-1} + \gamma_1 \mathbf{P}_s + \gamma_2 \mathbf{Y} + \gamma_3 \mathbf{W}} \cdot \mathbf{P}_s^{\beta_2} \cdot \mathbf{Y}^{\beta_3} \cdot \mathbf{W}^{\beta_4} . \qquad (3.10)$$

The resultant elasticities are as follow:

Own price

$$\eta_P = \beta_1 \cdot \mathbf{P} \cdot (1 + \ln \mathbf{P}) + \gamma_1 \cdot \mathbf{P}_s + \gamma_2 \cdot \mathbf{Y} + \gamma_3 \cdot \mathbf{W} \qquad (3.11)$$

Cross price

$$\eta \mathbf{P}_s = \beta_2 + \gamma_1 \cdot \mathbf{P}_s \cdot \ln \mathbf{P} \qquad (3.12)$$

Income

$$\eta_Y = \beta_3 + \gamma_2 \cdot \mathbf{Y} \cdot \ln \mathbf{P} \qquad (3.13)$$

Weather

$$\eta_W = \beta_4 + \gamma_3 \cdot \mathbf{W} \cdot \ln \mathbf{P} . \qquad (3.14)$$

From Equations 3.11 to 3.14, it can be seen that the elasticities are indeed functions of the levels of the various explanatory variables, but they are not necessarily so, systematically, as Betancourt described. For example, using the ACT data and disregarding the effects of variables other than the price of electricity itself—when the electricity price is measured in cents per kilowatt-hour—the own-price elasticity is -1.39, measured at the data means. If the same model is estimated using the price data, stated in dollars per kilowatt-hour, the elasticity is measured as -0.94. Now, while it may be accepted that the elasticities vary around these mean values, it is not desirable that they be dependent on the original units of measurement as these are. The difficulty is easily corrected through normalizing the input data, which can be accomplished by dividing each series by its standard deviation—although this is not a normal procedure in model development. But this transformation ensures, then, that the resultant elasticities will remain invariant to any scalar changes in the raw data. Also, the model will provide what might be termed an "instantaneous" elasticity, $\beta_1 \cdot \mathbf{P}_{-1}$, and an "equilibrium" elasticity, which would be $\beta_1 \cdot \mathbf{P}(1 + \ln \mathbf{P})$. Insomuch as the normalized variable is not a fraction, the equilibrium elasticity will be larger in absolute value than the instantaneous one.

Additionally, the level of collinearity that would result from the inclusion in Equation 3.9 of all of the additional variables in the exponent of the price term makes the identification of individual parameters impossible, for when the HDD data are included in that exponent, the FG R^2 increases from 0.72 to 0.99, while the coefficient of multiple determination remains unchanged. Thus, the inclusion of

Table 3.8. Betancourt and Box-Cox Elasticities (1963/64 to 1981/82), Australian Capital Territory Electricity Variable Elasticity of Demand Models

Fiscal year	Instantaneous	Equilibrium	Box-Cox	Fiscal year	Instantaneous	Equilibrium	Box Cox
1963/64	-0.45	-1.23	-1.22	1973/74	-0.28	-0.64	-0.77
1964/65	-0.43	-1.16	-1.13	1974/75	-0.26	-0.57	-0.69
1965/66	-0.41	-1.08	-1.09	1975/76	-0.24	-0.52	-0.65
1966/67	-0.40	-1.06	-1.04	1976/77	-0.23	-0.48	-0.60
1967/68	-0.39	-1.02	-1.03	1977/78	-0.23	-0.49	-0.59
1968/69	-0.38	-1.00	-0.97	1978/79	-0.23	-0.47	-0.57
1969/70	-0.37	-0.96	-0.96	1979/80	-0.22	-0.46	-0.55
1970/71	-0.34	-0.85	-0.90	1980/81	-0.24	-0.50	-0.55
1971/72	-0.32	-0.79	-0.85	1981/82	-0.28	-0.63	-0.59
1972/73	-0.31	-0.73	-0.83	at the means	-0.32	-0.76	-0.79

Source: Compiled by the author.

extra variables would not yield more information about the behavior of the own-price elasticity. Furthermore, comparing the normalized Betancourt model results with the BC results suggests that the latter may be a more appropriate formulation. Analysis of the results presented in Table 3.8 suggests that whereas the Betancourt model formulation might provide some insights, the simpler BC form provides similar elasticities, and so the maxim of parsimony recommends the latter. All of these results indicate that further analysis of the ACT demand behavior is warranted.

TASMANIA

The demand for electricity in Tasmania is of major contemporary interst in Australia because of the environmental implications of alternative forecasts of the future requirements of that state for electricity.[18] As mentioned earlier, HEC, a state government entity, is responsible for supplying electricity there, virtually all of which is derived from hydroelectric dams (95.2 percent of the total generated in 1981/82). HEC has as its working hypothesis that the availability of cheap electricity generates regional employment and provides that island state with a comparative advantage over the much larger and dominant economy of the Australian mainland.[19] Much of the discussion of this philosophy of economic growth has centered on the appropriate modeling strategy to adopt for preparing long-range electricity forecasts, with HEC favoring trend extrapolation. In 1980 Saddler et al. studied the costs and benefits of a proposal by HEC to develop a hydroelectric dam on the lower portion of the Gordon River below the Franklin River. Subsequently, Donnelly and Saddler (DS) (1982, 1984) reviewed Tasmania's future need for electricity for testimony before a Senate Select Committee of the Commonwealth Parliament and prepared several alternative scenarios for the growth in retail electricity demand there.[20] This section draws upon that work and extends the analysis to study the residential, commercial, and industrial demand. The retail electricity demand elasticities estimated from studies are presented in Table 3.9.

The Tasmanian electricity consumption data in the Saddler and Donnelly (SD) (1982, 1983) work are divided into two categories: retail and bulk users. The bulk users comprise 17 customers and

Table 3.9. Retail Electricity Demand Elasticities for Tasmania

	Saddler et al. (1980)		Hydro-electric Commission (1983)		Donnelly and Saddler (1984)			Saddler and Donnelly (1983)		
	Coefficient	t	Coefficient	t	Coefficient	t	FG R^2	Coefficient	t	FG R^2
P_E	-0.54	(-2.98)	-0.48	(-2.93)	-0.56	(-3.58)	0.88	-0.61	(-3.24)	0.88
P_S	0.30	(2.74)	0.24	(6.30)	0.31	(4.32)	0.25	0.29	(5.45)	0.16
Y	0.81	(5.83)	0.99	(7.88)	1.13	(8.53)	0.89	1.05	(7.02)	0.87
Constant	-1.29	(-3.38)	3.95	NR	7.15	(8.50)	–	6.49	(7.33)	–
Years	1961–77		1962–81		1961–80			1961–82		
Number of observations	17		20		20			22		
R^{-2}	0.988		0.986		0.985			0.978		
DW	1.32		1.39		1.696			1.665		

Note: NR, not reported.

Sources: H. D. W. Saddler, J. Bennett, I. Reynolds, and B. Smith, *Public Choice in Tasmania: Aspects of the Lower Gordon River Hydro-electric Development Proposal,* Centre for Resource and Environmental Studies Monograph 2 (Canberra: Australian National University, 1980); Tasmanian Hydro-electric Commission, "Load Forecast," Hobart (1983), p. 23; W. A. Donnelly and H. D. W. Saddler, "The Retail Demand for Electricity in Tasmania," *Australian Economic Papers,* 23 (1984):54. H. D. W. Saddler and W. A. Donnelly, "Electricity in Tasmania: Pricing, Demand, Compensation," Centre for Resource and Environmental Studies Working Paper No. 1983/84 (Canberra: Australian National University), 1983, p. 40.

represent the major industrial load in the state, accounting for approximately two-thirds of total consumption; the retail, or general load, customers account for the balance of demand. The retail total sales of electricity are obtained from the statistical section of HEC's annual report.[21] The per capita sales figure Q_E is calculated using the ABS annual estimates of the population of Tasmania. The price of electricity variable is defined as retail average revenue per kilowatt-hour deflated by the Hobart CPI—the base year is 1966/67—and since detailed tariffs are unavailable, the ex post average price is used in this model instead of some measure of marginal price.[22] The price of heating oil series is obtained from the Australian Institute of Petroleum (AIP) (1960–74) and the Prices Justification Tribunal (post 1974) published statistics, and are stated in "constant 1966/67 dollar" terms in the price of the substitute good variable constructed P_S. As the model's income variable, the ABS series for male average weekly earnings (AWE) is used. Since a state-level AWE became available only in 1966, the earlier years of data are derived from the national figures, using the average ratio of Tasmanian to national AWE series over the period 1966/67 to 1979/80. The income variable Y represents the CPI-deflated AWE series thus developed. Finally, as a weather variable W, the average minimum temperature for the five coldest months of the year (May to September) for the two major Tasmanian cities, Hobart and Launceston, is included so as to attempt to ascertain the influence of climate on the retail consumption of electricity there.[23]

A static demand model formulation is adopted and estimated from the data in Table 3.10. The retail demand for electricity is posited to be a function of the price of electricity, the price of substitute fuels, income, and temperature. The general functional relationship is of the form of Equation 3.1. Both linear and log-linear formulations were estimated; in neither is the temperature variable statistically significant, and it has been omitted from the reported results of the double logarithmic model, whose findings are judged to be the superior ones.

Equation 3.15 presents them below. As in the ACT model results, the respective t statistics appear in parentheses below the parameter estimates, and all of the parameters in this preferred model are statistically different from 0 and have the appropriate algebraic signs. Collinearity, again measured by the FG R^2, does not appear to be a problem as all of those values are considerably smaller

Table 3.10. Tasmanian Retail Electricity Dataset (1960/61 to 1981/82)

Fiscal year	Electricity: Kilowatt hours per capita	Electricity: Average price (cents per kilowatt-hour)	Heating oil price (cents per liter)	Average weekly earnings (dollars)	Monthly average minimum temperature (degrees centigrade)	Hobart: Consumer price index	Hobart: Other fuel index
1960/61	2,245.5	1.472	4.80	44.50	4.67	90.3	17.0
1961/62	2,342.0	1.592	4.80	45.70	5.02	90.7	16.8
1962/63	2,484.8	1.574	4.73	48.20	3.68	90.7	16.8
1963/64	2,648.7	1.554	4.84	52.30	4.56	91.7	17.1
1964/65	2,780.6	1.553	5.04	54.10	4.88	94.6	17.8
1965/66	2,898.3	1.554	5.18	58.50	4.67	97.9	18.2
1966/67	3,112.9	1.525	5.32	58.50	4.97	100.0	18.7
1967/68	3,035.5	1.609	5.50	62.00	4.52	104.6	19.3
1968/69	3,262.6	1.619	5.46	65.70	4.67	106.1	19.1
1969/70	3,435.2	1.615	5.46	70.90	4.77	108.5	19.1
1970/71	3,548.3	1.628	5.63	78.50	4.26	112.6	19.7
1971/72	3,813.8	1.766	5.65	87.50	3.67	119.9	20.8
1972/73	3,918.3	1.858	5.65	95.20	4.84	126.7	19.8
1973/74	4,095.8	1.852	5.70	110.50	4.89	142.5	22.2
1974/75	4,409.7	2.042	6.82	140.20	4.57	166.7	27.0
1975/76	4,607.1	2.428	9.10	157.80	4.85	190.0	33.5
1976/77	5,243.3	2.696	11.50	199.00	4.77	239.1	42.2
1978/79	5,564.6	2.984	15.50	211.60	4.25	257.7	53.3
1979/80	5,732.5	3.128	21.30	237.90	5.12	284.0	79.9
1980/81	5,810.9	3.637	30.50	271.00	NA	310.0	100.1
1981/82	5,804.2	4.241	NA	266.60	NA	341.1	115.7

Note: NA, not available.
Source: Compiled by the author.

than the R^{-2} for the equation.[24] It is to be anticipated that collinearity will be less severe in static demand models such as this one and those that Table 3.7 depicts. The DW statistic value indicates that first-order serial correlation is not a problem.

$$\ln \mathbf{Q}_E = 7.15 - 0.56 \ln \mathbf{P}_E + 0.31 \ln \mathbf{P}_S + 1.13 \ln \mathbf{Y}, \quad (3.15)$$

t	(8.497)	(−3.580)	(4.322)	(8.531)
FG R^2		0.88	0.25	0.89

$\bar{R}^2 = 0.985$
DW = 1.696
RMSE = 3.2%
Theil's U = 0.562.

These results suggest that a 1 percent increase in the price of electricity will result in about a 0.6 percent decline in the per capita consumption of electricity. There is a 95 percent probability that the elasticity estimates are within the following ranges:

$$-0.84 < \eta_{P_E} < -0.29$$
$$0.18 < \eta_{P_S} < 0.43$$
$$0.90 < \eta_Y < 1.36.$$

While these ranges are large, it is clear that retail electricity consumption in Tasmania is price inelastic, that heating oil as included in the model does not represent a "close" substitute for electricity, and that the commodity is generally considered to be a luxury good in Tasmania, as evidenced by its income-elastic response.[25] The RMSE in this equation is 3.2 percent of the per capita retail consumption, while the calculated Theil inequality coefficient, Theil's *U* statistic, suggests that the results from Equation 3.15 forecast better than a naive no change model, Equation 2.1—a judgmental forecasting model. Confidence limits can be placed upon the equation estimates, with the range of probability being a minimum at the mean values of the explanatory variables.[26] The 95 percent confidence interval, measured at the means, amounts to ±8 percent of the predicted consumption Q_E of 3,600 kilowatt-hours per capita. This same confidence limit placed on the remaining estimates differs across the observation set and is ±9 percent at the fiscal year 1960/61 values, with $Q_E = 2{,}400$ kilowatt-hours per capita, and ±10 percent at the 1979/80 values, with $Q_E = 6{,}000$ kilowatt-hours per capita. The confidence range for several scenarios that were constructed for forecasting purposes is calculated directly in the regression routine using the dummy variable approach suggested by Salkever (1976). He demonstrates that when extra data points are added to the original dataset, these taking on the scenario values for the independent variables and combining with dummy variables for those observations, then the "coefficients for each of these dummy variables will equal the prediction error for the corresponding data point, while the variances of these coefficients will equal the prediction error variances" (Salkever 1976, p. 394).

The results in Table 3.9 may be compared with the earlier Saddler et al. (1980) and Australian Department of National Development and Energy (DNDE) (1981) national electricity elasticities illustrated in Table 3.11. The former database is based upon calendar year information, and these data are used to replicate that model's results while adjusting for the first-order correlation present in the Saddler et al. results. All of the elasticities are similar to the current findings, and this outcome is to be expected.[27] The correction for serial correlation, which takes the form of an MA adjustment as defined by Equation 2.31, with the estimate MA $\theta = 1.00$, increases the income elasticity to 0.93, reduces the cross-price elasticity, and changes the own-price elasticity to -0.49. When these revised results are compared with the estimates in Equation 3.15, no statistically significant difference in the individual parameter estimates is indicated; however, when the parameters are taken collectively, the null hypothesis that the results derive from the same population is rejected.[28]

The DNDE estimates (1981) are not entirely comparable, since the definitions of the variables differ from those used in both of the cited analyses of electricity demand in Tasmania. These elasticities are presented solely to illustrate results of another study of the de-

Table 3.11. Previous Demand Elasticities

	Tasmania Saddler et al. (1980)		Australian Department of National Development and Energy	
Easticity	Original	Replicated	Substitute fuel-heating oil	Substitute fuel-crude oil
P_E	-0.54	-0.49	-0.86	-0.19[a]
P_S	0.30	0.21	0.25	0.11
Y	0.81	0.93	0.78	1.27
DW	1.32	1.771	NR	NR
Years	1961-77		1967/68 to 1978/79	
Number of observations	17		12	

Note: NR, not reported.

[a] Not significant at the 0.1 level.

Sources: H. D. W. Saddler, J. Bennett, I. Reynolds, and B. Smith, *Public Choice in Tasmania: Aspects of the Lower Gordon River Hydro-electric Development Proposal,* Centre for Resource and Environmental Studies Monograph 2 (Canberra: Australian National University, 1980); Australian Department of National Development and Energy, *Forecasts of Energy Demand and Supply: Primary and Secondary Fuels, Australia: 1980–81 to 1989–90* (Canberra: Australian Government Publishing Service, 1981), p. 149.

mand behavior of Australian electricity consumers. The differences in the data include the couching of the DNDE dependent variable in terms of national total public electricity demand rather than state-level retail per capita demand, because DNDE uses gross domestic product and not AWE. Generally, the per capita measures are more appropriate, since such a specification eliminates the "spurious correlation" that exists when estimates are developed using the absolute levels of the data (Maddala 1977, p. 266). Finally, DNDE reports as its preferred "most suitable equation" the specification that includes the price of crude oil, even though in that model the own-price elasticity estimate is not significantly different from 0.

In reaction to the Saddler et al. results, HEC in 1979 argued (in private communication) that a change in the underlying electricity demand relationship occurred during the observation period, that is, in 1972. That hypothesis is statistically tested by Donnelly and Saddler (1982) using an F statistic, and is rejected. To do this, the dataset is divided into two subsamples and the collective parameter estimates tested, this being similar to the test done to compare the Saddler et al. and the DS model results. Because the data in the DS model are for fiscal years, the subsamples selected for testing are 1960/61 to 1970/71 and 1971/72 to 1979/80 in the first instance and 1960/61 to 1971/72 and 1972/73 to 1979/80 in the second. Also, the test is run to divide the data toward assessing the effect of the impact, if any, of the OPEC market actions between the years 1972/73 and 1973/74. The null hypothesis F statistic results appear in Table 3.12. The changed regime hypothesis is effectively rejected in all cases.

Table 3.12. Structural Change Tests (1960/61 to 1979/80)

Subsample	F statistic	Degrees of freedom
1960/61 to 1970/71 1971/72 to 1979/80	0.722	4, 12
1960/61 to 1971/72 1972/73 to 1979/80	0.793	4, 12
1960/61 to 1972/73 1973/74 to 1979/80	0.861	4, 12

Source: Compiled by the author.

A set of seven scenarios is defined using differing assumed rates of growth in the explanatory variables (see Table 3.13). It will be noted that scenario number 8 refers to the 1979 HEC projections, for which no information is available on the assumed economic conditions, these projections being based upon the trend extrapolation model of HEC, as was explained earlier. The DS document (1982) constructs a retail consumption series that reflects those 1979 projections, but since HEC does not actually report an analogous series for retail consumption, this constructed series represents an interpretation of the HEC figures and therefore must be construed accordingly.[29]

The conditional forecasts derived by the application of the model using the Salkever dummy variable technique are presented as Equation 3.15 and appear in Table 3.14. The conditional forecasted annual rates of growth in retail electricity demand through the year 2000 obtained from the various scenarios range between 1.6 percent in the case of the "expensive oil" scenario up to 4.9 percent for the "continuation of historical patterns" scenario. The scenario dubbed as the "most likely" suggests a growth rate of less than one-half of the one forecasted by HEC: namely, 1.8 versus 4.2 percent per annum. The confidence limits on the scenario number 1 results indicate that there is a 95 percent chance of the 1999/2000 value for the average retail consumer to fall in the range between 6,200 and 11,500 kilowatt-hours. This upper limit of the most likely scenario

Table 3.13. Tasmanian Growth Rate Scenarios

Scenario	Percentage growth rate		
	P_E	P_S	Y
(1) Most likely	2	2	2
(2) Cheap electricity	0	2	2
(3) High growth rate	0	2	3
(4) Maximum income	0	0	3
(5) Expensive oil	2	5	1
(6) Equal growth rates	3	3	3
(7) Historical rates	-2.1	1.8	2.8
(8) Hydro-electric Commission	NR	NR	NR

Note: NR, not reported.
Source: Compiled by the author

Table 3.14. Log-Linear Model Scenario Forecasts (thousands kilowatt-hours per capita)

	Scenario							
Fiscal year	(1) Most likely	(2) Cheap electricity	(3) High growth	(4) Maximum income	(5) Expensive oil	(6) Equal growth	(7) Historical rates	(8) Hydro-electric Commission
1984/85	6.5	6.9	7.3	7.1	6.5	6.8	7.6	7.3
1989/90	7.1	8.0	8.9	8.4	7.0	7.8	9.7	9.0
1994/95	7.8	9.2	10.9	9.9	7.5	8.9	12.4	11.0
1999/2000	8.5	10.6	13.3	11.8	8.1	10.1	15.9	13.4
95% confidence range								
1999/2000	±31	±20	±25	±20	±33	±44	±16	NR
Annual growth rates (%)	1.8	2.9	4.0	3.4	1.6	2.6	4.9	4.2

Note: NR, not reported.
Source: Compiled by the author.

is still considerably below the HEC projection. In Figure 3.9 the 1980/81 and 1981/82 actual consumption levels, which were not used in the modeling, are plotted, as well as the 95 percent confidence interval around the most likely scenario forecasts.

Subsequent to the Senate Select Committee hearings, various other forecasts of electricity demand for Tasmania have been developed, some of which are listed in Table 3.15.[30] All of these forecasts are in terms of average megawatts demanded, so that it is necessary to adjust the DS "most likely" scenario projections to conform with the others. To do this, the DS per capita conditional forecast taken from Table 3.14 must be multiplied by the expected population trends of Tasmania and divided by the number of hours available in a year for generating electricity, with an adjustment factor applied to account for HEC internal uses of electricity and for transmission losses. It should be pointed out that the DS forecasts are the only

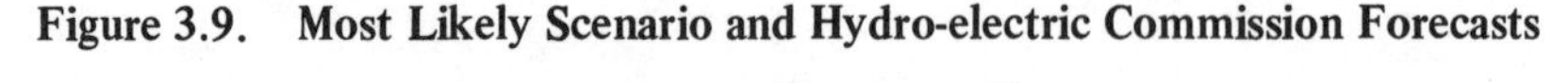

Figure 3.9. Most Likely Scenario and Hydro-electric Commission Forecasts

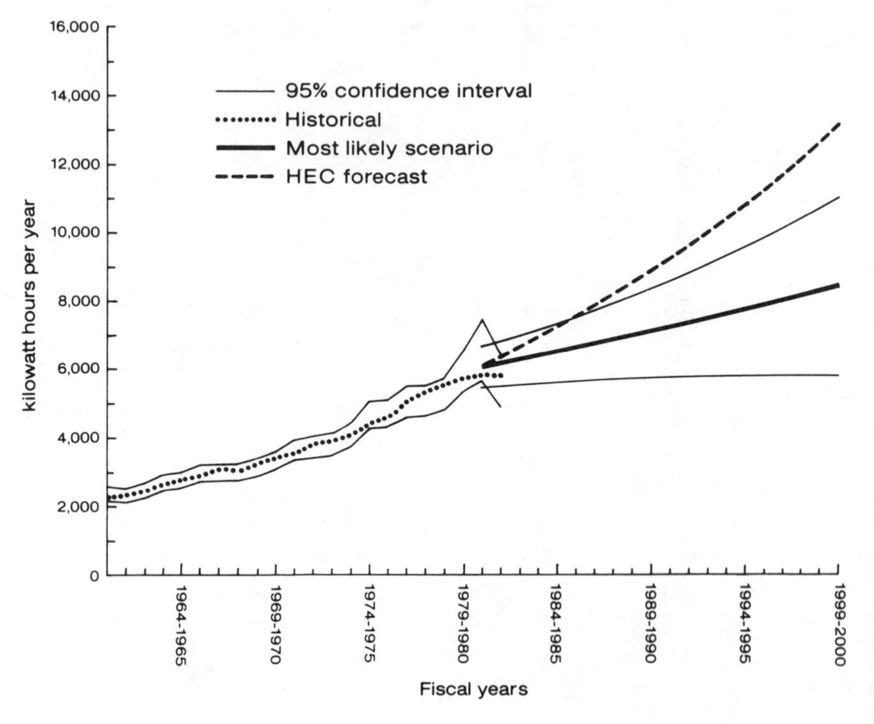

Table 3.15. Tasmanian Electricity Demand Forecasts (average megawatts)

Sector	Fiscal year			
	1981/82[a]	1989/90	1994/95	1999/2000
Major industrial	566			
HEC (1983)		698	762	831
Range		(672-725)	(717-807)	(782-881)
McLachlan (1982)		648	670	740
Range		(616-680)	(603-737)	(592-888)
General load	351			
HEC (1983)		540	667	818
Range		(515-586)	(623-722)	(742-884)
McLachlan (1982)		505	590	742
Range		(480-530)	(531-649)	(594-890)
Donnelly and Saddler (1982)	341[b]	401	449	496
95% confidence interval	(305-377)	(328-449)	(340-558)	(343-649)
South-West Tasmania Committee (New South Wales)		(375-429)	(441-536)	(510-666)
Directorate of Energy (1980)		533	NR	679
Range		(512-553)		(637-772)
Conservation case		503	NR	589
Range		(483-522)		(552-625)
Tasmanian Wilderness Society		428	497	579

Notes: NR, not reported.

[a] Tasmanian Hydro-electric Commission, "Report for the Year 1981–82" (Hobart: Parliament of Tasmania), 1982, p. 3, Table 1.

[b] Australian Bureau of Statistics, 1983, *Projections of Population of the States and Territories of Australia, 1981 to 2021.* Catalogue No. 3214.0, Canberra: Government Publishing Service.

Sources: Tasmanian Hydro-electric Commission, "Load Forecast," Hobart (1983), pp. 16 and 41, Tables 10 and 16; McLachlan, South-West Tasmania Committee (NSW) and Tasmanian Wilderness Society, in Senate Select Committee on South-West Tasmania, pp. 54, 75 (Table 4.8), 76; W. A. Donnelly and H. D. W. Saddler, "The Retail Demand for Electricity in Tasmania," Centre for Resource and Environmental Studies Working Paper No. R/WP-63, (Canberra: Australian National University, 1982), p. 13, Table 3 [HEC (1979) population figures used to convert "most likely" per capita conditional forecasts, adjusted by 1.12 for HEC use and transmission losses, to total megawatts]; Tasmanian Directorate of Energy, "Report to the Co-ordination Committee on Future Power Development" (Hobart: Tasmanian government, 1980), p. 219, Table 10.16.

econometric ones currently available for Tasmania, and that the methodology used in the other projections represents some combination of judgmental forecasting, usually trend extrapolation. The great divergence of results counsels that additional work needs to be done in identifying the factors affecting electricity demand in Tasmania.

As might be inferred from the description of the cooperation provided by HEC to "outside" analysts, the development of reliable sectoral demand for electricity equations requires substantial ingenuity and imagination and some heroic assumptions. The aggregate statistics that HEC supplies annually to ESAA indicate that in the fiscal year that ended in June 1982, there were 16,176 industrial customers. The 13 largest major users in 1982 accounted for 88.8 percent of the reported industrial consumption and approximately two-thirds of total electricity consumption; because of this discrepancy in the published ESAA figures, which greatly overstate the number of industrial users, those data must be adjusted so that the category reflects only the "bulk" consumers, with the balance of that ESAA classification added to the commercial sector. The resulting data are presented in Table 3.16. Now, whereas Donnelly and Saddler modeled the residential and commercial sectors as one, retail demand, the results obtained from modeling them separately suggest that the two components comprising that retail demand actually exhibit distinctly different behavioral responses.

The sectoral demand models are estimated using the general demand model specified in Equation 3.1; the dependent variable is average consumption per customer, and the ABS price index for "other fuels" in Hobart is employed as the substitute goods variable. The income variables for the respective sectors are household disposable income per residential customer, retail sales per commercial customer, and value added in manufacturing per bulk user. Sectoral revenue figures divided by the level of consumption provide the average price of electricity series, all of the monetary dataseries being deflated by the overall Hobart CPI. The weather variable is the Australian Bureau of Meteorology pseudo-HDD series for Hobart, measured from a base temperature of 12 degrees centigrade.[31] This variable is found to be insignificant in the commercial and industrial sectors and thus is omitted from those demand relationships. However, it is significant in the residential model, in contrast to the DS results with the temperature variable used in the retail demand specification model, and is included in the results reported in Table 3.17.

Table 3.16. Tasmanian Sectoral Model Dataset

Fiscal year	Residential			Commercial and minor industrial			Major industrial		Household income millions dollars	Australian disposal income ratio	Retail sales millions dollars	Value added millions dollars	Consumer price index	Other fuels price index	Heating degree days
	Megawatt-hours	Customers	Revenues thousands dollars	Megawatt-hours	Customers	Revenues thousands dollars	Megawatt-hours	Revenues thousands dollars							
1960/61	522,673	101,365	6,814	250,186	19,022	4,479	1,491,760	4,239	347	0.8840	163.5	124	90.3	17.0	501.4
1961/62	548,797	101,019	7,816	265,976	21,983	5,175	1,563,350	4,641	364	0.8843	166.1	127	90.7	16.8	399.6
1962/63	586,437	103,241	8,290	290,704	22,226	5,523	1,885,370	6,251	380	0.8890	172.9	141	90.7	16.8	646.8
1963/64	616,220	105,701	8,694	318,834	22,836	5,896	2,035,400	7,392	416	0.8835	180.2	151	91.7	17.1	522.6
1964/65	652,884	108,030	9,188	344,586	23,457	6,270	2,362,260	8,834	447	0.8726	191.0	165	94.6	17.8	515.9
1965/66	677,369	110,411	9,545	378,259	24,159	6,718	2,404,400	9,297	472	0.8681	198.3	173	97.9	18.2	486.8
1966/67	712,497	111,396	10,030	422,003	24,436	7,277	2,474,780	9,952	520	0.8683	215.0	192	100.0	18.7	447.5
1967/68	710,575	114,645	10,579	428,569	25,133	7,562	2,085,610	8,676	538	0.8588	227.9	195	104.6	19.3	515.6
1968/69	751,570	117,615	11,456	474,221	25,828	8,393	2,888,770	12,986	589	0.8616	240.9	194	106.1	19.1	498.2
1969/70	785,404	120,555	11,989	516,387	26,299	9,113	3,236,660	15,233	638	0.8521	256.0	223	108.5	19.1	439.9
1970/71	900.673	124,953	13,053	466,095	24,855	9,103	3,450,340	17,168	703	0.8533	272.1	230	112.6	19.7	512.3
1971/72	951,527	123,394	15,225	514,190	29,425	10,751	3,663,420	18,549	794	0.8479	291.3	240	119.9	20.8	499.7
1972/73	965,744	126,212	16,199	552,410	30,236	12,053	3,730,830	20,053	918	0.8546	342.6	278	126.7	19.8	414.7
1973/74	1,016,253	129,185	17,027	581,109	31,000	12,675	3,757,910	21,581	1,106	0.8428	402.3	333	142.5	22.2	403.3
1974/75	1,126,411	132,181	21,067	602,675	31,176	14,727	3,639,750	23,382	1,394	0.8218	485.4	395	166.7	27.0	443.5
1975/76	1,182,471	135,391	26,175	657,310	31,999	18,628	3,496,550	24,380	1,627	0.8255	532.3	445	190.0	33.5	369.6
1976/77	1,295,075	139,132	29,380	708,927	32,574	20,045	4,079,010	28,814	1,929	0.8184	618.0	520	217.7	36.9	566.8
1977/78	1,346,767	142,729	34,521	774,494	33,141	24,275	4,298,300	31,899	2,113	0.8182	685.4	480	239.1	42.2	433.4
1978/79	1,429,185	145,780	39,602	805,088	33,945	26,634	4,713,010	38,172	2,339	0.8278	745.6	549	257.7	53.3	459.3
1979/80	1,504,882	148,777	41,668	887,366	34,689	29,214	4,765,120	41,826	2,609	0.8212	807.7	654	284.0	79.9	485.3
1980/81	1,535,454	151,615	49,561	902,576	35,300	34,872	4,721,040	45,129	2,943	0.8172	885.7	713	310.0	100.1	485.3
1981/82	1,622,079	153,942	60,091	836,975	35,609	42,549	4,905,610	52,066	3,346	0.8070	968.1	731	341.1	115.7	485.3

Source: Compiled by the author.

Table 3.17. Box-Cox Estimates of Sectoral Demand Elasticities (1960/61 to 1981/82)

Sector	Elasticities				
	P_E	P_S	Y	W	
Residential	-0.16	0.19	1.32	0.10	
t	(-1.926)	(4.584)	(16.764)	(2.176)	
FG R^2	0.90	0.31	0.89	0.21	$\bar{R}^2$ = 0.986
					DW = 1.6360
					BCE λ = -1.71
					LLF = 6.786
Commercial	-0.83	0.18	0.30	NI	
t	(-7.074)	(4.037)	(2.249)		
FG R^2	0.80	0.17	0.78		$\bar{R}^2$ = 0.967
					DW = 1.8830
					AR ρ = 0.43
					(2.234)
					BCE λ = 1.06
					LLF = 28.714
Industrial	0.21	0.09	0.22	NI	
t	(0.518)	(0.500)	(0.779)		
FG R^2	0.34	0.37	0.04		$\bar{R}^2$ = 0.913
					DW = 1.5539
					AR ρ = 0.97
					(18.715)
					BCE λ = 1.18
					LLF = -5.546

Notes: NI, not included. Elasticities measured at the means of the variables.
Source: Compiled by the author.

The BC procedure is used on all of the sectors in an attempt to identify the appropriate function form. The parameter estimates are converted to elasticities, measured at the means of the variables, and these are what appear in the table.

In the residential sector, it can be seen from the value of the λ parameter estimate, −1.71, that neither the linear nor the log-linear functional form is appropriate, whereas in the commercial sector, the λ value approaches that that might be expected from the linear model. Serial correlation is a problem in both the commercial and the industrial sectors, as is evidenced by the DW statistics for these models, but an iterative Cochrane-Orcutt first-order AR adjustment based on Equation 2.30 was successful only in eliminating the serial correlation evidenced in the commercial sector. In the industrial sector, as representing only the bulk users, none of the parameter estimates is statistically significant, tending to support the DS-expressed hypothesis: "Because of the negotiated nature of these bulk contracts [i.e., the 13 to 17 major industrial consumers] that component of demand is not amenable to econometric modelling" (Donnelly and Saddler 1982, p. 3).

Once again, the residential demand for electricity is found to be highly price inelastic and income elastic and to have little potential for interfuel substitution. While this definition for the weather is statistically significant in the residential sector, its effect on consumption is small. The consumption of electricity in the commercial sector is more responsive to its price, but again little affected by the price of potential substitute fuels. The level of activity variable, retail sales, exhibits itself as an inelastic response in this sector's demand for electricity.

SUMMARY

Table 3.18 summarizes the various model results presented in this chapter and provides for comparison among them. It can be seen that the own-price elasticities for the residential sector vary substantially among the three regions for which estimates are available. Residential electricity consumption is income inelastic in the ACT, a high-income region; income elastic in Tasmania, a low-income region; and roughly of unitary elasticity in New South Wales, the state with the largest population. The cross-price elasticities in this sector differ by a factor of two between the ACT and Tasmania, and that elasticity is insignificant in New South Wales. (The mild climate in New South Wales relative to the other two regions may be one explanation for this response.) As might be expected, the commercial demand is more price elastic than is the residential demand, and it is income inelastic. These results provide evidence that the various states within Australia, as well as the individual consuming sectors, deserve to be modeled individually. Such research might determine if these apparent regional differences extend to the other states of Australia as well. As will be shown in Chapter 4, consumer behavior in the demand for motor gasoline differs regionally within Australia; it is reasonable to expect a similar result for electricity demand.

Table 3.18. Comparison of Australian Electricity Demand Elasticity Estimates

Elasticity	New South Wales and ACT Residential (1971) (Hawkins 1975)	ACT Residential (1963/64 to 1981/82)	Tasmania: Retail (Saddler and (Donnelly 1983)	Tasmania: Residential (1960/61 to 1981/82)	Tasmania: Commercial (1960/61 to 1981/82)
P_E	-0.55	-0.79	-0.61	-0.16	-0.83
P_S	NI	0.42	0.29	0.19	0.18
Y	0.93	0.27	1.05	1.32	0.30
W	NI	0.23	NI	0.10	NI
Number of observations	44	19	22	22	22
$\bar{R}^2$	0.988	0.977	0.978	0.986	0.967
Estimation procedure	OLS	BCE	OLS	BCE	BCE
Functional form	Linear	$\lambda = 0.59$	Log-linear	$\lambda = 1.71$	$\lambda = 1.06$

Notes: NI, not included. Elasticities measured at the means of the variables.

Sources: R. G. Hawkins, "The Demand for Electricity: A Cross-section Study of New South Wales and the Australian Capital Territory," *Economic Record,* 51(1975):10; H. D. W. Saddler, and W. A. Donnelly, "Electricity in Tasmania: Pricing, Demand, Compensation," Centre for Resource and Environmental Studies Working Paper No. 1983/24 (Canberra: Australian National University, 1983), p. 40; other data compiled by the author.

These Australian electricity demand estimates may be compared with studies done in the United States and the United Kingdom. Table 3.19 summarizes some of the more relevant studies. A wide diversity in the elasticity estimates is as evident in these studies as it is the Australian work. The distinct lack of robustness in all of these models argues strongly for consideration to be given to alternative modeling strategies, but often the data needed to sustain such approaches are not available. One approach that may warrant attention is the translog model specification, similar to the work to be discussed in Chapter 5. Such a strategy would substitute a set of demand equations for the single-equation models discussed in this chapter, and might thereby provide valuable insights into the substitution prospects among alternative energy sources. More temporal

Table 3.19. Selected Residential Electricity Demand Model Results

Study	Type	Elasticity: Own price	Elasticity: Cross price	Elasticity: Income	Elasticity: Weather	Data type	Period	Region	Other
FEA (1976)	LR	-0.15	–	0.10	–	CS TS A	1960–72	U.S. census	AP
Fisher and Kaysen (1962)	LR	-0.15	–	0.10	–	CS TS A	1946–57	U.S. states	AP
Griffin (1974)	LR	-0.27	–	0.53	–	TSA	1951–71	U.S. national	AP, DL
Halvorsen (1975)	LR	-1.00 to -1.21	0.04 to 0.08	0.47 to 0.54	-0.02 to -0.10	CS TS A	1961–69	U.S. states	AP
Houthakker (1951b)	LR	-0.89	0.21	1.17	–	CS	1937–38	U.K. towns	MP
Houthakker and Taylor (1970)	SR	-1.89	–	1.13	–	TS A	1947–64	U.S. national	AP
	LR	-1.89	–	1.93	–				
Houthakker, Verleger and Sheehan (1974)	SR	-0.09	–	0.13	–	CS TS A	1951–71	U.S. states	MP
	LR	-1.19	–	1.61	–				
Mount and Chapman (1979)	SR	-0.31	0.16	0.16	–	CS TS A	1963–72	U.S. states	AP
	LR	-1.17	0.61	0.61	–				
Mount, Chapman, and Tyrrell (1973)	SR	-0.14	0.02	0.02	–	CS TS A	1947–70	U.S. states	AP
	LR	-1.20	0.19	0.20	–				
Wilson (1971)	LR	-1.33	0.31	-0.46	0.04	CS	1966	U.S. cities	AP

Notes: LR, long run; SR, short run; CS, cross section; TS, time series; A, annual; M, monthly; AP, average price; MP, marginal price; DL, distributed lags.

Sources: FEA, *National Energy Outlook,* FEA-N75/713 (Washington, D.C.: U.S. Government Printing Office, 1976); F. M. Fisher and C. Kaysen, *The Demand for Electricity in the United States* (Amsterdam: North-Holland, 1962); J. M. Griffin, "The Effects of Higher Prices on Electricity Consumption," *Bell Journal of Economics and Management Science,* 5(1974): 515–39; R. Halvorsen, "Residential Demand for Electric Energy." *Review of Economics and Statistics,* 57(1975):12–18; H. S. Houthakker, "Some Calculations of Electricity Consumption in Great Britain," *Journal of the Royal Statistical Society,* A114 (1951b):359–71; H. S. Houthakker and L. D. Taylor, *Consumer Demand in the United States,* 2nd ed. (Cambridge, MA: Harvard University Press, 1970); H. S. Houthakker, P. K. Verleger, and D. P. Sheehan, "Dynamic Demand Analyses for Gasoline and Residential Electricity," *American Journal of Agricultural Economics,* 56(1974):412–18; T. D. Mount and L. D. Chapman, "Electricity Demand, Sulfur Emissions and Health: An Econometric Analysis of Power Generation in the United States," in *International Studies of the Demand for Energy,* W. D. Nordhaus, ed. (Amsterdam: North-Holland, 1979); T. D. Mount, L. D. Chapman, and T. J. Tyrrell, "Electricity Demand in the United States: An Econometric Analysis," ORNL-NSF-49 (Oak Ridge TN: Oak Ridge National Laboratory, 1973); J. W. Wilson, "Residential Demand for Electricity," *Quarterly Review of Economics and Business,* 11(1971):7–19.

disaggregation could serve better to identify the influence of the factors affecting electricity demand, since better measures of the marginal price and climatic variables can be developed for either monthly or quarterly datasets. Attention to diurnal variations in electricity consumption would provide the means to analyze the potential effects of the introduction of time-of-day tariffs, and such studies would concentrate on analysis of the factors affecting the demand for electric power in order to identify the appropriate mechanisms for reducing peak demand and smoothing the load curve. Hybrid models such as those introduced by Uri (1977, 1979) have proved to be a productive approach to the peak demand problem, but they represent a substantial increase in the data required for implementation. As an alternative to the Uri approach, econometric methods incorporating AR or MA error structures need to be explored too. However, as with all empirical work, such extensions must await the further development of the electricity database. The amount of modeling information to be derived is thus inevitably restricted by the quality of the published information and that that is released to interested analysts.

NOTES

1. Although this has changed somewhat since EC1 and EC2, individual forecasts within the industry still rely extensively upon judgmental and trend extrapolation forecasting procedures, as indicated in Chapter 2 [e.g., Bartels, Lopert, and Williamson (1985) on residential demand and the HEC load forecasts (1983)].

2. The Australian financial, or fiscal, year extends from July through June.

3. The ACT is the seat of the national capital, Canberra, and is similar in operation and responsibilities to the District of Columbia and Washington, D.C.

4. It should be noted that hydroelectric capacity in the United States is used to satisfy base-load requirements with peakload capacity provided by oil- or gas-fired turbines. However, the situation is reversed in Australia because of the uncertainty of rainfall and the need to maintain reservoir levels so that the primary irrigation function of the Snowy Scheme can be satisfied. Thus, base-load demand there is satisfied by coal-fired plants, with the Snowy hydro capacity being called upon only when the steam generation is insufficient to meet peak demands.

5. In addition to these studies and the ones mentioned in note 1, preliminary results of other Australian electricity demand studies by Stromback (1983) and Rushdi (1983) have recently been reported at professional meetings. Stromback models quarterly system consumption for Western Australia over the years

1962–82 and concludes that demand is highly price elastic in the long run, that is, –1.31, and income inelastic, that is, 0.19, as well. The model utilizes a polynomial lag structure in an attempt "better" to identify the "stock" and "flow" components that the author argues may be obscured in a simpler Koyck "lag" model. Conversely, Rushdi utilizes the Koyck lag structure to study the annual industrial electricity demand in South Australia over the years 1950–80. His findings are that industrial demand is price and output inelastic, that is –0.63 and 0.81, respectively, in the long run; and that neither oil nor coal is a close substitute for electricity, exhibiting cross-price elasticities of 0.20 and 0.13, respectively.

6. These are traditional definitions used by electricity authorities everywhere, and they are not without problems, because the consumption of a particular mode of usage is not always included in the proper generic class. For example, historically bulk-metered apartment complexes (multiunit housing with a single electricity meter) are usually included in the commercial category, because they are billed on that tariff rather than on a residential tariff. Similarly, large commercial users may be reported as industrial, once again because of the tariff arrangements. In some instances, adjustments have been made in present reporting that result in a closer agreement between the actual usage and the generic category in which the data are reported; but these modifications create an incompatibility of data between earlier and current years, so even adjustments designed to correct existing "errors in variables" will pose new challenges to the unwary modeler.

7. Hawkins (1975) states that various marginal prices were tested and those measures found not to be statistically significant. This issue of the most appropriate price variable to use in modeling will be discussed more later.

8. Portions of this section appear in Donnelly (1984, 1985), Donnelly and Leung (1983), and Donnelly and Diesendorf (1985), and are used here with permission.

9. Taylor (1975) raises even more fundamental theoretical questions based upon the possibility of multiple equilibria in the utility maximization caused by the consumer's nonlinear budget constraint implicit with a declining block rate tariff. Also, he points out that the resultant demand function would be discontinuous wherever the equilibrium switched from one segment of the budget constraint to another.

10. During 1977/78 the average ACT household consumed 9,456 kilowatt-hours, representing an average quarterly consumption of 2,364 kilowatt-hours. The average price per kilowatt-hour for that level of usage would be 2.32 cents, with the consumption falling in the lowest-priced final block of the tariff.

11. Berndt (1983) shows, "both analytically and empirically, that least squares estimates with the average intramarginal price variable excluded will typically differ very little from regression estimates with that variable included" (p. 14).

12. Roughly distributing the monthly system load similar to that depicted in Figure 3.1 would suggest quarterly consumption of perhaps 1,600, 2,000, 2,500, and 3,350 kilowatt-hours. The marginal price for these levels would be a constant 0.95 cents, and the average price per quarter would be 2.50, 2,39,

and 2.20 cents per kilowatt-hour, respectively, in Australian currency at 1966/67 values. Naturally, because of the constant marginal price, the average annual price throughout the year would not change from the 2.32 cents per kilowatt-hour that would result from a constant quarterly level of demand of 2.364 kilowatt-hours. Thus, the problem of identification of the demand curve may not really be an issue, at least with this current dataset.

13. Green and Salley (undated) have recently utilized household survey data for the United States to analyze the elasticities and substitutability of electricity and natural gas. Unfortunately, the data covering 120 standard metropolitan statistical areas are annual figures, and the authors do not utilize tariff information to calculate price variables yielding electricity price elasticities (incidentally, those appear to be excessively large, ranging from -0.38 to -0.97); but similar types of cross-sectional analyses in the future should provide valuable additional insights for policy analysts.

14. A degree day is defined as a 1-degree deviation of the average of the daily maximum and minimum temperatures from some arbitrary base temperature, such as a base temperature of 65 degrees Fahrenheit (the base used by the U.S. National Oceanic and Atmospheric Administration), or 18.3 degrees centigrade. These deviations are summed for all of the days in the period (month, quarter, or year). Negative (positive) deviations reflect heating (cooling) degree days. Models that include both HDD and CDD as explanatory variables, both calculated from the same base temperature, introduce high levels of multicollinearity, as simple logic demands a perfect negative collinearity between the two series. Tests run while the author was Chief of the Modeling and Forecasting Division at FEA (DOE) indicate a coefficient of multiple determination of between 0.90 and 0.99 for annual state-level values of the two variables. Models including both variables, as quite a few studies do, must suffer from problems of coefficient identification, although this issue has not been addressed in the literature. Therefore, in the present work, two separate bases are used: 12 degrees Celsius (53.6 degrees Fahrenheit) for HDD and 22 degrees Celsius (71.6 degrees Fahrenheit).

15. In private discussions with Pagan he suggests that more powerful LM tests may be derived and that when the null hypothesis is rejected by the particular form of the LM test as calculated in the SHAZAM version 4.5 used in this analysis, then the heteroscedasticity is likely to be acute.

16. Recall elasticity of demand reflects the percentage change in the quantity demanded for a given percentage change in one of the explanatory variables of the model, for example, its own price. Mathematically, this is

$$\sigma_{P_E} = \frac{\theta Q_E}{\theta P_E} \cdot \frac{P_E}{Q_E} ,$$

Or in this specific case,

$$\sigma_{P_E} = \left(\beta_1 \cdot \frac{Q_E}{P_E}\right) \cdot \frac{P_E}{Q_E} \equiv \beta_1 .$$

17. The long-run (LR) own-price elasticity is defined as

$$\sigma_{LR}P_E = \left(\frac{1}{1 - \beta_5}\right) \cdot \sigma_{SR}P_E \ .$$

18. Portions of this section appear in Donnelly and Saddler (1982, 1984) and Saddler and Donnelly (1982, 1983) and are reproduced with permission.

19. The efficacy of this position is widely debated there, and similar discussions have occurred in the United States concerning hydroelectric projects proposed and constructed by the U.S. Corps of Engineers.

20. It may be noted that some time after those Senate Select Committee Hearings were conducted, HEC included in a load forecasting report (HEC 1983) scenario results derived from an econometric model; while HEC still refuses to adopt that modeling strategy, it now, apparently, at least accepts its methodological validity.

21. Sales to King and Flinders Islands are excluded.

22. It may be noted that HEC maintains a largely defensive posture with respect to "external" analysts, including political members and civil service employees of both the state government of Tasmania and the Commonwealth government of Australia, as well as academics. This penchant for secrecy is not unique to HEC, permeating several of the other electricity-generating authorities (all government entities) in Australia, and this attitude thwarts many potentially useful analyses that might be undertaken. But the atmosphere is gradually changing, as those bodies recognize that improved analyses are not necessarily a threat to their existence and can actually benefit them.

23. Tasmania, as is so for the ACT, has a relatively harsher climate than the rest of Australia, but much of their residential space heating relies upon wood-burning stoves (the state is quite heavily forested). A deficiency of the retail electricity demand analysis is that insufficient information on this energy source is available to include it in the present model. Widespread usage of wood-burning stoves poses similar problems in studies in the United States as well.

24. See note 29 in Chapter 2.

25. Tasmania has a higher unemployment rate than the average in Australia, has experienced a recent history of out-migration, and exhibits the lowest rates of economic growth in that country, so this finding with respect to the effect of income changes on the demand for electricity there is not surprising.

26. The single-explanatory variable OLS model equation always passes through the means of the observations.

27. In the discussion of econometric methods presented in Chapter 2, it was indicated that the presence of serial correlation will result in inefficient, but unbiased and consistent, parameter estimates.

28. A t statistic is calculated to test the null hypothesis that the individual elasticity estimates in the replicated Saddler et al. and DS models come from the same population. The t values are -0.470, 1.391, and 1.515 for η_{P_E}, η_{P_S}, and η_Y, respectively. Taken collectively, the F statistic rejects the hypothesis $F_{(3, 16)} = 11.272$ (see Kmenta 1971, pp. 366–70).

29. HEC prepared a projection of general load (which is equal to retail sales less HEC's own internal uses and system losses) as part of its "Report on the Gordon River Power Development, Stage 2." The basis of its general load projection is explained in HEC (1979, p. 20). The growth rates used to reconstruct those HEC forecasts for retail sales appear in Donnelly and Saddler (1982, Appendix, Table F).

30. This information has been corrected and extended from that that appears in Australian Senate Select Committee (1982b).

31. This HDD series is one for Hobart that corresponds to the Canberra one used in the ACT model and described earlier under Australian Capital Territory.

REFERENCES

Anderson, K. P. 1973. "Residential Demand for Electricity: Econometric Estimates for California and the United States." *Journal of Business,* 46:526–53.

Australian Bureau of Statistics. 1983. "Projections of Population of the States and Territories of Australia, 1981 to 2021." Catalogue No. 3214.0. Canberra: Government Publishing Service.

Australian Bureau of Statistics, Quarterly, "Time-Series Data on Magnetic Tape and Microfiche," Catalogue No. 1311.0, Canberra: Government Publishing Service.

Australian Senate Select Committee on South-West Tasmania. 1982a. *Hearings, Thursday, 4 February 1982.* Parliament of the Commonwealth of Australia, *Official Hansard Report.* Canberra: Australian Government Publishing Service.

——. 1982b. *Report on Demand and Supply of Electricity for Tasmania and Other Matters.* Parliament of the Commonwealth of Australia, Canberra: Australian Government Publishing Service.

Bartels, R. 1983. "Modelling Household Demand for Electricity in N.S.W.: Literature Review and Suggestions." Department of Economic Statistics. Sydney: University of Sydney. Mimeo.

Bartels, R., P. Lopert, and S. Williamson. 1985. "The Residential Demand for Electricity in N.S.W." Sydney: Energy Authority of N.S.W.

Belsley, D. A., E. Kuh, and R. E. Welsh. 1980. *Regression Diagnostics: Identifying Differential Data and Sources of Collinearity.* New York: John Wiley & Sons.

Berndt, E. R. 1983. "Modelling the Aggregate Demand for Electricity: Simplicity vs. Virtuosity." Alfred P. Sloan School of Management Working Paper No. 1415-83. Cambridge, MA: Massachusetts Institute of Technology.

Betancourt, R. R. 1981. "An Econometric Analysis of Peak Electricity Demand in the Short Run." *Energy Economics,* 3:14–29.

Blattenberger, G. R., L. D. Taylor, and R. K. Rennhack. 1983. "Natural Gas Availability and the Residential Demand for Energy." *Energy Journal,* 4:23–45.

Box, G. E. P., and D. R. Cox. 1964. "An Analysis of Transformations." *Journal of the Royal Statistical Society,* B26:211–43.

Donnelly, W. A. 1985. "Electricity Demand Modeling." In *New Mathematical Advances in Economic Dynamics,* edited by D. F. Batten and P. F. Lesse, pp. 179–95. Sydney: Croom-Helm.

——. 1984. "Residential Electricity Demand Modeling in the Australian Capital Territory: Preliminary Results." *Energy Journal,* 5:119–31.

Donnelly, W. A., and M. Diesendorf. 1985. "Variable Elasticity of Demand Models for Electricity." *Energy Economics,* 7:159–62.

Donnelly, W. A., and E. S. Leung. 1983 "Residential Electricity Demand." *Search,* 14:206–11.

Donnelly, W. A., and H. D. W. Saddler. 1984. "The Retail Demand for Electricity in Tasmania." *Australian Economic Papers,* 23:53–59.

——. 1982. "The Retail Demand for Electricity in Tasmania." Centre for Resource and Environmental Studies Working Paper No. R/WP-63. Canberra: Australian National University.

Electricity Supply Association of Australia (ESAA). Annual. *The Electricity Supply Industry in Australia.* Melbourne: ESAA.

Fisher, F. M., and C. Kaysen. 1962. *The Demand for Electricity in the United States.* Amsterdam: North-Holland.

Green, R. D., and A. Salley. Undated. "Price Elasticities of Natural Gas and Electricity from Micro Data." In *Proceedings Ninth International Association of Science and Technology for Development International Symposium on Energy, Power and Environmental Systems,* edited by M. H. Hamza. pp. 92–95. Anaheim: ACTA Press.

Griffin, J. M. 1974. "The Effects of Higher Prices on Electricity Consumption." *Bell Journal of Economics and Management Science,* 5:515–39.

Halvorsen, R. 1975. "Residential Demand for Electric Energy." *Review of Economics and Statistics,* 57:12–18.

Hawkins, R. G. 1975. "The Demand for Electricity: A Cross-Section Study of New South Wales and the Australian Capital Territory." *Economic Record,* 51:1–18.

Houthakker, H. S. 1951a. "Electricity Tariffs in Theory and Practice." *Economic Journal,* 61:1–25.

——. 1951b. "Some Calculations of Electricity Consumption in Great Britain." *Journal of the Royal Statistical Society,* A114:359–71.

Houthakker, H. S., P. K. Verleger, and D. P. Sheehan. 1974. "Dynamic Demand Analyses for Gasoline and Residential Electricity." *American Journal of Agricultural Economics,* 56:412–18.

Houthakker, H. S., and L. D. Taylor. 1970. *Consumer Demand in the United States.* 2nd ed. Cambridge, MA: Harvard University Press.

Kmenta, J. 1971. *Elements of Econometrics.* New York: Macmillan.

Maddala, G. S. 1977. *Econometrics.* New York: McGraw-Hill.

Mount, T. D., and L. D. Chapman. 1979. "Electricity Demand, Sulfur Emissions and Health: An Econometric Analysis of Power Generation in the United States." In *International Studies of the Demand for Energy,* edited by W. D. Nordhaus, pp. 95–114. Amsterdam: North-Holland.

Mount, T. D., L. D. Chapman, and T. J. Tyrrell. 1973. "Electricity Demand in the United States: An Econometric Analysis." ORNL-NSF-49. Oak Ridge, TN; Oak Ridge National Laboratory.

Nordin, J. A. 1976. "A Proposed Modification of Taylor's Demand Analysis: Comment." *Bell Journal of Economics,* 7:719–21.

Pagan, A. R., and A. D. Hall. 1983. "Diagnostic Tests, as Residual Analysis (with comments)." *Econometric Reviews,* 2:158–254.

Pindyck, R. S. 1976. "International Comparisons of the Residential Demand for Energy: A Preliminary Analysis." MIT Energy Laboratory Working Paper No. MIT EL 176-032 WP. Cambridge, MA: Massachusetts Institute of Technology.

Pindyck, R. S., and D. L. Rubinfeld. 1981. *Econometric Models and Economic Forecasts.* 2nd ed. New York: McGraw-Hill.

Rushdi, A. A. 1983. "Industrial Demand for Electricity in South Australia." Paper presented at the Twelfth Conference of Economists, University of Tasmania, Hobart. Mimeo.

Saddler, H. D. W., J. Bennett, I. Reynolds, and B. Smith. 1980. *Public Choice in Tasmania: Aspects of the Lower Gordon River Hydro-electric Development Proposal.* Centre for Resource and Environmental Studies Monograph 2. Canberra: Australian National University.

Saddler, H. D. W., and W. A. Donnelly. 1983. "Electricity in Tasmania: Pricing, Demand, Compensation." Centre for Resource and Environmental Studies Working Paper No. 1983/24. Canberra: Australian National University.

——. 1982. "The Demand for Energy in Tasmania, with Particular Reference to Electricity." Senate Select Committee on South-West Tasmania, Official Hansard Report, pp. 150–212.

Salkever, D. S. 1976. "The Use of Dummy Variables to Compute Predictions, Prediction Errors and Confidence Limits." *Journal of Econometrics,* 4:393–97.

Savin, N. E., and K. J. White. 1978. "Estimation and Testing of Functional Form and Autocorrelation." *Journal of Econometrics,* 8:1–12.

Seaks, T. G., and S. K. Layson. 1983. "Box-Cox Estimation with Standard Econometric Problems." *Review of Economics and Statistics,* 65:160–64.

Stromback, C. T. 1983. "A Structural Dynamic Model of Electricity Demand in Western Australia." Paper presented at the First Annual Conference on Econometrics, Australian National University, Canberra. Mimeo.

Tasmanian Directorate of Energy. 1080. "Report to the Co-ordination Committee on Future Power Development." Hobart: Tasmanian government.

Tasmanian Hydro-electric Commission. 1983. "Load Forecast." Hobart.

——. "Report for the Year 1981–82," Hobart: Parliament of Tasmania.

Taylor, L. D. 1979. "Decreasing Block Pricing and the Residential Demand for Electricity." In *International Studies of the Demand for Energy,* edited by W. D. Nordhaus, pp. 65–79. Amsterdam: North-Holland.

——. 1975. "The Demand for Electricity: A Survey." *Bell Journal of Economics and Management Science,* 6:74–110.

Uri, N. D. 1979. "A Mixed Time-Series/Econometric Approach to Forecasting Peak System Load." *Journal of Econometrics,* 9:155–71.

——. 1977. "An Integrated Box-Jenkins/Econometric Model for Forecasting Time Series." *Proceedings of the American Statistical Association,* Part I: 404-7.

U.S. Federal Energy Administration. 1976. *National Energy Outlook.* FEA-N75/713. Washington, D.C.: U.S. Government Printing Office.

Wilson, J. W. 1971. "Residential Demand for Electricity." *Quarterly Review of Economics and Business,* 11:7-19.

Four

Gasoline Demand Modeling

OVERVIEW

The demand for gasoline is a derived demand with many similarities to the demand for electricity. Although the need to travel and the characteristics of the vehicle stock will affect this demand, the paucity of data on automotive stock and their usage patterns necessitates that a reduced form, dynamic demand model be estimated in order to capture any long-run effects of changing economic conditions. Three specific gasoline demand models are discussed and evaluated herein: the first is a national model, the second is a static model, and the last is a dynamic one.

INTRODUCTION

As is the case with electricity demand-modeling work, only a limited number of motor gasoline studies have been undertaken in Australia.[1,2] Folie (1977) has developed a set of state-level demand estimates, and the results suggest that consumer behavior differs across the six Australian states, but these results are not published. However, Donnelly (1982a), using more recent regional data, confirms the impression provided by Folie's single-equation OLS work in a set of restricted Zellner "seemingly unrelated equations" (SURE) estimates (1962). The other research to be analyzed is that of Schou and Johnson (SJ) (1979), but this work is done at the national rather than the state level. The contrasting results are illuminating, and are considered in detail, beginning with the national one and concluding with the pooled time-series cross-sectional model.

NATIONAL ESTIMATES

The national model specified by Schou and Johnson is the first econometric model published on Australian gasoline demand behavior, and it will be shown that, as in all national models for countries occupying large geographic territories such as Australia, Canada, and the United States, this one ignores important regional differences in the behavior of automobile and other vehicle drivers.[3] Sections of "outback" Australia are extremely isolated and are accessible only by airplane or long-distance road travel; thus, the response to changes in the price of gasoline and economic conditions in general of individuals living in such areas would be expected to differ from that for persons residing and working in metropolitan areas. It should, however, be recognized that only a small proportion of the Australian population lives outside the 11 major metropolitan centers in Australia.[4] The SJ estimation, which relies upon OLS, and the varying parameters approach of Cooley and Prescott (1973), which uses annual data for the years 1955–76, appear in Table 4.1 in terms of the modeling data. The results of several OLS model specifications are presented, and these include as explanatory variables price, income, stock of vehicles, and average fuel consumption. Statistical results of these models, and their derived elasticities are shown in Table 4.2: Regression I monetary variables are current dollars, while Regression II and III variables are represented in constant 1967 dollar terms. It can be seen that collinearity for the price, income, and stock of vehicles variables is high in the current dollar formulation; this contingency is not surprising, in that inflation might be expected to affect the price and income variables similarly. The income elasticity in this formulation is small; however, in terms of the level of collinearity present in the model, it is impossible to identify the precise value of that or any of the other parameter estimates used there.[5] The high collinearity among the price, the income, and the stock of vehicle variables suggests either that one or more of the

Table 4.1. Schou and Johnson Dataset

Year	Quantity of petrol (per capita	Price of petrol (cents per liter)	Stock of motor vehicles	Average petrol consumption (per vehicle)	Average weekly earnings (dollars)	Consumer price index
1955	445.5	7.8	212,970	1.947	34.70	74.0
1956	459.1	7.8	244,457	1.949	36.20	77.0
1957	457.3	7.8	236,189	1.886	38.10	81.5
1958	484.3	7.8	250,010	1.926	39.50	82.3
1959	510.5	7.8	263,158	1.971	40.70	83.6
1960	541.7	7.8	280,252	2.008	43.90	85.7
1961	546.6	7.6	296,865	1.952	46.00	89.2
1962	555.0	7.4	310,687	1.931	47.20	89.6
1963	611.1	7.5	329,257	2.045	48.40	89.8
1964	649.6	7.4	351,606	2.078	50.90	90.6
1965	675.4	7.9	374,224	2.071	55.30	94.0
1966	687.7	8.6	391,581	2.056	57.90	97.4
1967	711.9	9.0	410,663	2.068	61.90	100.0
1968	739.7	9.2	427,930	2.099	65.50	103.3
1969	774.6	9.3	451,040	2.130	70.40	106.0
1970	803.8	9.6	477,160	2.133	76.30	109.4
1971	829.4	10.3	503,920	2.124	84.80	114.6
1972	857.9	10.8	532,510	2.109	93.00	122.4
1973	911.1	10.8	563,410	2.145	101.50	129.8
1974	926.3	12.6	598,610	2.086	118.00	146.6
1975	952.8	14.6	633,300	2.046	148.20	171.1
1976	1,008.1	15.8	666,030	2.066	169.30	193.3

Source: K. Schou and L. W. Johnson, "The Short-run Price Elasticity of Demand for Petrol in Australia," *International Journal of Transport Economics,* 6(1979):364.

Table 4.2. Replicated Schou and Johnson Ordinary Least Squares Results

Regression		Elasticity				
		P_G	Y	S	C	
I		−0.07	0.05	0.63	0.94	$\bar{R}^2$ = 1.000
	t	(−2.956)	(2.999)	(34.951)	(15.250)	DW = 1.323
	FG R^2	0.98	0.99	0.98	0.85	LLF = −51.500
						Hall-Pagan $\chi^2(1)$ = 3.632
II		−0.02	−0.10	0.72	0.88	$\bar{R}^2$ = 1.000
	t	(−0.859)	(−1.537)	(16.733)	(23.798)	DW = 1.765
	FG R^2	0.69	1.00	1.00	0.60	LLF = −51.232
						Hall-Pagan $\chi^2(1)$ = 5.457
III		−0.05	NI	0.66	0.89	$\bar{R}^2$ = 1.000
	t	(−2.703)		(151.46)	(24.113)	DW = 1.498
	FG R^2	0.37		0.62	0.57	LLF = −52.664
						Hall-Pagan $\chi^2(1)$ = 6.280

Notes: Elasticities measured at the means of variables. NI, not included. Regression I is in nominal dollars; Regressions II and III are in constant dollars.

Source: Compiled by the author.

variables should be dropped from the model or that the equation is misspecified. The constant dollar OLS results fare somewhat better; however, at this juncture, the income and stock of vehicle variables exhibit virtually perfect collinearity. Furthermore, the income variable has a (theoretically) perverse algebraic sign that may be attributed to severe collinearity; so obviously one variable must be omitted if accurate estimates of any of the elasticities are to be obtained. The results preferred by SJ in this set of OLS specifications is Regression III, which (dropping the income variable) exhibits a distinct improvement in the collinearity and relates gasoline consumption per capita to the constant dollar price of gasoline, the number of registered vehicles, and the average consumption per vehicle. This relationship is specified in Equation 4.1, except that the income variable is omitted:

$$\mathbf{Q_G} = \alpha + \beta_1 \cdot \mathbf{P_G} + \beta_2 \cdot \mathbf{Y} + \beta_3 \cdot \mathbf{S} + \beta_4 \cdot \mathbf{C} + \mathbf{u}, \qquad (4.1)$$

where $\mathbf{Q_G}$ = gasoline consumption per capita
$\mathbf{P_G}$ = price in 1967 constant dollars
$\mathbf{Y}$ = average weekly earnings in 1967 constant dollars
$\mathbf{S}$ = number of registered vehicles
$\mathbf{C}$ = average consumption per vehicle.

It may be noted, as was to be expected, that omission of the income variable in the estimation has virtually no affect on the equation's coefficient of multiple determination $\bar{R}^2$. The choice between whether the income or the stock of vehicle variable is the more appropriate one to be dropped is an arbitrary decision of the modeler, but one that should not be subject merely to taste. The other results in regard to the DW statistic, which measures first-order serial correlation, and the LM χ^2 test for an homoscedastic error term, suggest potential problems of misspecification and heteroscedasticity.[6] Rather than pursuing the SJ modeling strategy, some alternative formulations to the Cooley-Prescott approach are considered. Instead, the model in Equation 4.1 is reestimated, using an MA serial correlation correction. Those results are provided as Model 1 in Table 4.3.

Table 4.3. Alternative National Gasoline Demand Model Specifications

Model	Estimation procedure	Elasticity P_G	Y	S	C	
1	MA	-0.05	NI	0.66	0.90	$\bar{R}^2$ = 1.000
	t	(-2.422)		(120.13)	(21.185)	DW = 1.896
	FG R^2	0.62		0.37	0.57	MAθ = 0.65 (3.820)
						LLF =-51.634
						Hall-Pagan $\chi^2(1)$ = 6.280
2	OLS	NI	NI	0.89	0.89	$\bar{R}^2$ = 0.999
	t			(61.095)	(19.044)	DW = 1.638
	FG R^2			0.88	0.88	LLF =-64.070
						Hall-Pagan $\chi^2(1)$ = 0.070
						BCE λ = 1.05
						Heteroscedasticity weighting = $1/\sqrt{Y}$
3	BCE AR	NI	NI	0.63	NI	$\bar{R}^2$ = 0.988
	t			(29.412)		DW = 1.642
						AR ρ = 0.36 (1.810)
						BCE λ =-0.31
						LLF =-90.281
						Hall-Pagan $\chi^2(1)$ = 1.225
						Heteroscedasticity weighting = $1/\sqrt{Y}$
4	BCE AR	-0.41	1.09	NI	NI	$\bar{R}^2$ = 0.990
	t	(-2.708)	(15.723)			DW = 1.964
	FG R^2	0.29	0.29			ρ = 0.69 (4.471)
						BCE λ = 0.39
						LLF =-92.805
						Hall-Pagan $\chi^2(1)$ = 1.206

Notes: NI, not indicated. Elasticities measured at the means of the variables.

Source: Compiled by the author.

While the MA adjustment in Model 1 eliminates the serial correlation in it, there remains evidence of heteroscedasticity, and therefore some additional formulations must be considered. The Model 2 and 3 results have the observations multiplied by the inverse of the square root of income as a simple adjustment for the problem, and the price variable is omitted because it is found to be statistically insignificant. Furthermore, the results of both of these models are indistinguishable from one another in terms of explanatory power, but only in the latter formulation do the two statistical problems seem to disappear. The Model 2 specification relies upon a BC transformation of the dependent and independent variables, which yields an estimate of λ of 1.05, suggesting that, as a maintained hypothesis, a linear model is acceptable; however, there is still evidence of collinearity being present. Dropping the average consumption per vehicle variable from the Model 2 form only slightly reduces the explanatory power of the regression, but reintroduces the serial correlation, which is successfully corrected by an AR adjustment for first-order serial correlation; and the BC estimate of λ indicates that a nonlinear specification for this model is the most appropriate. Finally, Model 4, which includes both the price and the income variables but omits the stock of vehicle and fuel efficiency variables, is tested. The presence of serial correlation in this form is also handled by an AR adjustment, inasmuch as the BC results suggest nonlinearity and heteroscedasticity is not an apparent problem. The adjusted R^2 is acceptable, and both the price and the income elasticities (which appear to be of reasonable magnitudes) are statistically significant and exhibit the appropriate algebraic signs. The appeal of this last model is its parsimony, its theoretical foundations, and its statistical results, although Schou and Johnson do not consider any of these adjustments, choosing instead to pursue the "varying parameters" framework.

The SJ implementation of the Cooley-Prescott model as an alternative to the models suggested here is applied to the Regression III specification, presumably because that model exhibits less collinearity than any of the other specifications on which Schou and Johnson report; but, as already noted, the decision to retain the vehicle stock variable in preference to the income variable is an ad hoc one. It might be argued that because of the derived demand nature of gasoline consumption that the stock of vehicle variable is to be preferred to an income variable on theoretical grounds; however, no rationale for this is provided by the authors. The time-varying parameters estimation method necessitates specification of a covariance structure, and Schon and Johnson consider four alternative ones, as described in Table 4.4. The results of the varying parameter model estimates appear in Table 4.5, presented in terms of the elasticities of the variables.

Whereas Schou and Johnson state that "the fairly high values of γ imply that permanent shifts in parameters over time are of greater relative magnitude than transitory shifts . . . [they conclude] that the parameters of [their] estimated demand function have been stable over the period 1955–1976" (p. 131). It can be seen that these elasticity values are similar to both the MA Model 1 results and to their own Regression I results. Can such apparent similarities be explained? Maddala (1977, p. 398) speculates that models misspecified

Table 4.4 Schou and Johnson Alternative Covariance Structures

Covariance matrix	Diagonal values
Σ_1	(1, 0, 0, 0)
Σ_2	(1, 1, 1, 1)
Σ_3	(OLS estimated standard errors of parameters)
Σ_4	(1, OLS estimated standard errors of slopes)

Source: K. Schou and L. W. Johnson, "The Short-Run Price Elasticity of Demand for Petrol in Australia," *International Journal of Transport Economics,* 6(1979):361.

Table 4.5. Schou and Johnson Varying Parameter Results

Model		Elasticity			γ
		P_G	S	C	parameter
Σ_1		-0.08	0.66	0.93	0.66
	t	(-2.48)	(45.04)	(18.14)	
Σ_2		-0.05	0.66	0.88	0.60
	t	(-2.20)	(67.4)	(23.3)	
Σ_3		-0.08	0.66	0.93	0.66
	t	(-2.74)	(46.57)	(18.53)	
Σ_4		-0.08	0.66	0.93	0.65
	t	(-2.71)	(47.17)	(18.54)	

Note: Elasticities measured at the means of the variable.

Source: K. Schou and L. W. Johnson, "The Short-Run Price Elasticity of Demand for Petrol in Australia," *International Journal of Transport Economics,* 6(1979):362, 364.

because of omitted variables would tend to exhibit problems of both autocorrelation and heteroscedasticity, and that in such circumstances the Cooley-Prescott varying parameters approach would appear to provide acceptable model results, such as is the case with the SJ estimates. The Model 3 specification, which explicitly addresses both the presence of first-order autocorrelation and that of heteroscedasticity, provides comparable explanatory power to the varying parameters results, but these latter suggest that the price elasticity falls within the range of –0.05 to –0.08, while the BC formulation, Model 3, rejects the price variable entirely in contrast to the Model 4 price elasticity point estimate of –0.41, with an income elasticity of roughly unity and a 0.95-to-1.24 confidence interval. A conclusion that might be drawn from this example is that sophistication may not provide additional insights into the behavior of demand relationships; underlying statistical problems evidenced by a model may be a more productive domain for inquiry.

STATE-LEVEL ESTIMATES

The first attempt in Australia at developing a state-level gasoline demand model was by Folie (1977). That model is specified as being linear in the logarithms of the data (log-linear) and of the form

$$\ln \mathbf{Q}_{Gj} = \alpha_j + \beta_{1j} \cdot \ln \mathbf{P}_{Gj} + \beta_{2j} \cdot \ln \mathbf{Y}_j + \mathbf{u}_j , \quad (4.2)$$

where $\mathbf{Y}$ = real per capita income
j = state subscript
ln = natural logarithm.

Unfortunately, a comprehensive analysis has never been published, and limited statistical information is presented in the conference paper that is available. Folie's dataset represents quarterly information covering the period from September 1958 through June 1974, but it is impossible to replicate or extend the Folie single-equation OLS results. Interesting comparisons should be possible between Folie's results (summarized in Table 4.6) and those of Donnelly (1982a).

It can be seen from the table that none of the price elasticities is significantly different from 0 at the 0.05 level, and only the South Australia price elasticity is significant at the 0.10 level. The New South Wales, Victoria, and Queensland price elasticities evidence similarities in magnitude, as do those for South Australia, Western Australia, and Tasmania. The income elasticities are all significant at the 0.05 level, and these appear to exhibit different patterns among the states. The Victoria and Queensland income elasticities are similar, as are those in South and Western Australia. Folie finds gasoline consumption to be generally income elastic in the long run, but still another pattern is suggested by the estimated values for the lagged dependent variable terms. However, since the original dataset

Table 4.6. Folie's State-Level Ordinary Least Squares Results

State	P_G			Y			Q_{t-1}		
	Short run	t	Long run	Short run	t	Long run	Parameter	t	R^2
New South Wales	−0.04	(−0.391)	−0.20	0.28	(2.350)	1.23	0.77	(8.862)	0.982
Victoria	−0.11	(−0.850)	−0.16	0.86	(4.806)	1.31	0.34	(2.522)	0.968
Queensland	−0.07	(−0.459)	−0.14	0.75	(4.016)	1.52	0.51	(2.691)	0.966
South Australia	−0.33	(−1.906)	−0.63	0.46	(4.070)	0.90	0.48	(4.330)	0.932
Western Australia	−0.24	(−1.510)	−0.64	0.41	(2.978)	1.11	0.63	(5.694)	0.980
Tasmania	−0.38	(−1.580)	−0.77	0.62	(3.655)	1.28	0.51	(4.146)	0.934

Source: M. Folie, "Competition in the Australian Retail Market for Petrol." Paper presented at the Sixth Conference of Economists, University of Tasmania, Hobart, 11-60 May, 1977, mimeo, calculated from Table 3.

is not available, these apparent similarities among regions cannot be tested statistically. Thus, in marked contrast to the SJ national results, Folie observes income to be a significant determinant of gasoline demand and also finds that the regional responses differ. The Model 4 results derived here from the SJ data yield price and income elasticity estimates of the same order of magnitude as are provided by the Folie long-run elasticities.

In a pooled time series of a state cross-sectional model of Australian gasoline demand, Donnelly (1982a)[7] defines these as a Zellner set (1962) of SURE, of state-level demand functions, and estimates these using an iterative procedure that converges to maximum likelihood estimates. The original intent of that research was to develop a multiequation model reflecting the derived demand nature of gasoline consumption, wherein the demand for gasoline is related to prices and the existing stock of motor vehicles. The stock of vehicles is made endogenous to the model, that is, a function of prices and income. Such an approach necessitates a reliable time series for the stock of motor vehicles, but only new vehicle registrations are available monthly by state in Australia, while vehicle censuses are taken only sporadically. Since the stock of registered vehicles at any point in time is determined by the number of vehicles on the registry at the last time period, plus new vehicle registrations, less vehicles scrapped, information on each of the components is required. This stock identity is

$$\sum_{i=1}^{t} \mathbf{S}_{it} \equiv \sum_{i=1}^{t-1} \mathbf{S}_{i\,t-1} + \sum_{k=1}^{m} \mathbf{N}_{k\,t-1} + \sum_{i=1}^{t} \sum_{k=1}^{m} \mathbf{R}_{ik}\,, \quad (4.3)$$

where **S** = registered vehicles
N = new vehicle registration
R = vehicles scrapped
i = year of manufacture
t = census year
k = reporting intervals of registrations.

The above equation will also include a regional, that is, *j* subscript, which further complicates the problem since regional transfers of vehicles must be accounted for as well, but the requisite information on scrappage and interstate transfers is not available. Motor vehicle censuses in Australia were taken in 1952, 1962, 1971, 1976, and 1979, but the data are not reported by year of manufacture for vehicles of an age greater than 11 years, leaving several alternative procedures for deriving scrapping rates. First, scrappage can be assumed to be proportional to registrations during intercensal periods. Adoption of such a procedure implies that changes in gasoline consumption respond to changes in new vehicle registrations only. This is not a valid assumption if the fuel efficiency of the stock of vehicles is thought to be changing, perhaps because of increased gasoline prices. Also, unless a first-in, first-out or some other arbitrary aging process is assumed, no vintage information on the stock of vehicles can be maintained. Even with such an ad hoc assumption, adjustments that force agreement with the census year data would be required, and such balancing would, by necessity, be arbitrary. A second alternative that could be considered is the use of some probability distribution for the scrappage rates, but while this approach allows the intercensal estimates to retain distinct vintages, it assumes that the scrappage rate is unrelated either to a particular vintage or to any economic impetus. Apocryphal stories abound among automobile buffs of "good" model years and are said to be supported by some empirical evidence, so the replacement component of new vehicle demand could be related to vintages. Additionally, incremental demand is theoretically related to economic conditions, and a related deterministic issue is the difficulty associated with selecting the appropriate probability distribution. [Sweeney (1979) in the United States and Filmer and Talbot (1974) in Australia choose to use a Poisson distribution.] Unfortunately, again, the selection criteria are arbitrary since, as has been mentioned, the lack of sufficient census detail precludes directly observing, and therefore estimating, all of the segments of the probability distribution. In particular, Hewitt and James (1975) report the average age of vehicles in Australia at scrappage to be between 11.2 and 12.3 years, while Donnelly (1982a) indicates 13.3 years might be the more recent incidence. However, it remains that the interval-reporting scheme used for the vehicle censuses makes it impossible to identify the actual distribution of scrappage rates. Finally, a polynomial survival rate function is estimated from the available motor vehicle census data; this approach does not prove to be any more fruitful than either of the other alternatives considered. Thus, because of the paucity of acceptable vehicle stock data, that multiequation approach is rejected in favor of a reduced form model of the Balestra and Nerlove dynamic demand formulation (1966), and includes the lagged dependent variable in the explanatory dataset. Verleger and Sheehan (1976) argue that such a specification provides an indirect measure of the influences of the unobservable stock of vehicles by relating gasoline demand both directly and indirectly to price and income changes. The indirect effect, operating through the lagged term, represents the "state" of demand, that is, changes in the stock of vehicles and their utilization rates, or, as Houthakker and Taylor (1970, p. 10) characterize this, a "psychological stock." Following the Verleger-Sheehan formulation, the desired consumption of gasoline, denoted as Q_G^*, is posited as a function of the price of gasoline and income:

$$Q_{G\,jt}^* = \gamma_{0j} + \gamma_{1j} \cdot P_{G\,jt} + \gamma_{2j} \cdot Y_{jt}, \tag{4.4}$$

where Q_G^* = desired demand
P_G = price of gasoline
Y = income
j = region
t = time period.

As can be seen, the variables have both spatial and temporal dimensions, and the unknown parameters, the γ terms, are permitted to vary spatially also. Although desired consumption is a function of price and income, adjustment to changes need not be instantaneous, with rigidities in the adjustment process being reflected by the influence of a slowly changing stock of vehicle variable, inasmuch as the stock of vehicles evolves and as a natural delay accrues in an environment of changing patterns of consumption. One means of taking such considerations into account is that of a simple proportional adjustment model, as in

$$Q_{G\,jt} - Q_{Gjt-n} = \Phi_j \cdot (Q_{G\,jt}^* - Q_{G\,jt-n}). \tag{4.5}$$

Here, the Φ_j parameter represents the proportional change that occurs in *n* time periods whenever a divergence between the desired level of consumption and the actual consumption occurs. Institutional factors such as the characteristics of the infrastructure and the effect of consumer behavior patterns mentioned earlier would determine the value of Φ, with a larger value representing a more rapid adjustment to a shock to the system. When the desired demand equation, Equation 4.4, is combined with the adjustment equation, Equation 4.5, and rewritten as a stochastic relationship that may be estimated, it becomes

$$Q_{G\,jt} = \alpha_j + \beta_{1j} \cdot P_{G\,jt} + \beta_{2j} \cdot Y_{jt} + \beta_{3j} \cdot Q_{G\,jt-n} + \epsilon_{jt}\,. \quad (4.6)$$

The direct, immediate effects that changes in price and income have on gasoline consumption are measured by the β_{1j} and β_{2j} parameters, and the indirect effect of these variables, which operates through the lagged term, is measured by γ_{1j} and γ_{2j}; or, as these are usually stated, $\beta_{1j}/(1 - \beta_{3j})$ and $\beta_{2j}/(1 - \beta_{3j})$, respectively. The estimated magnitude of the β_{3j} parameter provides an indication of the speed of the adjustment process, since that equals $(1 - \Phi_j)$, as has been shown. The proportion of adjustment toward the desired level of consumption achieved in *m* time periods after a shock to the system has occurred may be calculated as $(1 - \beta^m_{3j})$; thus, if $\beta_{3j} = 0.91$, approximately 25 time periods will be required to arrive at a 90 percent adjustment, and about 32 time periods to reach a 95 percent adjustment toward equilibrium.[8] In such an instance, the psychological stock "proxy" variable is important in determining consumption, because habits are slow to change. In addition, the value of the β_{3j} parameter provides a statistical means to choose between a "stock" and a "flow" adjustment interpretation of the model in Equation 4.6.

The data used to estimate the model are obtained from several sources, including ABS, the Department of Resources and Energy (DRE), and AIP. The dependent variable comes from the DRE monthly publication "Major Energy Statistics" and has been defined on a per diem, per capita basis. The per capita specification is used to eliminate any spurious correlation that might result if the variables were defined in terms of the actual levels. The gasoline price variable is the ABS gasoline price index (GPI); and as ABS did not construct a GPI prior to the September quarter of 1966, the data series up to this date is backcast using the AIP published prices (1973) for gasoline.[9] The regional income variable selected is the ABS male average weekly earnings (AWE) series; again, it is necessary to backcast the AWE figures for the September quarter of 1958 to the June quarter of 1966 inclusive. This is done using national wages and salaries data. The modeling datasets appear in Tables 4.7 through 4.11, both exogenous monetary variables being deflated by the respective regional CPIs so that the regressions appear in constant dollar terms.

An iterative Zellner SURE procedure is used to pool the time-series cross-sectional dataset, a technique that assumes that the equation residuals are spatially, but not serially, correlated (Kmenta 1971; Maddala 1977). The validity of this latter assumption is tested with the DW statistic and with the Durbin h statistic.[10] Spatial autocorrelation could result if some variable common to all equations were omitted in the specification; in such an instance, the SURE estimation technique results in a gain in efficiency over the single-equation OLS approach. As Kmenta (1971) explains: "This means that the gain in efficiency of the generalized least squares estimator [in this case, the iterative SURE estimates asymptotically approach the generalized least squares result] over the ordinary least squares estimator is greatest when the disturbances in the two equations are highly correlated and, at the same time, the explanatory variables are uncorrelated" (p. 534). Unfortunately, in the immediate case, the inclusion of the lagged dependent variable along with the concomitant high collinearity means that the efficiency gained, as referred to by Kmenta, will be somewhat attenuated. The χ^2 statistic of goodness-of-fit test for normality in the disturbance term is also presented in the reported results.[11]

Various cross-equation restrictions are tested, including, for example, the question of whether the regression parameters differ across regions. Mention should be made of inferences and their basic ramifications in regard to alternative hypotheses. The null hypothesis is described below, and Type I errors constitute the rejection of a null hypothesis when no significant differences exist. Alternatively, a researcher may accept erroneously a null hypothesis, thereby misjudging a difference as being *not* significant, when, in fact, a significant difference exists; this is denoted a Type II error.[12] The six sets of alternative national restrictions considered are as follow:[13]

1. No regional differences exist in the parameters:

$$\alpha_j = \alpha$$
$$\beta_{ij} = \beta_i, \text{ for } i = 1, \ldots, 3$$
$$j = 1, \ldots, 6.$$

Table 4.7. Gasoline Sales by State

Year	Quarter	New South Wales	Victoria	Queensland	South Australia	Western Australia	Tasmania	Northern Territory	Total Australia	Number of days
1958	3	406,258	357,765	170,756	121,535	87,362	38,437	3,132	1,185,245	92
1958	4	436,748	392,547	180,762	134,710	93,764	42,164	3,369	1,284,064	92
1959	1	425,719	389,420	163,818	129,922	92,277	43,552	2,518	1,247,226	90
1959	2	432,548	389,964	177,176	131,687	93,686	42,496	3,150	1,270,707	91
1959	3	439,476	385,900	184,340	131,610	93,937	41,110	3,537	1,279,910	92
1959	4	474,789	429,897	189,222	144,397	102,292	45,619	3,415	1,389,631	92
1960	1	471,408	422,829	178,966	133,451	100,447	47,461	2,700	1,357,262	91
1960	2	481,501	426,197	191,845	144,134	102,596	44,830	3,482	1,394,585	91
1960	3	489,211	426,302	194,023	137,432	100,828	43,447	4,064	1,395,307	92
1960	4	507,531	466,079	200,702	147,939	106,829	49,085	3,846	1,482,011	92
1961	1	508,668	445,154	187,377	145,989	106,033	48,262	3,001	1,444,484	90
1961	2	501,867	445,322	197,991	149,689	107,247	46,729	4,115	1,452,960	91
1961	3	503,898	434,571	201,265	144,974	106,538	44,857	4,619	1,440,722	92
1961	4	529,012	465,980	204,107	151,521	113,515	49,325	4,401	1,517,861	92
1962	1	541,041	469,362	200,169	154,439	115,175	51,593	3,442	1,535,221	90
1962	2	543,318	470,498	215,503	157,426	116,126	50,184	4,424	1,557,479	91
1962	3	550,556	462,180	218,972	154,849	114,548	48,975	4,901	1,554,981	92
1962	4	572,082	497,558	220,750	165,810	125,049	52,448	4,415	1,638,112	92
1963	1	574,146	499,848	210,057	163,478	123,612	55,158	3,887	1,630,186	90
1963	2	571,090	495,666	224,864	165,970	127,659	54,013	4,655	1,643,917	91
1963	3	593,098	500,717	236,625	166,297	125,404	51,717	5,296	1,679,154	92
1963	4	619,920	540,035	242,612	183,158	135,011	57,331	5,005	1,783,072	92
1964	1	624,948	535,121	238,389	177,730	135,091	59,085	3,942	1,774,306	91
1964	2	620,452	539,286	246,658	183,408	138,602	57,631	4,859	1,790,896	91
1964	3	639,968	536,499	260,173	183,807	137,719	56,194	6,596	1,820,956	92

1964	4	672,914	576,095	264,642	194,482	147,211	60,718	5,983	1,922,045	92
1965	1	657,939	564,835	254,200	189,741	143,693	63,781	4,391	1,878,580	90
1965	2	674,555	578,787	266,410	200,069	149,631	61,667	5,537	1,936,656	91
1965	3	685,593	580,760	282,754	208,411	157,126	61,200	6,005	1,981,849	92
1965	4	687,730	582,897	271,907	192,732	151,989	63,422	5,237	1,955,914	92
1966	1	685,289	590,452	264,415	197,478	156,795	65,041	4,856	1,964,326	90
1966	2	694,622	598,503	281,358	201,569	163,346	65,100	5,824	2,010,322	91
1966	3	695,621	585,892	281,217	199,542	157,159	60,995	6,538	1,986,964	92
1966	4	726,503	615,169	296,288	208,984	169,829	68,683	6,224	2,091,680	92
1967	1	738,351	630,207	287,082	210,267	170,188	71,555	4,974	2,112,624	90
1967	2	724,426	613,301	285,500	211,575	172,593	68,679	6,647	2,082,721	91
1967	3	747,010	628,271	305,584	213,690	175,708	65,932	8,001	2,144,196	92
1967	4	749,752	634,476	303,739	206,443	181,685	69,677	7,315	2,153,087	92
1968	1	766,737	638,632	292,019	212,476	186,177	72,851	6,223	2,175,115	91
1968	2	775,606	648,478	308,644	212,330	188,186	70,669	8,146	2,212,059	91
1968	3	781,375	658,420	321,910	217,604	192,459	70,178	10,197	2,252,143	92
1968	4	816,717	690,176	326,778	225,554	204,170	73,005	9,011	2,345,411	92
1969	1	809,193	676,769	315,631	223,850	204,306	77,033	8,938	2,315,720	90
1969	2	830,073	701,440	337,706	232,179	213,230	74,269	12,071	2,400,968	91
1969	3	831,549	685,647	344,581	228,605	213,167	72,542	13,278	2,389,369	92
1969	4	872,829	726,367	351,086	239,211	223,991	77,189	13,570	2,504,143	92
1970	1	877,548	721,885	341,557	236,807	220,908	79,356	11,784	2,489,845	90
1970	2	892,336	745,001	357,951	244,317	231,215	76,592	14,907	2,562,319	91
1970	3	895,845	723,175	365,220	243,248	234,743	76,056	17,421	2,555,708	92
1970	4	899,173	751,834	353,660	237,488	234,106	79,566	15,325	2,571,152	92
1971	1	917,667	755,330	351,527	240,989	235,406	82,630	13,388	2,596,937	90
1971	2	929,031	764,109	376,790	244,911	240,139	78,529	16,693	2,650,202	91
1971	3	954,663	778,864	386,600	255,051	249,622	79,793	19,098	2,723,691	92

(continued)

Table 4.7 continued

Year	Quarter	New South Wales	Victoria	Queensland	South Australia	Western Australia	Tasmania	Northern Territory	Total Australia	Number of days
1971	4	969,110	790,148	380,063	251,713	246,530	80,917	17,061	2,735,542	92
1972	1	984,036	811,474	382,540	260,450	250,508	86,048	13,947	2,789,003	91
1972	2	989,964	820,871	403,771	262,897	247,921	83,552	19,712	2,828,688	91
1972	3	927,436	793,344	395,056	238,680	254,099	80,656	19,890	2,709,161	92
1972	4	1,018,222	842,205	417,304	267,506	255,404	85,107	18,111	2,903,859	92
1973	1	1,023,827	846,583	425,256	270,534	263,600	91,445	15,307	2,936,552	90
1973	2	1,024,727	860,845	432,634	277,680	266,592	87,522	19,021	2,969,021	91
1973	3	1,055,000	872,893	465,889	290,165	284,323	88,689	22,553	3,079,512	92
1973	4	1,073,912	894,132	461,806	283,704	277,817	92,682	19,675	3,103,728	92
1974	1	1,032,416	876,411	418,842	288,695	278,676	95,659	17,234	3,007,933	90
1974	2	1,074,857	871,424	457,770	295,405	282,449	89,718	18,870	3,090,493	91
1974	3	1,081,789	900,395	483,470	301,755	297,205	91,134	24,459	3,180,207	92
1974	4	1,122,884	907,735	474,112	307,195	282,390	95,764	22,183	3,212,263	92
1975	1	1,085,384	912,544	468,326	302,468	293,384	96,108	17,607	3,175,821	90
1975	2	1,104,108	918,957	477,359	292,885	292,717	93,243	18,740	3,198,009	91
1975	3	1,167,778	947,724	525,384	315,534	315,258	95,755	24,769	3,392,202	92
1975	4	1,066,457	919,532	483,821	312,468	297,117	92,384	22,086	3,193,865	92
1976	1	1,134,694	965,844	483,195	318,432	312,226	100,109	19,083	3,333,583	91
1976	2	1,141,812	953,583	499,733	319,196	307,410	96,374	23,285	3,341,393	91
1976	3	1,159,417	967,092	528,411	318,745	318,427	96,430	27,470	3,415,992	92
1976	4	1,096,594	1,028,451	534,312	333,842	337,037	101,595	23,771	3,455,602	92
1977	1	1,185,039	990,199	505,561	337,333	327,609	105,209	19,307	3,470,257	90
1977	2	1,192,492	992,372	541,938	352,602	351,770	102,683	25,047	3,558,904	91
1977	3	1,215,207	1,027,468	571,824	340,407	350,390	102,430	29,176	3,636,902	92
1977	4	1,200,047	1,004,806	536,174	338,706	340,350	102,292	23,949	3,546,324	92

1978	1	1,205,680	1,020,817	540,087	341,834	346,017	107,214	20,894	3,582,543	90
1978	2	1,228,333	1,033,032	562,369	345,421	348,519	102,446	25,388	3,645,508	91
1978	3	1,241,781	1,071,576	597,675	360,184	382,928	106,301	28,387	3,788,832	92
1978	4	1,204,568	1,020,803	560,781	330,368	341,523	102,824	24,355	3,585,222	92
1979	1	1,260,888	1,053,679	570,683	353,643	356,180	110,535	21,502	3,727,110	90
1979	2	1,220,624	1,076,271	601,856	345,225	363,283	108,823	26,615	3,742,697	91
1979	3	1,247,772	1,023,763	598,613	331,637	354,376	100,682	30,858	3,687,701	92
1979	4	1,248,519	1,054,909	593,355	341,428	358,658	106,624	26,947	3,730,440	92
1980	1	1,240,237	1,033,772	579,729	332,774	353,274	109,250	22,838	3,671,874	91
1980	2	1,214,511	1,028,572	589,392	330,467	349,902	104,781	28,038	3,645,663	91
1980	3	1,249,108	1,002,537	596,352	318,368	351,056	100,637	31,570	3,649,628	92
1980	4	1,284,010	1,065,614	616,853	338,401	362,850	109,953	27,050	3,804,731	92
1981	1	1,237,706	1,020,096	581,252	319,402	345,368	103,600	23,968	3,631,392	90
1981	2	1,254,246	1,034,901	612,069	325,374	357,966	101,637	29,978	3,716,171	91

Source: Compiled by the author.

Table 4.8. Capital City Gasoline Price Index

Year	Quarter	Sydney	Melbourne	Brisbane	Adelaide	Perth	Hobart	Capital city average
1958	3	88.60	88.40	90.60	91.80	86.20	89.70	89.20
1958	4	88.90	88.70	91.00	92.10	86.10	90.10	89.70
1959	1	89.10	88.90	91.30	92.50	86.30	90.30	89.90
1959	2	89.30	89.10	91.40	92.60	86.90	90.30	90.10
1959	3	90.90	88.40	91.70	92.80	87.20	89.90	90.40
1959	4	90.50	88.60	91.90	93.10	87.10	90.10	90.80
1960	1	91.10	88.80	92.10	93.70	87.70	90.40	91.30
1960	2	92.10	89.50	92.30	94.80	89.00	92.70	92.30
1960	3	92.80	90.00	92.70	95.30	89.50	94.20	93.00
1960	4	93.70	89.40	92.80	95.70	90.30	94.40	93.50
1961	1	94.50	89.50	93.30	96.60	91.00	95.30	93.90
1961	2	95.10	89.00	92.70	96.70	91.50	95.20	94.40
1961	3	95.30	88.20	92.50	96.40	91.20	95.30	94.30
1961	4	95.00	88.20	92.60	95.90	90.90	95.00	94.10
1962	1	94.90	88.00	92.70	95.70	91.00	94.60	94.00
1962	2	94.80	89.50	92.60	94.50	91.20	93.70	93.90
1962	3	95.00	89.50	92.70	94.80	91.30	94.00	94.10
1962	4	95.40	89.50	92.70	94.50	91.20	94.00	94.20
1963	1	95.50	89.50	92.70	94.50	91.50	93.90	94.20
1963	2	95.70	89.60	92.80	95.20	91.90	94.30	94.40
1963	3	96.10	88.90	92.40	95.00	91.90	97.10	94.60
1963	4	96.20	88.80	92.40	94.60	92.20	96.90	94.50
1964	1	96.60	89.00	92.70	94.90	92.90	97.20	94.90
1964	2	97.80	88.60	92.60	96.10	93.80	97.70	95.60
1964	3	98.80	88.90	93.10	96.60	94.30	98.10	96.30
1964	4	99.50	89.40	93.50	97.40	94.30	98.80	97.10
1965	1	100.20	89.60	93.90	98.30	95.20	99.70	97.50
1965	2	101.00	89.20	94.20	99.10	96.20	99.30	98.10
1965	3	99.80	94.10	97.60	99.20	95.80	100.00	98.40
1965	4	100.40	95.20	98.50	100.00	96.70	100.70	99.20
1966	1	100.20	95.20	98.90	100.20	97.50	100.40	99.30
1966	2	100.60	95.40	99.00	100.90	99.10	101.00	99.90
1966	3	100.00	98.50	99.20	99.90	99.80	99.90	99.50
1966	4	100.00	100.50	99.20	100.00	99.80	99.90	100.00
1967	1	100.00	100.50	99.10	100.00	99.80	99.90	100.00
1967	2	100.00	100.50	102.40	100.00	100.50	100.20	100.50
1967	3	102.00	101.50	103.50	102.20	101.50	101.20	102.00
1967	4	104.50	104.10	106.70	105.70	104.10	104.00	104.70
1968	1	104.50	104.10	106.70	105.70	103.80	104.00	104.70
1968	2	104.50	104.10	106.70	105.70	103.80	104.00	104.70
1968	3	104.50	104.30	106.70	105.70	103.80	104.00	104.70
1968	4	103.40	103.00	105.40	104.20	102.70	104.60	103.60
1969	1	103.40	103.00	105.40	104.20	104.70	104.60	103.80
1969	2	103.40	103.00	105.40	104.20	104.70	104.60	103.80
1969	3	105.10	105.50	105.40	106.10	104.70	104.60	105.50
1969	4	105.10	105.50	105.40	106.10	104.70	104.60	105.50

(continued)

Table 4.8 continued

Year	Quar-ter	Sydney	Mel-bourne	Brisbane	Adelaide	Perth	Hobart	Capital city average
1970	1	105.10	103.80	105.40	106.10	104.70	104.60	104.90
1970	2	105.10	103.80	105.40	106.10	104.70	104.60	104.90
1970	3	107.70	106.30	109.30	108.70	107.30	107.10	107.60
1070	4	115.40	113.30	117.00	116.40	114.90	116.30	115.00
1971	1	113.90	111.10	115.70	115.10	113.40	115.30	113.50
1971	2	115.70	108.60	119.60	116.40	116.00	116.50	114.00
1971	3	115.70	99.10	119.60	116.40	116.00	116.50	111.00
1971	4	120.80	108.10	124.70	122.20	121.00	121.50	117.40
1972	1	120.80	107.90	124.70	122.20	121.00	121.50	117.30
1972	2	120.80	106.60	124.70	122.20	121.00	121.50	116.90
1972	3	120.80	110.90	124.70	122.20	121.00	121.50	118.30
1972	4	120.80	110.90	124.70	122.20	121.00	121.50	118.30
1973	1	120.80	109.90	124.70	122.20	121.00	121.50	117.90
1973	2	120.80	106.60	124.70	122.20	121.00	121.50	116.90
1973	3	120.80	120.40	124.70	122.20	121.00	121.50	121.40
1973	4	133.70	132.80	137.60	135.20	133.70	133.90	134.10
1974	1	136.90	132.80	141.70	135.20	133.70	136.40	135.80
1974	2	136.90	135.10	141.70	135.20	136.00	136.40	136.70
1974	3	144.80	141.60	145.20	142.90	138.50	144.20	143.20
1974	4	144.80	140.60	145.20	142.90	142.30	144.20	143.20
1975	1	170.00	147.90	156.20	169.60	151.30	154.10	159.60
1975	2	172.00	152.70	163.30	170.60	157.50	161.50	163.40
1975	3	171.80	149.00	164.20	169.10	159.10	162.30	162.20
1975	4	192.10	162.10	183.70	169.10	176.90	183.00	177.70
1976	1	204.60	161.70	188.70	170.60	179.30	186.30	183.10
1976	2	185.00	162.70	192.80	170.30	182.90	189.60	177.10
1976	3	188.40	165.20	197.10	170.70	186.80	195.70	180.10
1976	4	188.70	165.20	192.80	175.80	183.70	196.40	180.00
1977	1	196.30	174.60	196.40	184.20	184.90	206.00	187.20
1977	2	190.80	174.60	194.80	180.00	179.20	207.10	184.20
1977	3	187.80	180.70	198.60	188.20	183.00	215.20	187.10
1977	4	209.20	197.50	206.80	203.50	201.70	236.50	204.60
1978	1	204.50	197.50	205.30	199.60	200.90	236.50	202.40
1978	2	204.30	199.50	206.70	195.50	201.00	240.00	202.80
1978	3	210.10	195.20	205.70	200.70	199.20	208.90	203.00
1978	4	249.90	240.40	251.10	242.60	242.10	230.30	245.60
1979	1	264.30	244.30	255.80	244.40	243.00	247.70	252.80
1979	2	277.80	275.90	287.30	277.20	269.90	279.70	278.00
1979	3	293.40	311.60	299.50	290.20	295.00	317.60	301.20
1979	4	331.40	332.00	328.10	341.20	327.60	329.90	332.70
1980	1	317.30	326.80	320.50	336.70	324.40	328.70	324.60
1980	2	374.10	370.70	362.30	372.20	369.50	366.80	371.80
1980	3	358.10	356.20	346.60	351.40	351.40	379.20	356.00
1980	4	369.00	365.40	352.90	354.40	361.40	374.80	364.70
1981	1	370.50	372.70	354.30	363.60	366.10	377.40	369.10
1981	2	373.80	378.70	365.30	371.60	373.80	383.50	375.00

Source: Compiled by the author.

Table 4.9. Capital City Consumer Price Index

Year	Quarter	Sydney	Melbourne	Brisbane	Adelaide	Perth	Hobart	Capital city average
1958	3	84.20	81.70	81.00	82.90	83.00	83.40	82.80
1958	4	84.50	82.70	81.90	83.40	82.90	84.10	83.50
1959	1	84.70	83.20	82.60	84.00	83.20	84.40	83.80
1959	2	85.00	83.80	82.70	84.20	83.80	84.50	84.20
1959	3	85.30	84.00	83.50	85.00	84.00	84.80	84.50
1959	4	86.00	84.40	83.90	85.40	83.90	85.10	85.00
1960	1	86.70	85.20	84.40	86.40	84.90	85.60	85.80
1960	2	87.80	87.40	84.90	88.10	86.30	86.80	87.30
1960	3	88.60	88.80	85.80	88.80	86.90	89.10	88.30
1960	4	89.20	89.20	86.90	89.40	87.60	90.00	88.90
1961	1	89.90	89.60	88.00	90.10	88.40	90.90	89.50
1961	2	90.50	90.30	87.60	90.80	88.80	91.30	90.10
1961	3	903.0	901.0	88.20	90.20	88.30	91.40	89.90
1961	4	89.90	89.90	88.30	89.50	88.00	90.90	89.60
1962	1	89.80	89.50	88.70	89.20	88.10	90.30	89.40
1962	2	89.70	89.50	88.40	89.00	88.30	90.30	89.40
1962	3	90.00	89.70	88.50	89.00	88.50	90.40	89.60
1962	4	90.40	89.70	88.60	89.00	88.30	90.80	89.70
1963	1	90.50	89.60	88.70	89.00	88.70	90.70	89.70
1963	2	90.80	89.80	88.80	89.50	89.10	90.80	90.00
1963	3	90.80	90.10	89.20	89.70	89.00	91.20	90.20
1963	4	90.90	89.80	89.00	89.60	89.30	91.40	90.10
1964	1	91.40	90.30	89.70	90.20	90.10	91.90	90.70
1964	2	92.30	91.20	90.40	91.40	90.90	92.20	91.50
1964	3	93.40	92.10	91.60	92.70	91.80	93.30	92.60
1964	4	94.20	93.70	92.60	93.90	91.80	94.50	93.70
1965	1	94.70	94.50	93.50	94.20	92.80	94.90	94.30
1965	2	95.60	95.50	94.20	94.90	93.80	95.80	95.20
1965	3	96.70	96.40	96.00	95.50	94.30	97.00	96.20
1965	4	97.80	97.50	97.20	96.90	95.50	98.30	97.40
1966	1	97.80	97.50	98.20	97.20	96.30	97.80	97.60
1966	2	98.30	98.40	98.40	98.20	98.10	98.70	98.40
1966	3	98.80	98.90	99.00	98.60	98.80	98.60	98.80
1966	4	99.90	99.60	99.70	99.70	99.30	99.20	99.70
1967	1	100.10	100.00	100.30	100.20	100.30	100.60	100.10
1967	2	101.10	101.60	101.00	101.50	101.60	101.50	101.30
1967	3	102.50	102.90	102.80	102.60	102.00	104.30	102.70
1967	4	102.90	103.30	103.00	102.10	102.50	105.00	103.00
1968	1	103.20	103.80	103.70	102.60	103.10	104.60	103.40
1968	2	104.00	104.80	103.70	104.20	104.00	104.60	104.20
1968	3	104.40	105.10	104.70	104.20	104.30	105.00	104.60
1968	4	105.90	106.00	105.30	105.20	104.90	105.80	105.70
1969	1	106.70	106.60	105.80	105.50	105.60	106.50	106.40
1969	2	107.60	107.20	106.30	106.40	107.00	107.00	107.20
1969	3	108.40	107.60	107.20	106.90	107.70	107.40	107.80
1969	4	109.60	108.30	107.90	107.50	108.70	108.10	108.70

(continued)

Table 4.9 continued

Year	Quarter	Sydney	Melbourne	Brisbane	Adelaide	Perth	Hobart	Capital city average
1970	1	111.30	108.90	108.90	108.40	109.90	108.90	109.80
1970	2	112.90	110.10	109.70	110.00	111.40	109.60	111.20
1970	3	113.90	110.70	111.10	109.90	111.60	110.20	111.90
1970	4	116.20	112.60	113.30	111.80	113.50	112.40	114.00
1971	1	117.40	113.70	115.10	112.90	114.80	113.20	115.20
1971	2	119.80	115.20	117.20	115.40	116.40	114.60	117.20
1971	3	123.30	116.50	119.00	116.50	117.20	116.50	119.40
1971	4	126.00	119.70	121.30	119.10	120.50	120.30	122.20
1972	1	127.30	120.70	122.60	120.20	121.80	120.90	123.40
1972	2	128.50	121.90	123.60	121.10	123.10	122.00	124.50
1972	3	130.30	123.60	124.60	123.00	124.80	123.40	126.20
1972	4	132.00	125.00	126.60	124.30	125.30	125.10	127.70
1973	1	134.60	127.80	129.40	127.00	127.80	127.50	130.40
1973	2	138.80	132.30	133.90	131.60	131.40	130.80	134.70
1973	3	144.10	136.80	139.40	136.50	134.40	135.20	139.60
1973	4	149.40	141.90	144.00	141.90	138.60	141.10	144.60
1974	1	152.80	145.20	147.80	145.40	142.10	144.00	148.10
1974	2	158.70	151.90	153.10	151.70	147.10	149.90	154.10
1974	3	167.00	159.40	161.40	159.40	154.30	157.40	162.00
1974	4	173.00	164.60	166.60	166.40	163.60	165.50	168.10
1975	1	178.80	170.80	171.10	173.80	169.80	169.30	174.10
1975	2	185.50	176.70	175.60	179.30	176.70	174.60	180.20
1975	3	188.20	177.70	178.40	178.20	176.50	175.90	181.60
1975	4	196.90	188.40	189.60	188.60	187.70	189.10	191.70
1976	1	203.10	193.40	195.60	194.60	193.90	194.20	197.40
1976	2	207.60	198.30	200.10	200.40	200.40	200.60	202.40
1976	3	211.10	203.20	205.10	205.50	205.40	205.50	206.90
1976	4	222.80	215.30	217.80	220.00	219.00	217.20	219.30
1977	1	227.40	221.00	222.50	224.70	224.00	221.40	224.30
1977	2	232.30	227.00	226.60	230.10	229.30	226.70	229.60
1977	3	236.00	231.60	231.30	235.60	234.90	232.00	234.10
1977	4	241.80	236.60	236.30	241.00	242.50	238.10	239.60
1978	1	245.10	239.20	240.60	243.10	245.30	240.80	242.70
1978	2	249.80	245.20	245.20	247.60	249.80	245.50	247.70
1978	3	255.00	249.60	249.80	252.60	254.70	249.30	252.50
1978	4	261.60	253.90	257.60	256.90	260.80	254.50	258.20
1979	1	266.70	258.40	259.30	261.30	264.30	260.00	262.60
1979	2	274.30	265.30	265.20	267.90	271.40	267.00	269.60
1979	3	280.40	272.20	271.20	273.30	276.20	273.60	275.80
1979	4	288.50	279.70	279.70	283.60	285.40	281.20	284.10
1980	1	296.20	284.60	287.10	288.90	290.40	287.00	290.30
1980	2	304.40	293.30	293.90	297.30	297.70	294.10	298.30
1980	3	310.00	299.20	298.90	301.30	304.50	300.80	303.90
1980	4	316.80	305.40	305.60	307.40	309.00	306.20	310.30
1981	1	324.90	312.30	312.00	315.90	315.10	313.10	317.70
1981	2	331.50	319.30	320.40	323.60	322.10	320.10	324.70

Source: Compiled by the author.

Table 4.10. Average Weekly Earnings by State (dollars)

Year	Quarter	Sydney	Melbourne	Brisbane	Adelaide	Perth	Hobart	Capital city average
1958	3	40.60	41.80	34.30	35.80	39.90	36.30	40.00
1958	4	42.00	43.20	35.70	37.20	41.30	37.70	41.40
1959	1	39.60	40.80	33.40	35.00	39.00	35.40	39.00
1959	2	41.40	42.60	35.10	36.60	40.70	37.10	40.80
1959	3	42.40	43.50	36.00	37.50	41.60	38.00	41.70
1959	4	43.90	45.00	37.60	38.90	43.10	39.40	43.20
1960	1	42.60	43.70	36.20	37.70	41.80	38.20	41.90
1960	2	44.70	45.70	38.30	39.60	43.80	40.20	44.00
1960	3	45.20	46.20	38.80	40.10	44.30	40.60	44.50
1960	4	46.80	47.80	40.30	41.50	45.80	42.10	46.00
1961	1	44.30	45.30	37.90	39.20	43.40	39.70	43.60
1961	2	45.60	46.60	39.20	40.40	44.70	41.00	44.80
1961	3	45.50	46.50	39.00	40.30	44.60	40.90	44.70
1961	4	47.40	48.30	40.90	42.10	46.40	42.60	46.60
1962	1	45.30	46.30	38.90	40.20	44.40	40.70	44.60
1962	2	47.40	48.30	40.90	42.10	46.40	42.60	46.60
1962	3	47.00	47.90	40.50	41.70	46.00	42.30	46.20
1962	4	49.20	50.00	42.60	43.70	48.10	44.30	48.30
1963	1	46.90	47.80	40.40	41.60	45.90	42.10	46.10
1963	2	49.30	50.20	42.70	43.80	48.20	44.40	48.40
1963	3	49.30	50.10	42.70	43.80	48.20	44.40	48.40
1963	4	52.70	53.50	46.00	47.00	51.50	47.50	51.80
1964	1	49.40	50.30	42.90	43.90	48.30	44.50	48.60
1964	2	52.20	53.00	45.60	46.50	51.00	47.10	51.30
1964	3	53.60	54.30	46.90	47.80	52.30	48.40	52.60
1964	4	56.50	57.20	49.70	50.50	55.20	51.10	55.50
1965	1	53.60	54.30	46.90	47.80	52.30	48.40	52.60
1965	2	56.50	57.10	49.60	50.40	55.10	51.00	55.40
1965	3	57.50	58.10	50.70	51.40	56.10	52.00	56.40
1965	4	59.40	60.00	52.50	53.10	57.90	53.80	58.30
1966	1	56.60	57.20	49.80	50.50	55.20	51.20	55.60
1966	2	59.10	59.60	52.20	52.80	57.60	53.50	58.00
1966	3	62.20	63.50	56.40	56.90	59.30	56.70	61.10
1966	4	64.50	65.70	59.10	58.20	60.60	60.40	63.30
1967	1	61.00	60.90	54.70	55.60	57.70	56.80	59.50
1967	2	65.30	66.30	58.80	58.30	60.10	60.10	63.70
1967	3	65.10	67.00	59.20	59.80	62.40	60.30	64.40
1967	4	68.00	70.10	62.50	61.60	64.90	64.40	67.10
1968	1	64.80	64.60	57.50	59.10	62.50	60.20	63.20
1968	2	68.90	69.60	62.00	61.90	66.60	63.20	67.30
1968	3	69.50	69.80	62.00	63.20	67.10	63.00	67.80
1968	4	75.20	74.20	66.30	66.30	70.10	68.50	72.50
1969	1	69.70	70.60	62.10	63.40	67.40	63.60	68.50
1969	2	74.90	74.80	67.50	66.30	71.30	67.70	72.80
1969	3	75.80	76.30	67.80	68.60	73.50	69.00	74.10
1969	4	81.30	81.10	71.70	72.20	78.20	74.10	79.00

(continued)

Table 4.10 continued

Year	Quarter	Sydney	Melbourne	Brisbane	Adelaide	Perth	Hobart	Capital city average
1970	1	74.70	74.60	66.10	67.40	72.90	66.30	72.70
1970	2	82.00	81.40	72.10	72.80	78.20	74.30	79.40
1970	3	83.10	82.40	74.20	75.30	80.70	74.40	80.90
1970	4	89.40	87.80	79.30	77.30	85.90	80.40	86.30
1971	1	84.30	83.70	75.90	75.30	82.90	74.50	82.00
1971	2	92.40	91.50	82.70	80.90	90.00	84.80	89.80
1971	3	92.80	91.60	83.60	84.50	92.50	82.70	90.50
1971	4	99.70	96.80	89.80	88.00	96.30	91.20	96.50
1972	1	92.60	89.10	84.20	82.50	90.70	84.20	89.50
1972	2	100.20	97.90	91.80	88.80	94.70	92.00	97.20
1972	3	99.00	99.30	92.90	88.70	95.00	90.80	97.10
1972	4	107.70	105.90	99.20	95.00	99.00	99.30	104.40
1973	1	100.40	97.80	92.80	90.30	96.30	88.70	97.40
1973	2	111.40	108.20	103.60	99.50	105.00	102.00	108.20
1973	3	112.30	112.80	105.50	104.20	108.70	102.40	110.60
1973	4	123.40	121.20	115.60	110.40	115.00	117.00	120.40
1974	1	117.10	112.90	108.90	106.10	111.10	101.60	113.30
1974	2	132.10	128.30	124.00	120.80	125.30	120.60	128.70
1974	3	141.50	139.10	133.00	129.80	135.10	127.30	138.40
1974	4	158.70	155.20	150.00	141.80	147.40	146.70	154.40
1975	1	147.30	142.80	136.50	137.40	143.90	135.30	143.80
1975	2	160.70	154.20	150.70	145.30	156.00	151.40	156.40
1975	3	160.60	157.70	150.80	148.10	159.00	150.30	157.60
1975	4	178.20	178.00	169.90	163.70	173.70	164.70	175.70
1976	1	168.50	165.50	158.20	154.40	164.80	151.20	165.30
1976	2	183.60	180.70	175.00	167.60	178.80	165.10	179.80
1976	3	187.30	183.00	182.00	175.00	184.00	175.60	184.80
1976	4	197.90	197.40	189.80	182.80	194.50	188.70	195.40
1977	1	187.20	183.50	174.00	172.80	186.50	175.10	183.80
1977	2	201.80	200.40	192.30	187.40	198.20	185.50	198.70
1977	3	207.30	202.50	195.60	195.90	206.10	195.90	204.10
1977	4	217.10	211.90	206.80	199.40	211.10	204.80	212.50
1978	1	209.80	204.80	195.30	190.20	205.70	194.00	205.10
1978	2	219.40	217.90	211.20	203.90	213.90	201.10	216.30
1978	3	223.10	218.60	209.20	206.90	218.60	205.50	218.90
1978	4	232.80	231.30	219.70	211.70	223.50	214.50	228.20
1979	1	228.50	220.80	213.70	207.20	222.90	208.50	222.80
1979	2	236.30	235.70	228.70	215.50	228.30	217.80	232.80
1979	3	243.70	238.70	228.00	223.50	233.70	229.30	238.30
1979	4	253.60	252.30	235.10	230.10	242.20	239.70	248.50
1980	1	251.80	245.20	231.90	229.50	243.70	235.40	245.70
1980	2	265.70	256.90	248.10	238.90	258.60	247.30	258.20
1980	3	276.20	268.40	252.20	252.00	266.10	262.10	268.90
1980	4	295.60	288.80	282.00	271.30	283.10	288.00	290.00
1981	1	276.20	269.30	260.70	253.00	273.50	255.30	270.60
1981	2	301.60	291.60	281.60	271.20	293.50	280.70	293.10

Source: Compiled by the author.

Table 4.11. Population by State (thousands)

Year	Quarter	New South Wales	Victoria	Queensland	South Australia	Western Australia	Tasmania	Total Australia
1958	3	3,826	2,750	1,491	937	714	335	10,054
1958	4	3,845	2,764	1,495	943	718	336	10,101
1959	1	3,867	2,788	1,501	950	719	337	10,163
1959	2	3,880	2,805	1,515	958	723	339	10,219
1959	3	3,901	2,819	1,520	965	726	340	10,272
1959	4	3,919	2,830	1,524	971	730	341	10,316
1960	1	3,943	2,862	1,529	978	730	343	10,384
1960	2	3,960	2,877	1,543	984	733	344	10,442
1960	3	3,981	2,894	1,547	988	737	346	10,495
1960	4	4,009	2,908	1,550	995	742	347	10,552
1961	1	4,033	2,934	1,555	1,004	744	349	10,620
1961	2	4,054	2,950	1,576	1,030	758	350	10,719
1961	3	4,073	2,961	1,582	1,034	762	352	10,765
1961	4	4,092	2,975	1,589	1,039	767	354	10,817
1962	1	4,113	2,993	1,595	1,043	773	355	10,874
1962	2	4,131	3,003	1,600	1,048	778	356	10,918
1962	3	4,150	3,016	1,606	1,053	783	357	10,966
1962	4	4,171	3,031	1,613	1,049	789	358	11,013
1963	1	4,190	3,049	1,620	1,067	796	360	11,083
1963	2	4,203	3,061	1,628	1,074	800	361	11,129
1963	3	4,217	3,075	1,637	1,080	806	362	11,179
1963	4	4,236	3,092	1,650	1,087	811	363	11,240
1964	1	4,256	3,114	1,655	1,097	817	364	11,304
1964	2	4,269	3,126	1,662	1,104	821	365	11,349
1964	3	4,287	3,140	1,669	1,111	824	365	11,399
1964	4	4,309	3,159	1,678	1,120	831	367	11,465
1965	1	4,331	3,175	1,688	1,129	834	368	11,527
1965	2	4,347	3,186	1,697	1,137	838	368	11,574
1965	3	4,365	3,199	1,705	1,144	843	368	11,626
1965	4	4,387	3,217	1,712	1,154	851	370	11,694
1966	1	4,408	3,232	1,721	1,162	856	371	11,753
1966	2	4,418	3,242	1,728	1,167	861	372	11,790
1966	3	4,431	3,254	1,734	1,173	867	372	11,832
1966	4	4,451	3,272	1,740	1,179	877	374	11,894
1967	1	4,469	3,291	1,748	1,184	885	375	11,962
1967	2	4,482	3,296	1,753	1,188	892	375	11,987
1967	3	4,499	3,308	1,761	1,192	900	376	12,038
1967	4	4,520	3,325	1,769	1,196	910	378	12,099
1968	1	4,542	3,340	1,776	1,201	919	379	12,158
1968	2	4,553	3,345	1,782	1,206	928	380	12,194
1968	3	4,576	3,360	1,792	1,212	938	381	12,260
1968	4	4,599	3,377	1,801	1,219	950	383	12,331
1969	1	4,627	3,395	1,810	1,224	959	385	12,402
1969	2	4,644	3,405	1,817	1,229	968	385	12,447
1969	3	4,673	3,422	1,826	1,235	977	386	12,520
1969	4	4,699	3,440	1,834	1,242	989	387	12,591

(continued)

Table 4.11 continued

Year	Quarter	New South Wales	Victoria	Queensland	South Australia	Western Australia	Tasmania	Total Australia
1970	1	4,726	3,460	1,843	1,249	999	388	12,665
1970	2	4,734	3,464	1,847	1,253	1,004	338	12,690
1970	3	4,763	3,481	1,857	1,260	1,016	388	12,766
1970	4	4,792	3,501	1,867	1,270	1,027	390	12,846
1971	1	4,821	3,522	1,877	1,276	1,037	391	12,924
1971	2	4,825	3,520	1,881	1,277	1,043	390	12,937
1971	3	4,849	3,534	1,893	1,284	1,050	391	13,001
1971	4	4,873	3,549	1,907	1,290	1,060	392	13,070
1972	1	4,890	3,565	1,920	1,294	1,066	392	13,127
1972	2	4,905	3,577	1,932	1,299	1,071	392	13,177
1972	3	4,920	3,590	1,947	1,303	1,076	392	13,228
1972	4	4,935	3,604	1,962	1,308	1,081	394	13,284
1973	1	4,949	3,616	1,976	1,313	1,085	395	13,334
1973	2	4,961	3,628	1,987	1,318	1,090	396	13,380
1973	3	4,974	3,637	2,002	1,323	1,096	396	13,429
1973	4	4,990	3,651	2,019	1,329	1,103	397	13,491
1974	1	5,006	3,665	2,032	1,334	1,111	398	13,546
1974	2	5,019	3,677	2,046	1,341	1,117	399	13,599
1974	3	5,034	3,688	2,056	1,346	1,124	401	13,649
1974	4	5,063	3,706	2,074	1,326	1,138	402	13,709
1975	1	5,067	3,712	2,077	1,339	1,141	403	13,741
1975	2	5,075	3,719	2,084	1,341	1,147	405	13,771
1975	3	5,084	3,725	2,091	1,349	1,152	405	13,806
1975	4	5,096	3,730	2,102	1,355	1,159	407	13,849
1976	1	5,109	3,741	2,105	1,360	1,164	407	13,886
1976	2	5,118	3,746	2,112	1,363	1,170	407	13,915
1976	3	5,125	3,754	2,115	1,367	1,176	408	13,946
1976	4	5,139	3,765	2,122	1,373	1,184	409	13,991
1977	1	5,153	3,774	2,130	1,378	1,190	410	14,036
1977	2	5,165	3,782	2,137	1,382	1,197	411	14,074
1977	3	5,178	3,789	2,143	1,388	1,204	411	14,113
1977	4	5,192	3,799	2,155	1,394	1,211	412	14,163
1978	1	5,209	3,810	2,161	1,398	1,217	413	14,208
1978	2	5,227	3,818	2,167	1,400	1,222	414	14,249
1978	3	5,246	3,826	2,172	1,402	1,227	414	14,287
1978	4	5,263	3,836	2,180	1,405	1,232	416	14,331
1979	1	5,281	3,847	2,190	1,407	1,237	416	14,380
1979	2	5,301	3,854	2,197	1,410	1,243	418	14,424
1979	3	5,316	3,860	2,209	1,412	1,248	419	14,465
1979	4	5,336	3,869	2,223	1,415	1,254	420	14,516
1980	1	5,355	3,879	2,234	1,417	1,259	422	14,567
1980	2	5,373	3,887	2,248	1,420	1,265	423	14,616
1980	3	5,392	3,896	2,260	1,424	1,270	424	14,665
1980	4	5,413	3,908	2,276	1,428	1,277	425	14,727
1981	1	5,433	3,917	2,289	1,432	1,282	426	14,780
1981	2	5,453	3,927	2,303	1,435	1,288	427	14,833

Source: Compiled by the author.

2. Only a scale difference exists across regions:

$$\beta_{ij} = \beta_i, \text{ for } i = 1, \ldots, 3 \quad j = 1, \ldots, 6.$$

3. Only the adjustment rate differs:

$$\alpha_j = \alpha$$
$$\beta_{ij} = \beta_i, \text{ for } i = 1,2 \quad j = 1, \ldots, 6.$$

4. Both the adjustment rate and the scale differ:

$$\beta_{ij} = \beta_i, \text{ for } i = 1,2 \quad j = 1, \ldots, 6.$$

5. Only the price response is invariant across regions:

$$\beta_{ij} = \beta_1, \text{ for } j = 1, \ldots 6.$$

6. Only the income elasticity is invariant across regions:

$$\beta_{2j} = \beta_2, \text{ for } j = 1, \ldots, 6.$$

Equation 4.6 is estimated in both linear and log-linear form, and as the statistical differences obtained are small, only the constant elasticity results are reported.[14] The results of the six restriction tests are given in Table 4.12.

Table 4.12. National Parameter Restriction Tests

	df_1	df_2	F
Price, income, lagged dependent variable, and intercept	20	504	6.954
Price, income, and lagged dependent variable	15	504	4.359
Price, income, and intercept	15	504	4.529
Price and income	10	504	3.849
Price	5	504	3.538
Income	5	504	3.832

$F_{0.95(5,400)} = 2.23, F_{0.99(5,400)} = 3.06$

Source: Compiled by the author.

These national cross-equation restrictions are all rejected at the 0.01 level of significance; thus, regional differences appear to exist in both the immediate and the long-term behavioral response to system shocks. This means that the results obtained from a national gasoline demand function, which implicitly assumes identical responses among regions, may be misleading to decision makers. Tables 4.13 and 4.14 provide for a comparison of the single-equation OLS model estimates and those derived from an iterative SURE set of estimates. The latter generally exhibit smaller values for the parameter of the lagged dependent variable, implying that regional demand has less dependence on previous consumption levels than would be suggested by the single-equation OLS model formulation; therefore, the OLS results would tend to overestimate the potential benefits to be achieved from changes in the nature and composition of the automobile stock and from adjustments in consumer behavior patterns. Overall, also, the Zellner estimates indicate more inelastic price responses and slightly higher income elasticities; these differences may be attributed to some form of spatial autocorrelation, but no tests are made to identify its form.

In the SURE estimates, some systematic behavior appears to occur across states; for example, the short- and long-run price elasticities in New South Wales, Victoria, and Queensland resemble each other's, and the short-run price elasticities in South Australia, Western Australia, and Tasmania do also. These similarities in the parameter estimates suggest that some form of cross-equation restrictions might be appropriate to the modeling strategy. The following set of state-level restrictions is tested and are not statistically rejected; the F statistic is 0.42:

1. Price, income, and adjustment rate parameters identical in New South Wales and Victoria
2. Short-run price elasticity identical in South Australia, Western Australia, and Tasmania
3. Short-run income elasticity identical in Western Australia and Tasmania
4. Adjustment rate identical in Western Australia and Tasmania

The results from applying Zellner's SURE estimating technique to a set of state-level gasoline demand equations, and from imposing

Table 4.13. Single-Equation Ordinary Least Squares Gasoline Demand Elasticities (1958Q3–1981Q2)

Region	Price elasticity			Income elasticity						Goodness-of-fit	
	Short run	t	Long run	Short run	t	Long run	R^{-2}	DW	Durbin's h	normality $\chi^2(4)$	SEE (%)
New South Wales	-0.10	(-3.062)	-0.48	0.14	(2.339)	0.68	0.989	2.031	-0.252	13.284	2.4
Victoria	-0.11	(-4.570)	-0.55	0.14	(2.947)	0.71	0.989	1.920	0.380	10.471	2.2
Queensland	-0.13	(-2.803)	-0.67	0.15	(2.395)	0.81	0.988	1.810	1.049	10.439	2.9
South Australia	-0.18	(-4.640)	-0.45	0.29	(5.061)	0.72	0.975	1.972	0.152	9.232	3.1
Western Australia	-0.18	(-4.924)	-1.19	0.08	(1.416)	0.55	0.987	1.959	0.141	8.717	2.8
Tasmania	-0.17	(-4.544)	-1.52	0.04	(0.967)	0.34	0.986	1.723	1.355	3.127	2.6
Australia	-0.12	(-4.416)	-0.67	0.12	(5.333)	0.66	0.991	1.839	0.797	16.264	2.1

Source: Compiled by the author.

Table 4.14. Unrestricted Iterative Seemingly Unrelated Equations Gasoline Demand Elasticities (1958Q3–1981Q2)

Region	Price elasticity			Income elasticity						Goodness-of-fit	
	Short run	t	Long run	Short run	t	Long run	$\bar{R}^2$	DW	Durbin's h	normality $\chi^2(4)$	SEE (%)
New South Wales	-0.09	(-3.490)	-0.35	0.20	(4.689)	0.78	0.988	2.046	-0.338	8.380	2.3
Victoria	-0.09	(-5.253)	-0.38	0.19	(6.012)	0.78	0.989	1.932	0.285	10.944	2.1
Queensland	-0.09	(-2.580)	-0.37	0.21	(4.745)	0.88	0.987	1.794	1.033	13.189	2.8
South Australia	-0.18	(-5.830)	-0.49	0.26	(6.570)	0.70	0.975	1.998	-0.005	6.816	3.1
Western Australia	-0.16	(-5.371)	-0.87	0.13	(2.910)	0.68	0.987	1.910	0.374	7.890	2.7
Tasmania	-0.15	(-4.968)	-0.89	0.09	(3.076)	0.55	0.986	1.726	1.258	10.841	2.6
Implied Australia	-0.10	NA	-0.40	0.19	NA	0.75	NA	NA	NA	NA	NA
System $R^2 = 0.9996$, $\chi^2(18) = 692.50$											

Note: NA, not applicable.
Source: Compiled by the author.

appropriate cross-equation restrictions, suggest that consumers in the two most populous Australian states, New South Wales and Victoria, respond similarly to price changes in gasoline and to income changes in both the short and the long run; while consumers in South Australia, Western Australia, and Tasmania evidence similar short-run price elsaticities. The latter outcome implies that only the latter two states have the same long-run response to price shocks. Further, Western Australia and Tasmania indicate identical income elasticities in both the short and the long run. An explanation for some of the contrasts and similarities might be the largely urban nature of gasoline consumption in New South Wales and Victoria, whereas rural drivers covering long distances in the other states may be reacting to other conditions. Queensland, because of the rapid development of its retirement population and the attraction of the "Gold Coast" ocean resorts, might reasonably be expected to display unique responses, as it, in fact, appears to do. The implied national elasticities, derived as weighted averages of these SURE state-level estimates, reveal a long-run income elasticity that is approximately one and three-quarters of the corresponding price elasticity, substantially different from the single-equation OLS national results that appear in Table 4.13, where the price and income responses are equivalent.

In a dynamic demand model formulation, a choice must be made between the "stock" adjustment and alternative "flow" adjustment interpretations of the modeling results. The former interpretation implies that the β_{3j} parameter, the coefficient of the lagged endogenous variable, reflects a measure of the rate of vehicle depreciation, that being $(1 - \Phi_j)$, where Φ is the rate of vehicle depreciation. Recall that in Equation 4.6 the β_{3j} term derives from the adjustment Equation 4.5, which assumes a fixed consumption of gasoline per vehicle. Thus, current consumption levels are determined by the existing vehicle stock; and this includes gasoline consumption accounted for by new vehicles, plus the consumption of the remaining "old," or, that is, those registered in the previous period. If a scrappage rate of between 12 and 13 years is assumed, the corresponding depreciation rate would be about 0.08. Table 4.16 presents the F statistic calculated for testing $H_0 : \beta_{3j} = 0.92$, the null hypothesis. This null hypothesis is rejected for all regions individually and taken collectively, and suggests the appropriateness of the flow adjustment interpretation.[15] The table also presents the results of testing the hypothesis in regard to the question of whether or not the

Table 4.15. Restricted Iterative Seemingly Unrelated Equations Gasoline Demand Elasticities (1958Q3–1981Q2)

Region	Price elasticity			Income elasticity			$\bar{R}^2$	DW	Durbin's	Goodness-of-fit normality	SEE
	Short run	t	Long run	Short run	t	Long run			h	$\chi^2(4)$	(%)
New South Wales	-0.09	(-5.508)	-0.38	0.19	(6.998)	0.77	0.988	2.048	-0.321	8.380	2.3
Victoria	-0.09	(-5.508)	-0.38	0.19	(6.998)	0.77	0.989	1.926	0.304	9.689	2.1
Queensland	-0.09	(-2.690)	-0.39	0.20	(4.627)	0.86	0.987	1.796	1.021	6.546	2.8
South Australia	-0.16	(-7.068)	-0.46	0.25	(6.753)	0.70	0.975	2.003	-0.020	5.510	3.1
Western Australia	-0.16	(-7.068)	-1.03	0.09	(3.606)	0.54	0.987	1.935	0.230	10.746	2.7
Tasmania	-0.16	(-7.068)	-1.03	0.09	(3.606)	0.54	0.986	1.722	1.266	5.545	2.6
Implied Australia	-0.10	NA	-0.42	0.18	NA	0.73	NA	NA	NA	NA	NA

System R^2 = 0.9996, $\chi^2(11)$ = 689.73, $F(7,504)$ = 0.420

Restrictions imposed: Short-run price and income elasticities in New South Wales and Victoria are (respectively) equal.
Short-run price elasticities in South and Western Australia and Tasmania are equal.
Short-run income elasticities in Western Australia and Tasmania are equal.
Lagged response is equal in New South Wales and Victoria and in Western Australia and Tasmania.

Pooled cross-sectionally correlated and time-series autoregressive model

										$\chi^2(24)$	
Australia	-0.10	(-5.225)	-0.84	0.05	(3.266)	0.41	0.985	1.994	0.033	49.325	2.8

Source: Compiled by the author.

Table 4.16. Stock Adjustment and Long-Run Income Elasticity Hypothesis Tests

Region	Q_{t-4} parameter estimates	$H_0 : \beta_{3i} = 0.92$ F	$H_0 : \eta$ income, long run = 1.0 F	*df*
New South Wales and Victoria	0.75	-7.005	-5.228	1,511
Queensland	0.77	-3.895	-2.459	1,511
South Australia	0.64	-7.021	-6.779	1,511
Western Australia and Tasmania	0.84	-3.807	16.613	1,511
All states collectively	–	17.982	16.371	4,511

Source: Compiled by the author.

individual states exhibit long-run income elasticities of unity, an assumption that may be found to be implicitly biased toward the positive in many judgmental forecasting models. These assumptions of unitary income elasticities are all rejected.[16]

A problem already discussed with respect to single-equation models is that of collinearity, and this difficulty may be exacerbated in a system of equations where "high" collinearity exists among the various independent variables across equations, such as the price of gasoline and incomes in the various states. While the impact of collinearity on the parameter estimates is difficult to assess (see Belsley, Kuh, and Welsh 1980), the relatively simple FG test, used previously in electricity modeling, confirms that in this pooled analysis income and the lagged dependent variables are those most severely affected. Table 4.17 presents these FG R^2 measures, and the parameter estimates are further tested using a series of "sliding window" estimates in an attempt to ascertain their stability.

The 60-quarter sliding window estimates are developed using an unrestricted SURE estimation.[17] Those results appear in Tables 4.18 through 4.23, and Figure 4.1 provides a graph of the parameter estimates. The variation in parameter estimates among states appears to decline appreciably over time for the price elasticity and for the adjustment rate variable. The mean value for each state's sliding window estimates is quite similar to the unrestricted SURE results derived from using the entire dataset. In most instances the means

Table 4.17. Farrar-Glauber R^2

	Q_{t-4}	P	Y
New South Wales	0.957	0.252	0.958
Victoria	0.937	0.140	0.940
Queensland	0.960	0.239	0.963
South Australia	0.930	0.243	0.933
Western Australia	0.939	0.168	0.941
Tasmania	0.911	0.062	0.912
Australia	0.956	0.255	0.959

Source: Compiled by the author.

derived from the sliding window estimates are within 2 standard deviations of the results obtained by using the entire period. In general, New South Wales and Victoria exhibit more temporal stability in parameters than do the other states. The lagged dependent variable is consistently the most stable parameter, with the income elasticity generally more stable than the price elasticity, although the latter converges more toward a single national value. Figure 4.2 illustrates the temporal history for the standard errors of the estimates, and, as can be seen, the standard errors for each state have increased slightly over time. No formal tests for constancy of regression relationships such as those suggested by Brown, Durbin, and Evans (1975), based upon the cumulative sums of the residuals, and Hackl (1979), based upon the moving sums of the regression residuals, have been made. Thus, only the subjective opinion that the restricted SURE parameter estimates as presented in Table 4.15 appear to be reasonable is offered.

Therefore, using those restricted SURE estimates and assuming no future shocks to the system, that is, assuming that real price and income remain constant at the levels of the June quarter of 1981, national per capita consumption could be expected to decline by approximately 2.4 percent from the current 2.75 liters per day to 2.68 liters per day. Table 4.24 shows the changes that would be expected across the various states. It is interesting to note that in equilibrium the model anticipates that the variance in the existing regional consumption rates will decline by about 20 percent.

Table 4.18. Sixty-Quarter Sliding Window Estimates–New South Wales

Observation period	Parameters								Durbin's
	P	*t*	Y	*t*	Q_{t-1}	*t*	$\bar{R}^2$	DW	*h*
59Q3–74Q2	-0.12	(-2.502)	0.26	(5.804)	0.69	(19.641)	0.984	1.724	0.954
59Q4–74Q3	-0.14	(-2.959)	0.27	(6.181)	0.67	(19.164)	0.984	1.712	1.031
60Q1–74Q4	-0.16	(-3.277)	0.27	(6.318)	0.66	(18.754)	0.983	1.632	1.389
60Q2–75Q1	-0.16	(-3.455)	0.26	(6.261)	0.66	(18.700)	0.983	1.692	1.206
60Q3–75Q2	-0.17	(-3.554)	0.26	(6.293)	0.66	(17.831)	0.983	1.705	1.177
60Q4–75Q3	-0.14	(-2.887)	0.24	(5.915)	0.69	(18.170)	0.982	1.718	0.929
61Q1–75Q4	-0.24	(-5.260)	0.27	(5.947)	0.62	(14.387)	0.978	1.732	0.635
61Q2–76Q1	-0.17	(-3.858)	0.28	(5.883)	0.63	(14.124)	0.977	1.988	-0.243
61Q3–76Q2	-0.18	(-4.369)	0.28	(5.967)	0.62	(13.967)	0.977	2.095	-0.751
61Q4–76Q3	-0.19	(-4.479)	0.25	(5.548)	0.63	(14.403)	0.978	2.209	-1.246
62Q1–76Q4	-0.16	(-3.365)	0.19	(3.921)	0.70	(15.116)	0.979	2.179	-0.890
62Q2–77Q1	-0.17	(-3.481)	0.18	(3.812)	0.70	(15.216)	0.978	2.236	-1.183
62Q3–77Q2	-0.16	(-3.430)	0.18	(3.737)	0.70	(15.006)	0.977	2.259	-1.084
62Q4–77Q3	-0.15	(-3.308)	0.17	(3.516)	0.72	(14.951)	0.977	2.245	-1.033
63Q1–77Q4	-0.16	(-3.226)	0.25	(4.938)	0.63	(12.639)	0.974	2.280	-1.351
63Q2–78Q1	-0.16	(-3.104)	0.25	(4.898)	0.63	(12.421)	0.974	2.287	-1.321
63Q3–78Q2	-0.17	(-3.235)	0.25	(4.845)	0.63	(12.296)	0.974	2.323	-1.368
63Q4–78Q3	-0.16	(-2.993)	0.25	(4.904)	0.62	(12.085)	0.973	2.314	-1.333
64Q1–78Q4	-0.18	(-3.511)	0.24	(4.511)	0.62	(11.668)	0.972	2.309	-1.336
64Q2–79Q1	-0.16	(-3.376)	0.22	(4.038)	0.66	(12.720)	0.970	2.238	-1.447
64Q3–79Q2	-0.19	(-3.925)	0.21	(3.650)	0.67	(12.656)	0.969	2.301	-1.289
64Q4–79Q3	-0.16	(-3.538)	0.20	(3.539)	0.69	(13.255)	0.968	2.300	-1.331
65Q1–79Q4	-0.14	(-3.388)	0.21	(3.670)	0.69	(13.266)	0.966	2.254	-1.337
65Q2–80Q1	-0.14	(-3.487)	0.22	(3.903)	0.68	(13.531)	0.965	2.325	-1.418
65Q3–80Q2	-0.12	(-3.522)	0.24	(4.229)	0.67	(13.404)	0.964	2.372	-1.594
65Q4–80Q3	-0.11	(-3.503)	0.26	(4.638)	0.66	(13.270)	0.962	2.342	-1.540
66Q1–80Q4	-0.10	(-3.237)	0.28	(5.007)	0.64	(12.923)	0.960	2.356	-1.530
66Q2–81Q1	-0.09	(-3.108)	0.27	(4.733)	0.65	(13.107)	0.958	2.329	-1.471
66Q3–81Q2	-0.09	(-3.060)	0.27	(4.796)	0.65	(13.180)	0.956	2.313	-1.577
Mean	-0.15		0.24		0.66				
Standard deviation	0.03		0.03		0.03				
Entire observation set									
59Q3–81Q2	-0.09	(-3.490)	0.20	(4.689)	0.74	(20.069)	0.988	2.046	-0.338

Source: Compiled by the author.

Table 4.19. Sixty-Quarter Sliding Window Estimates–Victoria

Observation period	Parameters								
	P	t	Y	t	Q_{t-1}	t	$\bar{R}^2$	DW	Durbin's h
59Q3–74Q2	-0.08	(-2.204)	0.24	(6.704)	0.70	(18.946)	0.996	1.724	0.895
59Q4–74Q3	-0.08	(-2.213)	0.25	(6.684)	0.69	(17.843)	0.996	1.688	1.221
60Q1–74Q4	-0.09	(-2.311)	0.22	(5.791)	0.72	(18.198)	0.995	1.527	1.736
60Q2–75Q1	-0.08	(-2.201)	0.22	(5.793)	0.72	(18.344)	0.995	1.631	1.491
60Q3–75Q2	-0.09	(-2.472)	0.21	(5.589)	0.72	(17.581)	0.995	1.628	1.509
60Q4–75Q3	-0.06	(-1.643)	0.19	(4.916)	0.76	(18.517)	0.995	1.619	1.534
61Q1–75Q4	-0.09	(-2.747)	0.21	(5.834)	0.73	(18.644)	0.995	1.608	1.455
61Q2–76Q1	-0.09	(-2.664)	0.21	(5.795)	0.73	(18.656)	0.995	1.687	1.167
61Q3–76Q2	-0.08	(-2.676)	0.21	(5.932)	0.72	(18.550)	0.995	1.704	0.804
61Q4–76Q3	-0.09	(-3.037)	0.19	(5.495)	0.73	(19.062)	0.995	1.832	0.059
62Q1–76Q4	-0.16	(-5.088)	0.16	(4.399)	0.74	(18.351)	0.996	1.963	-0.324
62Q2–77Q1	-0.15	(-4.721)	0.17	(4.871)	0.73	(18.743)	0.996	2.033	-0.196
62Q3–77Q2	-0.13	(-3.907)	0.17	(4.769)	0.73	(17.963)	0.996	2.047	-0.252
62Q4–77Q3	-0.14	(-4.162)	0.17	(3.516)	0.73	(17.686)	0.996	2.055	-0.322
63Q1–77Q4	-0.13	(-3.693)	0.21	(5.639)	0.69	(16.626)	0.995	2.059	-0.452
63Q2–78Q1	-0.12	(-3.455)	0.20	(5.590)	0.70	(17.577)	0.995	2.126	-0.639
63Q3–78Q2	-0.12	(-3.357)	0.20	(5.351)	0.70	(17.117)	0.995	2.124	-0.554
63Q4–78Q3	-0.11	(-3.107)	0.19	(5.151)	0.72	(17.373)	0.995	2.090	-0.454
64Q1–78Q4	-0.12	(-3.559)	0.19	(5.116)	0.72	(17.730)	0.995	2.119	-0.485
64Q2–79Q1	-0.11	(-3.604)	0.18	(4.849)	0.74	(19.519)	0.995	2.077	-0.546
64Q3–79Q2	-0.10	(-3.418)	0.17	(4.754)	0.76	(20.830)	0.994	1.986	-0.204
64Q4–79Q3	-0.12	(-4.557)	0.18	(4.892)	0.74	(20.837)	0.994	2.063	-0.308
65Q1–79Q4	-0.11	(-4.500)	0.19	(5.024)	0.74	(20.772)	0.993	2.092	-0.639
65Q2–80Q1	-0.11	(-4.754)	0.20	(5.330)	0.73	(20.966)	0.993	2.151	-0.619
65Q3–80Q2	-0.11	(-5.119)	0.22	(5.659)	0.71	(20.903)	0.993	2.134	-0.579
65Q4–80Q3	-0.11	(-5.536)	0.23	(5.916)	0.70	(20.416)	0.993	2.042	-0.328
66Q1–80Q4	-0.11	(-5.601)	0.24	(6.168)	0.69	(20.060)	0.992	2.055	-0.263
66Q2–81Q1	-0.10	(-5.202)	0.25	(6.139)	0.69	(19.589)	0.992	2.050	-0.258
66Q3–81Q2	-0.10	(-5.326)	0.25	(6.030)	0.68	(19.305)	0.992	1.998	-0.272
Mean	-0.11		0.20		0.72				
Standard deviation	0.02		0.03		0.02				
Entire observation set									
59Q3–81Q2	-0.09	(-3.490)	0.19	(6.012)	0.75	(25.965)	0.989	1.932	0.285

Source: Compiled by the author.

Table 4.20. Sixty-Quarter Sliding Window Estimates–Queensland

Observation period	Parameters						$\bar{R}^2$	DW	Durbin's
	P	*t*	Y	*t*	Q_{t-1}	*t*			*h*
59Q3–74Q2	-0.28	(-3.337)	0.37	(6.594)	0.58	(9.926)	0.996	1.710	1.174
59Q4–74Q3	-0.29	(-3.665)	0.36	(6.649)	0.57	(10.135)	0.996	1.622	1.471
60Q1–74Q4	-0.17	(-2.239)	0.32	(5.847)	0.62	(10.910)	0.995	1.448	2.098
60Q2–75Q1	-0.20	(-2.563)	0.35	(6.248)	0.58	(9.936)	0.994	1.656	1.444
60Q3–75Q2	-0.17	(-2.190)	0.35	(6.206)	0.58	(9.710)	0.994	1.679	1.266
60Q4–75Q3	-0.16	(-2.089)	0.34	(6.124)	0.60	(10.241)	0.994	1.757	0.936
61Q1–75Q4	-0.20	(-2.552)	0.39	(7.006)	0.53	(8.825)	0.994	1.816	0.567
61Q2–76Q1	-0.20	(-2.543)	0.40	(7.097)	0.50	(8.230)	0.994	1.865	0.419
61Q3–76Q2	-0.22	(-2.801)	0.40	(7.149)	0.49	(8.116)	0.994	1.920	0.186
61Q4–76Q3	-0.21	(-2.725)	0.38	(6.778)	0.51	(8.605)	0.994	1.955	-0.259
62Q1–76Q4	-0.24	(-3.222)	0.36	(6.465)	0.52	(8.766)	0.995	2.102	-0.647
62Q2–77Q1	-0.25	(-3.449)	0.36	(6.584)	0.52	(8.901)	0.995	2.087	-0.464
62Q3–77Q2	-0.29	(-4.218)	0.34	(6.292)	0.53	(9.048)	0.995	2.045	-0.266
62Q4–77Q3	-0.31	(-4.598)	0.31	(5.857)	0.56	(9.602)	0.995	1.969	-0.031
63Q1–77Q4	-0.29	(-4.265)	0.36	(6.505)	0.51	(8.414)	0.994	2.051	-0.469
63Q2–78Q1	-0.32	(-5.012)	0.35	(6.377)	0.51	(8.489)	0.994	2.148	-0.853
63Q3–78Q2	-0.32	(-5.112)	0.35	(6.473)	0.51	(8.456)	0.994	2.190	-0.832
63Q4–78Q3	-0.33	(-5.359)	0.33	(6.520)	0.52	(9.086)	0.994	2.168	-0.833
64Q1–78Q4	-0.29	(-4.485)	0.31	(5.656)	0.56	(9.186)	0.994	2.073	-0.529
64Q2–79Q1	-0.24	(-3.957)	0.27	(5.135)	0.62	(10.669)	0.993	1.963	-0.185
64Q3–79Q2	-0.16	(-2.894)	0.25	(4.568)	0.68	(11.935)	0.992	1.835	0.314
64Q4–79Q3	-0.12	(-2.442)	0.23	(4.327)	0.71	(13.628)	0.992	1.939	0.234
65Q1–79Q4	-0.11	(-2.343)	0.23	(4.407)	0.72	(13.880)	0.992	1.909	0.202
65Q2–80Q1	-0.10	(-2.194)	0.24	(4.639)	0.71	(14.453)	0.992	1.950	0.196
65Q3–80Q2	-0.10	(-2.396)	0.26	(4.993)	0.70	(14.416)	0.991	1.926	0.249
65Q4–80Q3	-0.10	(-2.382)	0.28	(5.318)	0.68	(14.224)	0.991	1.892	0.403
66Q1–80Q4	-0.09	(-2.249)	0.30	(5.699)	0.66	(13.788)	0.991	1.881	0.488
66Q2–81Q1	-0.09	(-2.169)	0.30	(5.508)	0.66	(13.683)	0.991	1.879	0.495
66Q3–81Q2	-0.08	(-2.000)	0.32	(5.753)	0.65	(13.300)	0.990	1.801	0.608
Mean	-0.20		0.32		0.59				
Standard deviation	0.08		0.05		0.07				
Entire observation set									
59Q3–81Q2	-0.09	(-2.580)	0.21	(4.745)	0.76	(18.966)	0.987	1.794	1.033

Source: Compiled by the author.

Table 4.21. Sixty-Quarter Sliding Window Estimates–South Australia

Observation period	Parameters								Durbin's
	P	*t*	Y	*t*	Q_{t-1}	*t*	$\bar{R}^2$	DW	*h*
59Q3-74Q2	-0.41	(-4.352)	0.34	(6.189)	0.44	(6.780)	0.990	1.812	0.773
59Q4-74Q3	-0.43	(-4.840)	0.32	(5.415)	0.45	(6.733)	0.990	1.799	0.886
60Q1-74Q4	-0.52	(-6.098)	0.32	(5.879)	0.43	(6.347)	0.990	1.716	1.048
60Q2-75Q1	-0.47	(-5.708)	0.31	(5.733)	0.45	(6.856)	0.990	1.740	1.036
60Q3-75Q2	-0.47	(-5.666)	0.31	(5.741)	0.45	(6.926)	0.990	1.750	0.935
60Q4-75Q3	-0.46	(-5.916)	0.30	(5.972)	0.46	(7.373)	0.990	1.831	0.612
61Q1-75Q4	-0.55	(-7.370)	0.32	(6.240)	0.39	(6.146)	0.991	0.847	0.521
61Q2-76Q1	-0.47	(-6.540)	0.35	(6.800)	0.39	(6.240)	0.991	1.857	0.578
61Q3-76Q2	-0.47	(-7.157)	0.35	(6.880)	0.38	(5.966)	0.991	1.868	0.393
61Q4-76Q3	-0.45	(-7.231)	0.34	(6.680)	0.39	(6.199)	0.991	1.901	0.021
62Q1-76Q4	-0.42	(-7.223)	0.30	(5.934)	0.44	(6.801)	0.991	1.970	0.112
62Q2-77Q1	-0.42	(-7.437)	0.30	(6.259)	0.44	(6.946)	0.991	1.975	-0.008
62Q3-77Q2	-0.44	(-8.190)	0.29	(6.062)	0.45	(7.041)	0.991	1.959	0.065
62Q4-77Q3	-0.43	(-8.101)	0.29	(6.160)	0.44	(6.869)	0.991	1.970	0.127
63Q1-77Q4	-0.43	(-8.014)	0.34	(7.458)	0.38	(6.174)	0.991	1.965	0.152
63Q2-78Q1	-0.43	(-7.974)	0.33	(7.603)	0.39	(6.509)	0.991	1.961	0.128
63Q3-78Q2	-0.41	(-7.593)	0.35	(8.080)	0.37	(6.378)	0.991	1.960	0.084
63Q4-78Q3	-0.42	(-7.671)	0.34	(7.847)	0.38	(6.451)	0.991	1.983	-0.006
64Q1-78Q4	-0.42	(-7.785)	0.33	(7.702)	0.39	(6.945)	0.991	1.995	0.003
64Q2-79Q1	-0.37	(-6.990)	0.31	(6.935)	0.45	(8.163)	0.989	1.889	-0.034
64Q3-79Q2	-0.30	(-6.160)	0.30	(6.615)	0.52	(10.032)	0.988	1.824	0.452
64Q4-79Q3	-0.29	(-6.285)	0.29	(6.521)	0.53	(11.112)	0.988	1.916	0.339
65Q1-79Q4	-0.22	(-5.330)	0.31	(6.461)	0.56	(11.444)	0.987	1.880	-0.000
65Q2-80Q1	-0.21	(-5.488)	0.32	(6.780)	0.55	(12.034)	0.986	1.999	-0.127
65Q3-80Q2	-0.18	(-5.170)	0.35	(7.284)	0.55	(12.378)	0.987	2.072	-0.691
65Q4-80Q3	-0.19	(-5.775)	0.37	(8.184)	0.54	(12.634)	0.987	2.052	-0.345
66Q1-80Q4	-0.19	(-6.001)	0.38	(8.088)	0.53	(12.211)	0.987	2.017	-0.103
66Q2-81Q1	-0.18	(-6.027)	0.39	(8.193)	0.52	(11.948)	0.987	2.022	-0.093
66Q3-81Q2	-0.18	(-6.143)	0.41	(8.009)	0.50	(11.353)	0.986	1.935	-0.018
Mean	-0.37		0.33		0.45				
Standard deviation	0.12		0.03		0.06				
Entire observation set									
59Q3-81Q2	-0.18	(-5.830)	0.26	(6.570)	0.63	(15.249)	0.975	1.998	-0.005

Source: Compiled by the author.

Table 4.22. Sixty-Quarter Sliding Window Estimates–Western Australia

Observation period	Parameters								Durbin's
	P	t	Y	t	Q_{t-1}	t	$\bar{R}^2$	DW	h
59Q3–74Q2	-0.36	(-3.236)	0.07	(1.439)	0.86	(22.374)	0.997	1.668	1.277
59Q4–74Q3	-0.33	(-3.075)	0.06	(1.289)	0.86	(22.017)	0.997	1.602	1.521
60Q1–74Q4	-0.22	(-2.237)	0.03	(0.633)	0.89	(22.078)	0.996	1.472	1.885
60Q2–75Q1	-0.15	(-1.565)	0.04	(0.777)	0.89	(21.391)	0.996	1.529	1.905
60Q3–75Q2	-0.12	(-1.216)	0.03	0.572)	0.91	(21.494)	0.996	1.520	1.914
60Q4–75Q3	-0.10	(-1.022)	0.00	(0.048)	0.93	(21.430)	0.996	1.578	1.624
61Q1–75Q4	-0.11	(-1.173)	0.06	(1.186)	0.88	(20.701)	0.996	1.531	1.900
61Q2–76Q1	-0.13	(-1.338)	0.06	(1.196)	0.87	(20.590)	0.996	1.493	1.789
61Q3–76Q2	-0.15	(-1.618)	0.05	(1.058)	0.87	(21.186)	0.996	1.595	1.384
61Q4–76Q3	-0.13	(-1.442)	0.04	(0.841)	0.87	(20.887)	0.996	1.675	1.033
62Q1–76Q4	-0.16	(-2.035)	0.05	(1.147)	0.85	(20.054)	0.996	1.800	0.409
62Q2–77Q1	-0.15	(-2.049)	0.06	(1.301)	0.84	(19.903)	0.996	1.908	0.338
62Q3–77Q2	-0.26	(-3.988)	0.03	(0.705)	0.84	(18.783)	0.995	1.911	0.017
62Q4–77Q3	-0.25	(-4.260)	0.05	(0.930)	0.83	(18.098)	0.995	1.953	0.184
63Q1–77Q4	-0.22	(-3.766)	0.05	(0.974)	0.83	(17.654)	0.995	1.891	0.146
63Q2–78Q1	-0.22	(-3.835)	0.06	(1.327)	0.81	(17.277)	0.995	1.928	0.289
63Q3–78Q2	-0.16	(-2.706)	0.10	(1.999)	0.78	15.813)	0.994	1.823	0.496
63Q4–78Q3	-0.19	(-3.182)	0.09	(1.827)	0.78	(15.199)	0.994	1.976	-0.109
64Q1–78Q4	-0.22	(-3.865)	0.14	(2.685)	0.73	(14.047)	0.994	2.019	-0.134
64Q2–79Q1	-0.24	(-4.301)	0.13	(2.741)	0.72	(14.718)	0.994	2.037	-0.236
64Q3–79Q2	-0.22	(-4.399)	0.13	(2.796)	0.73	(15.916)	0.993	1.971	-0.118
64Q4–79Q3	-0.26	(-5.651)	0.16	(3.332)	0.69	(16.230)	0.993	2.024	-0.169
65Q1–79Q4	-0.20	(-4.820)	0.17	(3.398)	0.70	(15.872)	0.991	2.066	-0.778
65Q2–80Q1	-0.19	(-4.916)	0.17	(3.633)	0.70	(16.519)	0.991	2.122	-0.533
65Q3–80Q2	-0.18	(-5.188)	0.19	(4.097)	0.68	(16.541)	0.990	2.059	-0.313
65Q4–80Q3	-0.17	(-5.165)	0.22	(4.594)	0.66	(15.634)	0.990	1.911	0.087
66Q1–80Q4	-0.16	(-5.203)	0.24	(5.051)	0.64	(15.364)	0.990	1.909	0.343
66Q2–81Q1	-0.16	(-5.193)	0.25	(5.190)	0.64	(15.080)	0.989	1.897	0.417
66Q3–81Q2	-0.15	(-5.098)	0.26	(5.519)	0.62	(14.529)	0.988	1.730	0.608
Mean	-0.19		0.10		0.79				
Standard deviation	0.06		0.07		0.09				
Entire observation set									
59Q3–81Q2	-0.16	(-5.371)	0.13	(2.910)	0.82	(24.134)	0.987	1.910	0.374

Source: Compiled by the author.

Table 4.23. Sixty-Quarter Sliding Window Estimates–Tasmania

Observation period	Parameters P	t	Y	t	Q_{t-1}	t	$\bar{R}^2$	DW	Durbin's h
59Q3–74Q2	-0.08	(-0.887)	0.10	(2.809)	0.86	(26.087)	0.996	1.611	1.479
59Q4–74Q3	-0.07	(-0.785)	0.09	(2.698)	0.86	(25.241)	0.996	1.565	1.705
60Q1–74Q4	-0.10	(-1.259)	0.08	(2.199)	0.87	(25.340)	0.996	1.498	1.962
60Q2–75Q1	-0.05	(-0.553)	0.09	(2.623)	0.86	(23.987)	0.996	1.463	2.060
60Q3–75Q2	-0.03	(-0.389)	0.10	(2.813)	0.85	(24.066)	0.995	1.481	1.984
60Q4–75Q3	-0.04	(-0.449)	0.09	(2.705)	0.85	(23.408)	0.995	1.487	2.069
61Q1–75Q4	-0.10	(-1.188)	0.11	(2.888)	0.81	(21.094)	0.994	1.381	2.179
61Q2–76Q1	-0.10	(-1.264)	0.10	(2.782)	0.81	(22.117)	0.995	1.573	1.579
61Q3–76Q2	-0.11	(-1.426)	0.10	(2.716)	0.81	(21.925)	0.995	1.606	1.241
61Q4–76Q3	-0.12	(-1.580)	0.08	(2.360)	0.81	(21.812)	0.995	1.721	0.524
62Q1–76Q4	-0.15	(-1.929)	0.07	(2.070)	0.81	(21.802)	0.995	1.845	0.314
62Q2–77Q1	-0.15	(-1.899)	0.07	(1.887)	0.82	(22.067)	0.995	1.746	0.838
62Q3–77Q2	-0.17	(-2.013)	0.06	(1.692)	0.83	(21.616)	0.995	0.700	1.134
62Q4–77Q3	-0.16	(-1.880)	0.06	(1.618)	0.83	(20.930)	0.994	1.667	1.267
63Q1–77Q4	-0.12	(-1.590)	0.08	(2.354)	0.81	(20.389)	0.994	1.727	1.104
63Q2–78Q1	-0.12	(-1.588)	0.08	(2.305)	0.81	(21.443)	0.994	1.735	1.051
63Q3–78Q2	-0.11	(-1.483)	0.09	(2.473)	0.80	(21.581)	0.993	1.742	0.939
63Q4–78Q3	-0.08	(-1.280)	0.10	(2.696)	0.79	(21.019)	0.993	1.750	0.998
64Q1–78Q4	-0.08	(-1.185)	0.10	(2.738)	0.79	(20.366)	0.993	1.759	0.945
64Q2–79Q1	-0.08	(-1.267)	0.10	(2.655)	0.80	(21.283)	0.993	1.787	0.801
64Q3–79Q2	-0.06	(-0.903)	0.09	(2.639)	0.81	(22.244)	0.992	1.722	0.969
64Q4–79Q3	-0.15	(-2.791)	0.09	(2.394)	0.79	(20.289)	0.991	1.804	0.655
65Q1–79Q4	-0.14	(-3.093)	0.09	(2.510)	0.79	(20.395)	0.990	1.858	0.308
65Q2–80Q1	-0.15	(-3.450)	0.11	(3.025)	0.78	(21.110)	0.991	1.964	0.134
65Q3–80Q2	-0.16	(-4.055)	0.12	(3.280)	0.77	(20.829)	0.990	1.959	0.127
65Q4–80Q3	-0.15	(-4.212)	0.14	(3.815)	0.75	(20.206)	0.989	1.979	0.076
66Q1–80Q4	-0.14	(-4.027)	0.16	(4.523)	0.73	(19.578)	0.988	1.990	-0.040
66Q2–81Q1	-0.14	(-4.213)	0.16	(4.466)	0.72	(19.327)	0.988	2.006	-0.040
66Q3–81Q2	-0.16	(-4.750)	0.17	(4.357)	0.72	(18.874)	0.987	1.857	0.181
Mean	-0.11		0.10		0.81				
Standard deviation	0.04		0.03		0.04				
Entire observation set									
59Q3–81Q2	-0.15	(-4.968)	0.09	(3.076)	0.83	(28.200)	0.986	1.726	1.258

Source: Compiled by the author.

Figure 4.1. Sixty-Quarter Sliding Window Parameter Estimates (Unrestricted Seemingly Unrelated Equations) *All States*

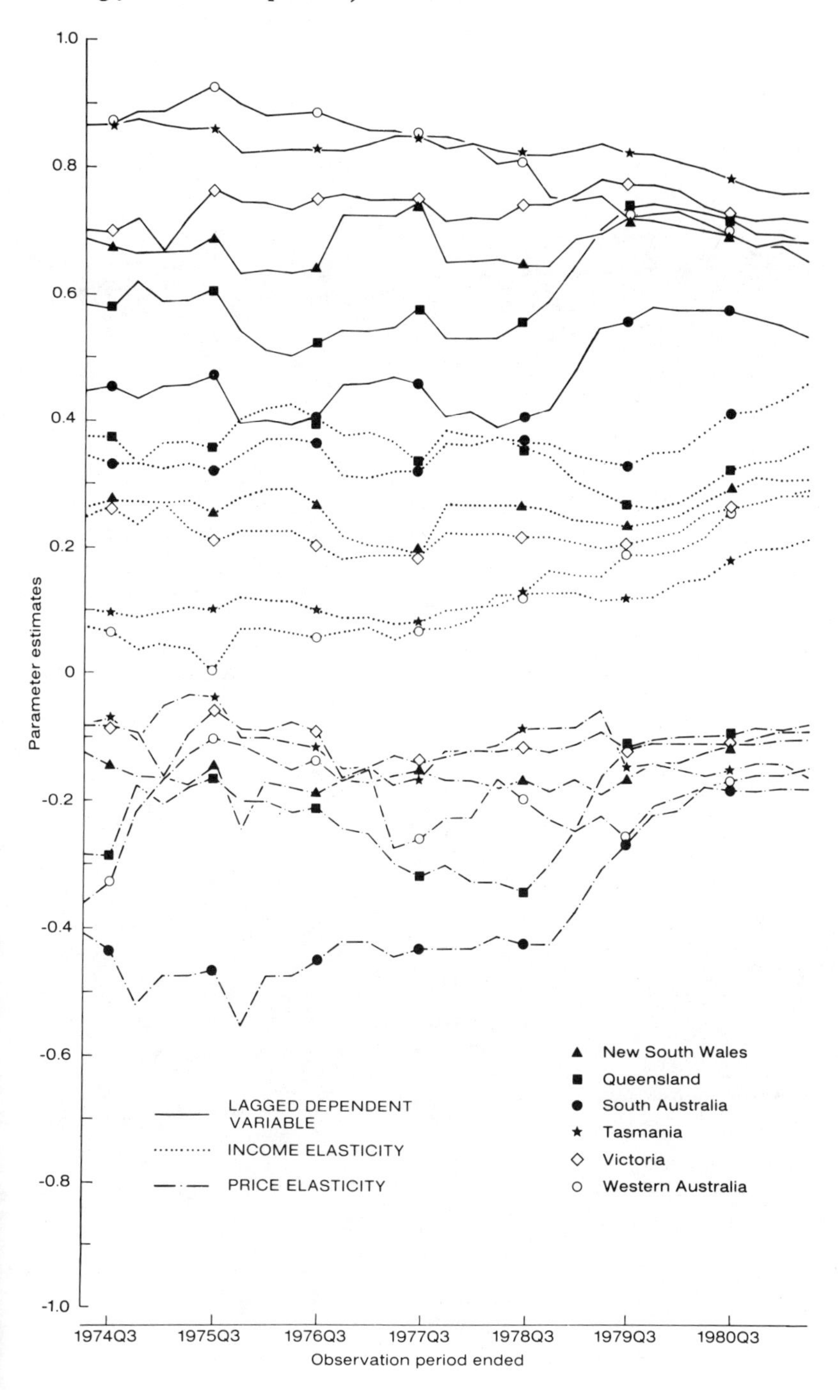

Figure 4.2. Sixty-Quarter Sliding Window Standard Errors of the Estimates (Unrestricted Seemingly Unrelated Equations) *All States*

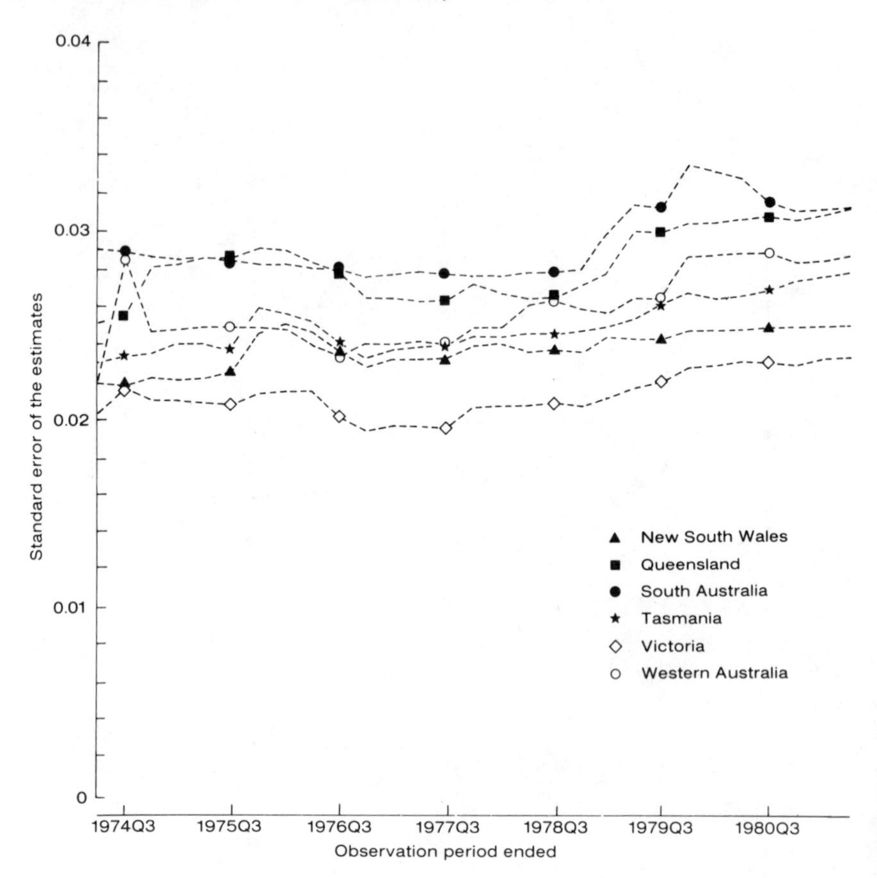

Table 4.24. No Shock Conditional Forecasts

	June quarter 1981	Equilibrium	Percentage change
New South Wales	2.54	2.54	0.1
Victoria	2.89	2.79	-3.4
Queensland	2.89	2.89	0.2
South Australia	2.71	2.61	-3.8
Western Australia	3.03	2.74	-9.8
Tasmania	2.68	2.47	-8.0
Australia	2.75	2.68	-2.4

Source: Compiled by the author.

In summary, the iterative SURE results, with appropriate cross-equation restrictions imposed, provide short-run price elasticities that are similar in magnitude to price elasticities reported for the United States as revealed in Table 4.25. The short-run income elasticities are, however, substantially smaller than those for the United States, but this result may be attributable to reliance upon AWE in the present study as the proxy for income instead of the use of a measure of household disposable income, as many of the U.S. studies employ. Also, the long-run price elasticities are generally larger in this quarterly model than is the case for the annual or cross-sectional studies done elsewhere.

The price response in Western Australia and Tasmania, while differing substantially from that in the other Australian states, is not unique among gasoline demand model results, as it is similar in size to the values that Drollas (1984) presents for Belgium, Denmark, and Italy. In addition, the long-run income elasticities calculated here are significantly less than unity in contrast to the highly income-elastic response that Folie (1977) reports and that Phlips (1972) and DRI (1973) reveal for the United States and that Drollas (1984) cites for Austria, Belgium, Sweden, and the United Kingdom. It should be noted, though, that the Drollas study is the only single-equation study in the table that uses post–oil embargo data, and a question is raised as to whether the effect of spatial autocorrelation may influence these results, since Baltagi and Griffin (1983) and Dahl (1982) find that in pooled time-series cross-sectional analyses, the long-run income elasticities are less than unity, as is established in the Australian case. Overall, the present results suggest that attempts at modeling national energy demand responses may provide misleading information, since important regional differences in response to price and income changes do exist in gasoline demand, and spatial correlation effects may be important. Various alternative national aggregation schemes are tested and rejected, and a statistically determined aggregation scheme is selected that indicates some similarities in responses across regions: Namely, gasoline demand is found to be price inelastic, except in Western Australia and Tasmania, where the long-run price elasticity is roughly unity and the gasoline is consistently found to be income inelastic.

Additional research could identify and incorporate some of the missing variables that may account for the observed differences in regional responsiveness, and further analysis of the stability of the

Table 4.25. Dynamic Demand Model Gasoline Elasticities

	Price elasticity		Income elasticity			
	Short run	Long run	Short run	Long run	Time span	Notes
Australia						
Brain and Schuyers (1981)	−0.11	−0.22	0.72	NR	NR–1974	A N
Donnelly (1982a)	−0.09 to −0.18	−0.48 to −1.52	0.04 to 0.29	0.34 to 0.81	58Q3–81Q2	Q C S
Folie (1977)	−0.04 to −0.38	−0.14 to −0.77	0.28 to 0.86	0.90 to 1.52	58Q3–74Q2	Q S
Austria						
Drollas (1984)	−0.52 to −0.57	−0.82 to −0.85	0.42 to 0.75	1.37 to 1.39	1950–80	A N
Belgium						
Drollas (1984)	−0.48 to −0.50	−1.17 to −1.62	0.38 to 0.57	1.27 to 1.34	1950–80	A N
Canada						
Reza and Spiro (1979)	−0.21	−0.33	0.60	1.44	69Q1–76Q3	Q N
Denmark						
Drollas (1984)	−0.31 to −0.38	−1.07 to −1.27	0.14 to 0.31	0.57 to 0.86	1950–80	A N
France						
Drollas (1984)	−0.44	−0.58	0.26	1.08	1950–80	A N
Italy						
Drollas (1984)	−0.38 to −0.41	−1.08 to −1.44	0.31 to 0.39	1.34 to 1.46	1950–80	A N
New Zealand						
Baas et al. (1982)	−0.09	−0.12	0.63	0.84	1962–81	A N
Hughes (1980)	−0.11	−0.14	0.57	0.79	69Q1–79Q4	Q N
Sweden						
Drollas (1984)	−0.16 to −0.17	−0.37 to −0.52	0.46 to 0.65	1.46 to 1.50	1950–80	A N

West Germany						
Donnelly (1982b)	−0.17	−0.21	0.99	1.22	1970–80	A N
Drollas (1984)	−0.41 to −0.53	−0.82 to −1.20	0.31 to 0.42	1.07 to 1.15	1950–80	A N
United Kingdom						
Drollas (1984)	−0.24 to −0.28	−0.55 to −0.62	0.06 to 0.22	0.90 to 1.21	1950–80	A N
United States						
Data Resources Inc. (1973)	−0.07 to −0.14	−0.24 to −0.32	0.28 to 0.45	0.94 to 1.03	63Q1–72Q4	Q S
Drollas (1984)	−0.32 to −0.35	−0.70 to −0.73	0.63 to 0.75	0.55 to 0.93	1950–80	A N
Houthakker and Taylor (1970)	−0.16	−0.45	NR	NR	1929–41, 1946–61	A N
Houthakker, Verleger, and Sheehan	−0.07	−0.25	0.30	1.00	63Q1–73Q4	Q N
Kraft and Rodekohr (1980)	−0.20	−0.72	0.14	0.49	1954–72	A S
Nordhaus (1979)	−0.22	−0.76	0.39	0.84	1959–72	A N
Phlips (1972)	−0.11	−0.68	NR	1.34	1929–41, 1946–67	A N
Sweeney (1979)	−0.12	−0.73	0.85	0.78	1953–73	A N
12 European Countries plus the U.S.						
Kennedy (1974)	−0.46	−0.83	0.74	1.33	1962–72	A C N
18 OECD Countries						
Galtagi and Griffin (1983)	−0.10 to −0.16	−0.64 to −0.92	0.13 to 0.16	0.61 to 0.84	1960–78	A C N
38 Countries						
Dahl (1982)	−0.20	−0.98	0.10	0.50	1970–78	A C N

Notes: NR, not reported; A, annual; Q, quarterly; C, cross section; N, national; S, state level.
Source: Compiled by the author.

elasticity estimates using different time intervals could shed light on the issue of whether or not the elasticities are constant. Perhaps a usable stock of vehicle series, improved information on utilization rates, and fuel efficiency measures might allow for the estimation of a vintage capital stock model of the form Sweeney (1979) suggests, although Drollas (1984) argues quite persuasively that "one need not resort to particularly elaborate equation systems based upon such data to explain a high proportion of the variation in gasoline consumption" (p. 74). The results presented in this chapter tend to support the robustness of the Koyck lag–type dynamic demand model utilized here.

NOTES

1. Gasoline (motor gasoline) is an aggregate of regular gasoline (standard petrol) and super gasoline (super petrol) as reported in "Major Energy Statistics" published monthly by DRE.

2. For a comprehensive survey of gasoline demand models, see Dahl (1984).

3. Parts of this section are reproduced from Donnelly (1984) with permission of the editor of the *International Journal of Transport Economics*.

4. These include the capital cities of Canberra, Sydney, Melbourne, Brisbane, Adelaide, Perth, Hobart, and Darwin and the regional centers of Geolong, Newcastle, and Woolongong.

5. As in the previous chapter, the measure of collinearity will be the FG R^2.

6. In the light of Pagan's caveat with respect to the power of the LM statistic as it is derived in the version of the SHAZAM regression package, and as mentioned in note 15 in Chapter 3, and which is used to replicate the SJ analysis: The indication of heteroscedasticity that Regression III exhibits must be taken seriously and should be explicitly addressed by the modeler.

7. Part of the material here is reproduced from Donnelly (1982a) with permission of the editor of *Economic Record*.

8. The mean and median lags for this form of geometric lag model are calculated as $\beta_{3j} / (1 - \beta_{3j})$ and $\ln 0.5 / \ln \beta_{3j}$, respectively (see Pindyck and Rubinfeld 1981, pp. 232–34). This implies about 2.5 years for the mean lag and just over 1.66 years for the median lag with this value for β_{3j}.

9. Gasoline is referred to as petrol in Australia so the GPI is actually the petrol price index.

10. Durbin (1970) specifies the *h* statistic as an alternative to the DW statistic for use with equations that include lagged dependent variables. McNowan and Hunter (1980) run simulations to test the power of Durbin's *h* statistic and conclude that it may be less powerful than the traditional DW statistic, and, as Guilkey (1975) indicates, with a system of equations, such as is used here, neither

the DW nor Durbin's *h* statistic is the more appropriate test of first-order serial correlation. However, because of the simplicity of calculation these are the statistics reported. In addition, as the model is specified using quarterly data, a higher-order serial correlation might be anticipated, for example, a four-quarter autoregressive or moving-average process rather than the first-order process; and, furthermore, such higher-order processes are not identified by these statistics whenever they are calculated in the usual manner.

11. The assumption that the disturbances term is normally distributed is not necessary for the parameter estimates to be BLUE; however, the estimates "can no longer claim to be efficient because, without the specification of the distribution form, the Cramer-Rao lower bound of their variances is not known" (Kmenta 1971, pp. 248–49).

12. The null hypotheses being tested with a F statistic on this system of state-level equations concern various combinations of intercept, price, income, and adjustment rate national equalities, and as will be shown, these are rejected, thereby indicating a possible Type I error.

13. Many combinations of restrictions are possible and could be tested, but generally only national summations are applied in analyses of gasoline demand relationships, and these seem to be the most interesting ones for evaluation.

14. In Donnelly (1981) both the linear and the log-linear results appear as estimated over the interval from the September quarter of 1958 to the December quarter of 1980 inclusive and from the September quarter of 1966 through the December quarter of 1980.

15. Verleger and Sheehan (1976) reached a similar conclusion that the flow adjustment model should be accepted using U.S. data.

16. This "hypothesis test" on the long-run income elasticity is implemented by rewriting the long-run (LR) elasticity definitional equation, namely, $\eta_{Y,LR} \equiv \beta_{2j} / (1 - \beta_{3j}) = 1.0$ or $\beta_{2j} = 1.0 - \beta_{3j}$. This is the relationship tested.

17. Sliding window estimates involve dividing the observation set into overlapping subsamples. In this case, the 92 quarters provide 29 60-quarter samples, which are used to derive the parameter estimates.

REFERENCES

Alt, C., A. Bopp, and G. Lady. 1979. "Econometric Analysis of the 1974/75 Decline in Petroleum Consumption in the USA: Some Policy Implications." *Energy Economics,* 1:27–32.

Australian Bureau of Statistics. Quarterly. "Time-Series Data on Magnetic Tape and Microfiche." Data Extracted from the Australian National University Department of Statistics Economic Time Series Databank, Canberra.

Australian Department of Resources and Energy. Monthly. "Major Energy Statistics" and "Sales of Petroleum Products by State Marketing Area," Canberra.

Australian Institute of Petroleum. 1973. "List Prices of Petrol, Australian Capital Cities, 1945–1973." Melbourne: Petroleum Information Bureau.

Baas, H. J., W. R. Hughes, and C. G. Treloar, 1982, "Elasticities of Demand for Petrol and Aviation Fuel in New Zealand 1961 to 1981," Report prepared for the Ministry of Energy, Wellington. Deparrment of Economics Technical Publication No. 20, Hamilton, New Zealand: University of Waikato.

Balestra, P., and M. Nerlove. 1966. "Pooling Cross Section and Time Series Data in the Estimation of a Dynamic Model: The Demand for Natural Gas." *Econometrica,* 34:585–612.

Baltagi, B. H., and J. M. Griffin. 1983. "Gasoline Demand in the OECD: An Application of Pooling and Testing Procedures." *European Economic Review,* 22:117–37.

Belsley, D. A., E. Kuh, and R. E. Welsh. 1980. *Regression Diagnostics: Identifying Differential Data and Sources of Collinearity.* New York: John Wiley & Sons.

Brain, P., and G. S. Schuyers. 1981. *Energy and the Australian Economy.* Melbourne: Longman Cheshire.

Brown, R. L., J. Durbin, and J. M. Evans. 1975. "Techniques for Testing the Constancy of Regression Relationships." *Journal of the Royal Statistical Society,* B37:149–63.

Cooley, T. T., and E. C. Prescott. 1973. "Systematic (Non-random) Variation Models Varying Parameter Regression: A Theory and Some Applications." *Annals of Economic and Social Measurement,* 2:463–73.

Dahl, C. A. 1984. "Gasoline Demand Survey." Department of Economics. Baton Rouge: Louisiana State University. Mimeo draft.

——. 1982 "Do Gasoline Demand Elasticities Vary?" *Land Economics,* 58:373–82.

Data Resources Inc. 1973. "A Study of the Quarterly Demand for Gasoline and Impacts of Alternative Gasoline Taxes." Interim Report to the U.S. Environmental Protection Agency and the President's Council of Environmental Quality. Lexington, MA.

Donnelly, W. A. 1985. "A State-Level, Variable Elasticity of Demand for Gasoline Model." *International Journal of Transport Economics,* 12:193–202.

——. 1984. "The Australian Demand for Petrol." *International Journal of Transport Economics,* 11:189–205.

——. 1982a. "The Regional Demand for Petrol in Australia." *Economic Record,* 58:317–27.

——. 1982b. Petrol and Diesel Fuel Demand Elasticities for Germany and Belgium." Report to the Australian Department of Transport, Canberra. Mimeo.

——. 1981. "The Demand for Petrol in Australia." Centre for Resource and Environmental Studies Working Paper No. R/WP-61. Canberra: Australian National University.

——. 1979. "A State-Level Monthly Demand for Gasoline Specification: Kentucky." Unpublished.

Drollas, L. P. 1984. "The Demand for Gasoline: Further Evidence." *Energy Economics,* 6:71–82.

Durbin, J. 1970. "Testing for Serial Correlation in Least-Squares Regression When Some of the Regressors Are Lagged Dependent Variables." *Econometrica,* 38:410–21.

Filmer, R. J., and S. Talbot. 1974. "Demand for Passenger Motor Vehicles." Paper presented at the Australian National University, Canberra, July 19. Mimeo.

Folie, M. 1977. "Competition in the Australian Retail Market for Petrol." Paper presented at the Sixth Conference of Economists, Economic Society of Australia and New Zealand, University of Tasmania, Hobart. Mimeo.

Goodwin, P. B., and M. J. H. Mogridge. 1981. "Hypotheses for a Fully Dynamic Model of Car Ownership." *International Journal of Transport Economics,* 313–26.

Guilkey, D. K. 1975. "A Test for the Presence of First-Order Vector Autoregressive Errors When Lagged Endogenous Variables Are Present." *Econometrica,* 43:711–17.

Hackl, P. 1979. "Moving Sums of Residuals: A Tool for Testing the Constancy of Regression Relationships over Time." In *Models and Decision Making in National Economies*, edited by J. M. L. Janseen, L. F. Pau, and D. Straszak, pp. 219–25. Amsterdam: North-Holland.

Hewitt, D. R., and P. N. James. 1975. "Forecasting Motor Spirit Demand." Petroleum Branch. Canberra: Australian Department of Minerals and Energy. Mimeo.

Houthakker, H. S., and L. D. Taylor. 1970. *Consumer Demand in the United States.* 2nd ed. Cambridge, MA: Harvard University Press.

Houthakker, H. S., P. K. Verleger, and D. P. Sheehan. 1974. "Dynamic Demand Analyses for Gasoline and Residential Electricity." *American Journal of Agricultural Economics,* 56:412–18.

Hughes, W. R. 1980. "Petrol Consumption in New Zealand 1969–79: Fixed and Time Varying Parameter Results." *New Zealand Economic Papers,* 14: 28–42.

Kennedy, M. 1974. "An Economic Model of the World Oil Market." *Bell Journal of Economics and Management Science,* 5:540–77.

Kmenta, J. 1971. *Elements of Econometrics.* New York: Macmillan.

Kraft, J., and M. Rodekohr. 1980. "A Temporal Cross Section Specification of the Demand for Gasoline Using a Random Coefficient Regression Model." *Energy,* 5:1193–202.

Maddala, G. S. 1977. *Econometrics.* New York: McGraw-Hill.

McNown, R. F., and K. R. Hunter. 1980. "A Test for Autocorrelation in Models with Lagged Dependent Variables." *Review of Economics and Statistics,* 62:313–17.

Mehta, J. S., G. V. L. Narasimham, and P. A. V. B. Swamy. 1978. "Estimation of a Dynamic Demand Function for Gasoline with Different Schemes of Parameter Variation." *Journal of Econometrics,* 7:263–79.

Nordhaus, W. D. 1976. "The Demand for Energy: An International Perspective." In *Proceedings of the Workshop on Energy Demand: May 22–23, 1975,* edited by W. D. Nordhaus, pp. 511–87. Laxenburg, Austria: International Institute for Applied Systems Analysis.

Ostro, B. D., and J. L. Naroff. 1980. "Decentralization and the Demand for Gasoline." *Land Economics,* 56:169–80.

Philips. L. 1972. "A Dynamic Version of the Linear Expenditure Model." *Review Of Economics and Statistics,* 54:450–88.

Pindyck, R. S., and D. L. Rubinfeld. 1981. *Econometric Models and Economic Forecasts.* 2nd ed. New York: McGraw-Hill.

Ramsey, J., R. Rasche, and B. Allen. 1975. "An Analysis of the Private and Commercial Demand for Gasoline." *Review of Economics and Statistics,* 57:502–7.

Reza, A. M., and M. H. Spiro. 1979. "The Demand for Passenger Car Transport Services and for Gasoline." *Journal of Transport and Economic Policy,* 13: 304–19.

Schou, K., and L. W. Johnson. 1979. "The Short-Run Price Elasticity of Demand for Petrol in Australia." *International Journal of Transport Economics,* 6:357–64.

Sweeney, J. L. 1979. "Effects of Federal Policies on Gasoline Consumption." *Resources and Energy,* 2:3–26.

Verleger, P. K., and D. P. Sheehan. 1976. "A Study of the Demand for Gasoline." In *Econometric Studies of U.S. Energy Policy,* edited by D. W. Jorgenson, pp. 179–241. Amsterdam: North-Holland.

White, K. J. 1978. "A Generalized Computer Program for Econometric Methods–SHAZAM." *Econometrica,* 46:239–40.

Zellner, A. 1962. "An Efficient Method of Estimating Seemingly Unrelated Regressions and Tests for Aggregation Bias." *Journal of the American Statistical Association,* 57:348–68.

Five

Energy Substitution Modeling

OVERVIEW

The production function approach to analyzing economic issues has a long tradition, beginning with the flow analysis of Quesnay in the *Tableau Economique* published in the year 1952,[2] and continuing on to the first parametric production function specified by Cobb and Douglas (1928; hereafter CD). Questions relating to energy substitution sparked a trend toward multifactor modeling, a search for more flexible functional forms, and the substantial work on the transcendental logarithmic (translog) form. Several models of sectoral energy demand, both for aggregate manufacturing energy use and for the iron and steel industry, are discussed.

INTRODUCTION

Economics, as the study of the efficient allocation of resources, analyzes business decision making under the rubric of the theory of the firm. Alternative paradigms have been suggested, but the dominant one remains that of "profit maximization" behavior: The firm may attempt to optimize in terms of profit maximization per se or in terms of cost minimization for a particular level of output.

Other production theories such as that of "satisficing" behavior, which involve desired market share or revenue growth, have not been as amenable to statistical evaluation as have the profit-maximizing ones, so the assumption is that firms are optimizers predominately.

In a survey article on production functions, Fuss, McFadden, and Mundlak (1978) provide five objectives to be served by such analysis: (1) explanation of the distribution of income shares among factors of production; (2) identification of returns to scale; (3) understanding of the degree of substitutability of the factors of production; (4) determination of the potential for either nested or additive decomposition of factors of production (separability); and (5) implications of technical change. These issues can be quantified and addressed in the context of production functions and their first and second derivatives. Using the Fuss notation, "Consider an n input, one output production function $y = f(v_i, \ldots, v_n)$, with partial derivatives $f_i = \partial f/\partial v_i$ and $f_{ij} = \partial^2 f/\partial v_i \partial v_j$" (Fuss, McFadden, and Mundlak 1978, p. 230).

Table 5.1 presents those "economic effects" that the authors designate as being the major concerns relating to production analyses.

Table 5.1. Economic Effects of Production

Economic effect	Formula	Number of distinct effects
Output level	$y = f(v)$	1
Returns to scale	$\mu = \left[\sum_{i=1}^{n} v_i \cdot f_i \right] / f$	1
Distributive share	$s_i = v_i \cdot f_i / \left[\sum_{j=1}^{n} v_j \cdot f_j \right]$	n − 1
"Own-price" elasticity	$\epsilon_i = v_i \cdot f_{ii} / f_i$	n
Elasticity of substitution	$\sigma_{ij} = \dfrac{-f_{ii} / f_i^2 + 2(f_{ij} / f_i f_j) - f_{ij} / f_j^2}{1 / v_i \cdot f_i + 1 / v_j \cdot f_j}$	$\dfrac{n(n-1)}{2}$

Source: M. A. Fuss, D. McFadden, and Y. Mundlak, "A Survey of Functional Forms in the Economic Analysis of Production," in *Production Economics: A Dual Approach to Theory and Applications*, vol. 1, M. Fuss and D. McFadden, Eds. (Amsterdam: North-Holland, 1978), p. 231. Reprinted with permission.

They conclude that "a necessary and sufficient condition for a function form to reproduce comparative statics effects *at a point* without imposing restrictions across these effects is that it have $(n + 1)(n + 2) / 2$ distinct parameters, such as would be provided by a Taylor's expansion to second-order" (p. 231). And they warn that "if a parsimonious flexible form is fitted to observations over an extensive domain, as is normally the case in econometric production analysis, then the fitted form will *not* in general be a second-order approximation to the true function at any chosen point [emphasis added]" (p. 234).

Fuss, McFadden, and Mundlak make the distinction between analytic studies and predictive models (p. 220). The first parametric production function estimated, the CD (1928), which was developed to test the hypothesis of constant returns to scale, is an example of the analytic genre of study, whereas the work of Hudson-Jorgenson (1974), designed for forecasting energy demand, is of the latter type. A problem remains, however, since modelers need to be certain that their statistical analysis is actually performed on the testable hypotheses specified and not on the maintained hypotheses that include the technological, behavioral, statistical, and regularity assumptions in regard to functional form that accrue within the definitional structure.

Let us consider some of the functional specifications that have been utilized in empirical work, beginning with the CD formulation (1928, p. 152), which relates production to the labor and capital inputs used. Equation 5.1 is the original specification wherein one maintained hypothesis is that of constant returns to scale:

$$\mathbf{P}' = b \cdot \mathbf{L}^{k} \cdot \mathbf{C}^{(1-k)}, \qquad (5.1)$$

where $\mathbf{P}'$ = index of production
$\mathbf{L}$ = index of labor
$\mathbf{C}$ = index of capital
$0 < k < 1$.

The formulation provides information on some of the "economic effects" of interest to economists but not on returns to scale, as these are always unity:[3]

$$\mu_{CD} = k + (1 - k) \equiv 1.$$

The distributive shares in the CD are k for labor and $(1 - k)$ for capital, and the own-price elasticities are $(k - 1)$ and $-k$ for labor and capital, respectively. The CD elasticity of substitution is always equal to 1. However, that original analysis was designed to test marginal productivity theory (that factor shares are equal to the marginal product of the factor) and to attempt to determine the returns to scale; to accomplish this, the exponents on the labor and capital inputs must be allowed to be independent, as depicted in Equation 5.2:

$$V = A \cdot L^{\alpha} \cdot K^{\beta}, \qquad (5.2)$$

where $\mathbf{V}$ = value added
$\mathbf{K}$ = capital
$0 < \alpha, \beta < 1$.

When this more general formulation is applied to aggregate manufacturing data for the United States covering the period 1899–1922, Douglas (1948, p. 12) estimates $\alpha = 0.73$, $\beta = 0.25$, and $\alpha + \beta \cong 1$. This appears to justify the maintained hypothesis that manufacturing production exhibits constant returns to scale; and since the α and β exponent values are similar in magnitudes to the factor shares actually paid to labor and capital, it appears that the marginal productivity theory is confirmed here as well. These time-series results, when interpreted in conjunction with the cross-sectional study of industries for the year 1919, seem to suggest that the restricted estimation defined by Equation 5.1 is justified; however, methodological problems raise questions as to whether the summation of $(\alpha + \beta)$ is biased toward unity because of the data definitions employed, and this consideration results in doubt as to whether the statistical results have or have not been applied validly to a testable hypothesis [see Cramer (1969, pp. 233–37) for a discussion of this problem].

The formulation of Equation 5.2 of the CD production function can represent increasing, constant, or decreasing returns to scale, depending upon whether the value of $(\alpha + \beta)$ is greater than, equal to, or less than 1. A maintained hypothesis of the functional form is that of declining marginal productivity, since α and β (and k in Equation 5.1) are assumed to be positive fractions. The marginal physical product of labor (MPPL), the first derivative of the production function taken with respect to labor, declines as the labor input is increased. By way of illustration, the evaluation of the derivative of the MPPL is

$$\begin{aligned} MPPL &= \partial V / \partial L \\ &= \alpha \cdot V / L \end{aligned} \qquad (5.3)$$

and

$$\begin{aligned} \partial(MPPL) / \partial L &= \partial^2 / \partial L^2 \\ &= \alpha \cdot (\alpha - 1) \cdot V / L^2 < 0, \text{ if } 0 < \alpha < 1. \end{aligned}$$

The elasticity of substitution between capital and labor can be evaluated based upon the relationship presented in Table 5.1 or may be viewed as a measure of the relative change in the optimal factor input ratio as it pertains to the relative change in the input prices, that is to say, in the immediate CD function instance:

$$\sigma_{LK} = \frac{d\,[\ln(\mathbf{L}/\mathbf{K})]}{d\,[\ln(\mathbf{P_L}/\mathbf{P_K})]} = \frac{\partial(\mathbf{L}/\mathbf{K}) / (\mathbf{L}/\mathbf{K})}{\partial\,(\mathbf{P_L}/\mathbf{P_k}) / (\mathbf{P_L}/\mathbf{P_K})} \equiv 1, \qquad (5.4)$$

where σ = elasticity of substitution
d = total derivative
ln = natural logarithm
$\mathbf{P_L}$ = wage rate
$\mathbf{P_K}$ = capital rental rate.

This result is easily demonstrated if we assume that the production function as depicted in Equation 5.2 applies, and the factor payments define a cost function that is minimized subject to some predetermined level of output. The proof follows accordingly:

If $\quad \mathbf{V = A \cdot L^{\alpha} \cdot K^{\beta}}$

Minimize: $\quad \mathbf{C = P_L \cdot L + P_K \cdot K}$

Subject to: $\quad \mathbf{V(L, K) = V_0}$.

Setting up the Lagrangian function,

$$\mathbf{Z = P_L \cdot L + P_K \cdot K + \lambda \cdot (V_0 - A \cdot L^{\alpha} \cdot K^{\beta})} ;$$

then

$$\partial Z/\partial \lambda = V_0 - A \cdot L^{\alpha} \cdot K^{\beta} = 0$$
$$\partial Z/\partial L = P_L - \lambda \cdot [\alpha \cdot (V/L)] = 0$$
$$\partial Z/\partial K = P_K - \lambda \cdot [\beta \cdot (V/K)] = 0.$$

Setting the partial derivatives of the Lagrangian function with respect to the labor and capital inputs equal to each other yields

$$P_L - \lambda \cdot [\alpha \cdot (V/L)] = P_K - \lambda \cdot [\beta \cdot (V/K)]$$

or

$$L/K = (\alpha/\beta) \cdot (P_K/P_L).$$

Thus, it may be seen that

$$\partial(L/K)/\partial(P_L/P_K) = \alpha/\beta.$$

Therefore,

$$\sigma_{CD} \equiv 1, \text{QED}.$$

Thus, the CD function represents the first parametric production function to be estimated that may provide empirical insights into some of the economic effects mentioned by Fuss, McFadden, and Mundlak (1978). The functional form used by Cobb and Douglas was in actuality specified earlier by Wicksell (1916) in a paper discussing the "critical point" of declining productivity in agriculture. Therein, a three-input production function, which includes a "working factor" of capital and agricultural land [and is "homogeneous and of degree one" (p. 287)] is specified and is

$$p = a^{\alpha} \cdot b^{\beta} \cdot c^{\gamma}, \qquad (5.5)$$

$$\text{where } \alpha + \beta + \gamma = 1.$$

Wicksell, however, describes only the general attributes of such a function–namely, if α, β, and γ all increase by 10 percent, then output p increases by 10 percent also—but he makes no attempt to estimate such a relationship.

The most widely applied generalization of the CD function is the CES function defined by Arrow et al. (1961), which, when specified as a cost function, is

$$\ln c(y, \mathbf{p}) = \alpha_0 + \alpha_y \cdot \ln y + \ln(\Sigma_i \alpha_i \cdot \mathbf{p}_i^\rho)^{(1/\rho)}. \qquad (5.6)$$

The elasticity of substitution for the CES production function is simply

$$\sigma_{CES} = 1/\rho.$$

Thus, the CD represents a special case of the CES function. This is where $\rho = 1$. As will be shown later, another commonly used functional form, the translog, also takes on the CD as a special case.

Let us consider a specific three-factor CES production function of the form

$$\mathbf{Q} = \gamma \cdot [\delta_1 \cdot \mathbf{K}^\rho + \delta_2 \cdot \mathbf{L}^\rho + (1 - \delta_1 - \delta_2) \cdot \mathbf{M}^\rho]^{(1/\rho)}, \qquad (5.7)$$

where $\mathbf{Q}$ = output
$\mathbf{M}$ = materials
γ = neutral efficiency parameter
ρ = substitution parameter
δ_1, δ_2 = distributional parameters.

As the name implies, the CES function exhibits a constant elasticity of substitution between pairs of factor inputs. Not only is this elasticity constant, but it is also identical for all pairwise combinations of factor inputs, being always $(1/\rho)$. The CES function can represent other commonly hypothesized production relationships, including the fixed coefficient I-O case: When $\rho = \infty$, the inputs are perfect complements, and, in the infinitely divisible input case, whenever $\rho \to 0$, the isoquants are straight lines and the inputs are perfect substitutes. It can be seen that the CES function is more versatile (flexible) than either the CD or the I-O functional forms, and this flexibility allows the statistical testing of the appropriate form based upon the estimated value of the ρ parameter. This interest in developing more general forms has led to the definition of an entire class of flexible functional forms, including, among others, the generalized BC (Berndt and Khaled 1979), a form that takes on the generalized square root quadratic, the generalized Leontief, and the generalized translog as special cases; the Fourier (Gallant 1981); the minflex Laurent (Barnett, Lee, and Wolfe 1984); the generalized McFadden (Diewert and Wales 1984); and the CES translog (Pollak, et al. 1984). The underlying objective of the search for flexible functional forms is to impose the fewest constraints on the parameters of the model, and the form that has enjoyed the most widespread application to date is the translog. Christensen, Jorgenson, and Lau (1973) utilize the transcendental logarithmic model to

> represent the production frontier by functions that are quadratic in the logarithms of the quantities of inputs and output. These functions provide a local second-order approximation to any production frontier. The resulting frontiers permit a greater variety of substitution and transformation patterns than frontiers based upon constant elasticity of substitution and transformation. (p. 28)

In this chapter there are three examples of translog energy modeling: first, an energy translog model for 13 industry sectors; second, an aggregate manufacturing model; and last, a model of the iron and steel industry.

The question of factor substitutability and the demand for energy has been widely studied using historical data for aggregate manufacturing [e.g., Berndt and Christensen 1973a, 1974; Berndt and Wood 1975; Brain and Schuyers 1981; Fuss 1977; Griffin 1977a; Griffin and Gregory 1976; Magnus 1979; Pindyck 1977; Turnovsky, Folie, and Ulph (TFU) 1982]. Attention has also been devoted to analyzing the energy demands of individual manufacturing subsectors [e.g., Duncan and Binswanger 1976; Halvorsen and Ford 1979; Hawkins 1977, 1978; Hudson and Jorgenson 1974; Field and Grebenstein 1980; Fishelson and Long 1978; Humphrey and Moroney 1975; Magnus and Woodland 1980; Williams and Laumas 1981; Wills 1979; Turnovsky and Donnelly (TD) 1984]. Other researchers have used pseudodata derived from simulations of optimization models (Griffin 1977b; Kopp and Smith 1981); however, the latter approach of modeling does not yield any information not already contained in the mathematical programming structures on which they are based.

The majority of studies have been done on aggregate manufacturing, since longer, and qualitatively better, historical time series can be constructed for the aggregate sector than for individual sectors. In fact, few countries have yet begun the consistent collection of data in regard to energy consumption. Further, the problem of having a "reasonable" number of observations is exacerbated when the industrial classification scheme undergoes major revisions. In the immediate instance, the Australian Standard Industrial Classification (ASIC) sectoral definitions were significantly revised in 1967/68 to bring about consistency among the various periodic censuses and to ensure that the resulting series would be "broadly convertible" to the United Nations' International Standard Industrial Classification, but in 1970/71, no survey data were published because of financial exigencies.

Theoretical issues also arise concerning the appropriateness of various aggregations used in the modeling: for example, inter- and intraindustry and inter- and intrafactor and additionally the imposed regularity conditions of positivity, linear homogeneity, symmetry, and concavity.[4] Specifically, it may be questioned whether it is valid (1) to aggregate industries to estimate the production function for total manufacturing, (2) to aggregate firms to estimate an industry production function, or even (3) to aggregate processes to estimate a firm's production function. Other problems have to do with the issue of the existence of aggregate inputs; for example, do administrative employees and production workers form a consistent aggregate? Finally, are the inputs separable, allowing the estimation of a capital (**K**), labor (**L**), and energy (**E**), or **KLE**, production function, rather than the full **KLEM** model, which includes other materials (**M**) inputs? The models discussed here will address some of these questions.

THE TRANSCENDENTAL LOGARITHMIC PRODUCTION FUNCTION

As is customary in such modeling, a twice-differentiable monotonic quasi-concave production function is assumed to exist of the general form

$$\mathbf{Q} = f(\mathbf{K}, \mathbf{L}, \mathbf{E}, \mathbf{M}). \tag{5.8}$$

Total output is a function of the four inputs, and a second-order approximation for this function is specified by a flexible functional form, the translog being one such function. By applying duality theory and assuming that the regularity conditions of (1) positivity, (2) linear homogeneity, and (3) concavity are met by this production function, a corresponding translot unit cost function can then be derived (Diewert 1974). Burgess (1975) points out that the translog function is not self-dual except at the actual point of approximation, whereas the CD or CES specifications are both self-dual. As a result, the translog production and translog cost functions do not necessarily describe the same technology, and the two approaches often yield quite different estimates of the elasticities of substitution (Kuh 1976). Such results are rather unsettling, particularly since Christensen, Jorgenson, and Lau (1973) argue that "these frontiers can be regarded as alternative approximations to the same technology" (p. 38). This inherent discrepancy between the theoretical and the empirical results is one of the weaknesses of the translog functional form that deserves further investigation. The translog unit cost function is defined as

$$\ln \mathbf{C} = \ln \alpha_0 + \Sigma_i \alpha_i \cdot \ln \mathbf{P}_i + \tfrac{1}{2}\Sigma_i \Sigma_j \beta_{ij} \cdot \ln \mathbf{P}_i \cdot \ln \mathbf{P}_j, \quad \text{for } i = 1, \ldots, n; \; j = 1, \ldots, n. \tag{5.9}$$

This applies where $\mathbf{C}$ is the minimum cost of producing a given level of output and the $\mathbf{P}_i$ and $\mathbf{P}_j$ terms are the factor prices of the various n inputs. By logarithmically differentiating this function, a set of linear demand equations suitable for estimation purposes can be derived:

$$\mathbf{S}_i = \alpha_i + \Sigma_j \beta_{ij} \cdot \ln \mathbf{P}_j + \mathbf{u}_i, \; i = 1, \ldots, n. \tag{5.10}$$

Here, the $\mathbf{S}_i$ terms are the value shares of the input i in total cost and the $\mathbf{u}_i$ terms are the random errors in cost-minimizing behavior. Theil (1980) points out a shortcoming in the use of the translog, which may be described as follows: Since the translog is a second-order approximation, the differentiation process transforms the system of value-share equations into a first-order approximation only; hence, the quality of the approximation will deteriorate overall.

In the complete system of n demand equations, the value shares sum to 1, the disturbance covariance matrix is singular, and consequently one demand equation must be dropped from the system before estimation. Maximum likelihood estimates will be invariant to the omitted equation, and Kmenta and Gilbert (1968) show that iterative Zellner parameter estimates converge to maximum likelihood estimates. The three restrictions that must be imposed on this system of equations to satisfy the conditions of linear homogeneity and symmetry are

$$\Sigma_i \alpha_i = 1 \tag{5.11}$$

$$\Sigma_j \beta_{ij} = 0, i = 1, \ldots, n. \tag{5.12}$$

$$\beta_{ij} = \beta_{ji}., i = 1, \ldots, n \quad j = 1, \ldots, n \tag{5.13}$$

As the restrictions on the production function are actually imposed on the parameters of the system of value-share equations, Simmons and Weiserbs (1979) raise questions as to the validity of the symmetry tests. A well-behaved set of cost functions will be "concave in input prices with the predicted shares being nonnegative at each observation point" (Turnovsky, Folie, and Ulph 1982, p. 67), or, alternatively, each observation must be tested.

Finally, a proxy variable A_i may be added in the attempt to identify technical progress, be it Hicks neutral or some other form of factor-biased change. The approach proposed by Binswanger (1974) and Duncan and Binswanger (1976), which introduces time as a proxy variable for technical progress, is adopted in both the TFU and TD studies, wherein the variable is assumed to grow at an exponential rate and the set of demand equations are of the following form:

$$\mathbf{S}_i = \alpha_i + \Sigma_j \beta_{ij} \cdot \ln \mathbf{P}_j + \gamma_i \cdot \mathbf{A} + u_i, i = 1, \ldots, n \tag{5.14}$$

With allowances for technical progress, the additional restriction of Equation 5.15 is imposed:

$$\Sigma_i \gamma_i = 0. \tag{5.15}$$

The estimated parameters obtained from the set of input share functions in Equation 5.14 are used to calculate the own-elasticities and cross-elasticities of demand η_{ii} and η_{ij} and the partial elasticities of substitution σ_{ii} and σ_{ij}, as defined by Allen (1938) [see the discussion in Chapter 1 on the AES as a measure of the response of the function to changes in factor prices, where output is constant]. Those AES relationships that Berndt and Wood (1975, p. 261) derive for the translog cost function are

$$\sigma_{ii} = (\beta_{ii} - \mathbf{S}_i) / \mathbf{S}_i^2 + 1, \text{ for i}, \ldots, n \qquad (5.16)$$

and

$$\sigma_{ij} = \beta_{ij} / (\mathbf{S}_i \cdot \mathbf{S}_j) + 1, \text{ for } i = 1, \ldots, n \quad i \neq j.$$

The $\mathbf{S}_i$ and $\mathbf{S}_j$ terms are the cost shares for the *n* factor inputs as defined in Equation 5.14, and the β terms are the parameter estimates from those same relationships. The factor input price elasticities of demand are then defined as

$$\eta_{ij} = \mathbf{S}_j \cdot \sigma_{ij}, \text{ for } i, j = 1, \ldots, n. \qquad (5.17)$$

There are several forms of separability that can be tested, namely, global, strong, and weak. Global separability is equivalent to testing whether the coefficients of all of the cross-product terms are equal to 0, which implies that the translog function is of the CD form. Strong (or linear) separability requires that the cross-product terms of the individual inputs be equal to 0. A necessary and sufficient condition for two inputs to be weakly separable is that the marginal rate of substitution between them is independent of the quantities of the other inputs (see Leontief 1947, p. 364). Berndt and Christensen (1973b) argue that the weak separability restriction may be expressed as a set of equality restrictions on the AES; for example, two inputs, $\mathbf{X}_1$ and $\mathbf{X}_2$, are said to be weakly separable from the other inputs at a point if

$$\mathbf{Q} = f(\mathbf{X}_1, \mathbf{X}_2, \mathbf{X}_3, \mathbf{X}_4) = g[h(\mathbf{X}_1, \mathbf{X}_2), \mathbf{X}_3, \mathbf{X}_4] \qquad (5.18)$$

$$\text{If: } \sigma_{X_1 X_3} = \sigma_{X_2 X_3} \neq 1$$

$$\text{And: } \sigma_{X_1 X_4} = \sigma_{X_2 X_4} \neq 1.$$

They show that these restrictions reduce to three independent non-linear restrictions on the parameter estimates of Equation 2.11:

$$\alpha_1 \cdot \beta_{23} - \alpha_2 \cdot \beta_{13} = 0 \quad (5.19)$$
$$\alpha_1 \cdot \beta_{24} - \alpha_2 \cdot \beta_{14} = 0$$
$$\beta_{11} \cdot \beta_{22} = (\beta_{12})^2 .$$

Blackorby, Primont, and Russel (1977) question the validity of testing separability restrictions with the various flexible functional forms, and they demonstrate that the translog cannot model non-homothetic weakly separable functions. In fact, if the translog as an exact form is used, testing for weak separability with the three non-linear restrictions above is equivalent to testing for a mixture of strong and homothetic weak separability. One way around the difficulty, suggested by Woodland (1978), is to restructure the problem so that a variable profit function is estimated, since the translog can be used to test for the existence of a single separable group. An alternative approach, proposed by Denny and Fuss (1977), is to use approximate analysis to justify applying the less stringent set of restrictions and then to test for separability at the point of approximation only; this approach means that the last constraint can be dropped.

A function such as the general one specified as Equation 5.8 is weakly separable in fuels if the marginal rates of interfuel substitution are invariant for different levels of the other factor inputs. Weak separability allows the managerial decisions to be modeled as hierarchical, with management first determining the level of aggregate energy consumption from the KLEM function and then calculating the interfuel mix from a similarly defined energy submodel (see Fuss 1977). If weak separability between the aggregate fuel input and the other aggregate inputs is assumed, the general functional form is

$$\mathbf{Q} = \mathbf{g}\,[\mathbf{K}, \mathbf{L}, \mathbf{h}(\mathbf{s}, \mathbf{o}, \mathbf{e}, \mathbf{g}), \mathbf{M}], \quad (5.20)$$

where s = solid fuels
o = liquid fuels
e = electricity
g = gas.

The remainder of this chapter examines two different models of industrial energy demand in Australia: an aggregate manufacturing model and an iron- and steel-sector model.

SECTORAL ENERGY DEMAND

In 1976 Duncan and Binswanger published the first Australian translog energy demand model using a 14-industry classification system for the manufacturing sector.[5] The modeling data that cover the financial years 1948/49 through 1966/67 are not published. They accept as a maintained hypothesis that industrial energy demand is separable from the other factor inputs in the production function. Thus, they consider only the question of interfuel substitution wherein that substitution is unaffected by changes in the relative prices of the other factors of production; as is to be demonstrated, subsequent studies statistically test that separability hypothesis in aggregate manufacturing and in the iron and steel industry. In unpublished work Duncan and Binswanger (1974a, 1974b) test for a well-behaved cost function in each of these industries, but do not report on any tests for energy separability. With respect to the cost functions, they conclude that a CD formulation should be rejected in all cases except for the skins and leather industry,[6] homogeneity should be rejected in only 3 of the remaining 13 industry sectors tested,[7] and symmetry with homogeneity imposed is rejected in 7 of the 10 industries modeled. Only in the industry sectors for bricks, pottery, and glass, wood furniture, and rubber can the maintained hypotheses of homogeneity and symmetry be accepted, but unfortunately the convexity condition is violated in each.

In regard to the published results of the energy submodel that analyzes the four energy inputs of coal, fuel oil, electricity, and coal gas, the CD unitary elasticity of substitution model is consistently rejected. The price homogeneity restrictions are rejected in 5 of the 15 industries, and symmetry (with homogeneity imposed) is rejected in 7 of the remaining 11 industries. Concavity in the energy input is violated for all industries when evaluated at the data means, and none of the industry-specific cost functions or the energy submodel cost functions are well behaved in this work. These problems result in numerous positive own-price elasticity of demand estimates.[8] The authors consider, in addition, the type of technological factor bias suggested by the data in five of the industry sectors.[9] Electricity consistently exhibits a factor-using bias in these sectors; coal is factor saving in three of the four sectors in which it is used as an energy source and neutral in the remaining sawmills sector; fuel oil is factor using in the sawmills and the wood furniture sectors, factor saving in the rubber sectors, and insignificantly different from 0 in the skins and leather industry. Overall Duncan and Binswanger (1976) recognize that the "results are disappointing" (p. 299) and suggest that inadequacies in the data may be one explanation, but do not elucidate. However, the failure of empirical results to satisfy the various functional constraints required for well-behaved models is not unusual. Furthermore, the nature of the translog, in that it represents a local approximation to the underlying function solely, may suggest an imperfect fit of whatever data are available to the basic hypothetical structure of the model.

AGGREGATE MANUFACTURING

Turnovsky, Folie, and Ulph (1982) present results for Australian aggregate manufacturing using the translog functional form, with considerable effort being devoted to the construction of the requisite modeling data.[10] As was mentioned previously, ABS made substantial changes in the ASIC definitions between fiscal years 1967/68 and 1968/69. The classifications went from the original factory census basis to a definition according to manufacturing establishment, and ABS did not provide both series for the transition year, leaving accurate bridging tables for individual industries difficult, if not impossible, to construct. However, the aggregate manufacturing series and the iron and steel industry data are successfully linked.[11] The TFU model is estimated on the share equations as specified in Equation 5.14 using data covering the years 1946/47 to 1974/75, with series being aggregated by way of Divisia indexes in prices and quantities—see Diewert (1976, 1978). The price of capital is an aggregate of the implicit rental price for buildings and structures and the implicit rental price for plant and equipment, while administrative employees and production workers are aggregated; the "material price indices were readily available" (Turnovsky, Folie, and Ulph 1982, p. 65). Similar price and quantity aggregations are done for the fuels submodel, which analyzes the substitution among "solid fuels (black coal, brown coal, brown coal briquettes, coke and coke oven gas), oil, electricity, and gas (coal gas and natural gas)" (p. 65). The aggregate manufacturing and energy submodel value shares and price indexes modeling data appear in Tables 5.2 and 5.3.

The maintained hypotheses of symmetry (Equation 5.13) and homogeneity (Equations 5.11 and 5.12) are rejected by the aggregate manufacturing model data; however, this appears to be a common problem, and these hypotheses are rejected in most other studies as well. Concavity is evidenced at all data points.

These results notwithstanding, all of the hypotheses are imposed a priori by the authors for the subsequent analyses. Tables 5.4 and 5.5 allow the comparison of the elasticities of substitution and factor price elasticities for the TFU model and the Berndt and Wood results (1975). The TFU study concludes that capital and energy are substitutes and that labor and energy are complements in Australian manufacturing, whereas the Berndt-Wood results for the United States suggest the opposite behavior. The literature offers several alternative explanations for such divergent capital and energy substitution findings (see Berndt and Wood 1981; Griffin 1981a, 1981b; Hogan 1979; Kang and Brown 1981) but none suffices for the TFU results. The mean Australian σ_{KE} value of 2.26, derived as it is from time-series data, seems to refute "the conventional explanation for the sharply contrasting sets of results . . . that the time-series studies are reflecting short-term relationships while the cross-section studies, with their wider variation in relative input prices, are capturing the long-term effects" (Turnovsky, Folie, and Ulph 1982, p. 62).

Gibbons (1984), Vinals (1984), and Stapleton (1981) contribute to this continuing debate with no resulution, and the conflicting results remain unexplained. The labor-energy complementarity conclusion may raise a question concerning the appropriateness of aggregating administrative employees and production workers into a single labor input. Berndt and White (1979) find that nonproduction workers are complements to energy, whereas production workers are substitutes. This issue is explored further in the results of the iron and steel model in the next section. The capital-labor elasticity of substitution of 2 does not accord well with a CD aggregate value-added relationship that would imply a value close to unity, as has been discussed.

Finally, it might be noted from the information in Tables 5.4 and 5.5 that the U.S. model's elasticities appear to be slightly more

Table 5.2. Aggregate Manufacturing Model Dataset (1946/47 to 1974/75)

	KLEM model						
Fiscal year	Value shares			Price indexes			
	Capital	Labor	Energy	Capital	Labor	Energy	Material
1946/47	0.1669	0.2530	0.0232	100.00	100.00	100.00	100.00
1947/48	0.1651	0.2538	0.0222	111.25	113.13	103.80	111.42
1948/49	0.1590	0.2563	0.0229	122.50	127.93	117.44	126.17
1949/50	0.1657	0.2532	0.0240	134.96	141.99	129.72	143.63
1950/51	0.1623	0.2457	0.0240	161.65	169.93	148.67	177.19
1951/52	0.1549	0.2501	0.0287	201.65	208.61	177.80	215.43
1952/53	0.1619	0.2535	0.0291	236.71	225.88	199.26	234.91
1953/54	0.1648	0.2472	0.0281	254.52	237.41	199.52	222.81
1954/55	0.1655	0.2488	0.0270	262.11	252.21	194.87	221.48
1955/56	0.1663	0.2506	0.0266	282.22	267.70	196.75	231.54
1956/57	0.1768	0.2471	0.0273	290.44	278.38	202.31	246.31
1957/58	0.1826	0.2427	0.0273	298.15	287.83	198.29	238.27
1958/59	0.1879	0.2411	0.0270	303.83	296.16	197.99	228.18
1959/60	0.1907	0.2406	0.0257	308.41	318.39	193.95	232.90
1960/61	0.1902	0.2438	0.0264	318.44	330.63	196.84	232.22
1961/62	0.1955	0.2417	0.0266	303.46	336.50	190.84	228.18
1962/63	0.1966	0.2362	0.0262	283.97	346.46	189.49	225.51
1963/64	0.1997	0.2316	0.0256	278.16	361.30	182.38	227.53
1964/65	0.1998	0.2345	0.0250	294.83	388.17	174.97	231.54
1965/66	0.2025	0.2371	0.0247	308.53	402.08	172.81	238.27
1966/67	0.2103	0.2361	0.0247	314.41	427.42	173.73	242.95
1967/68	0.2121	0.2366	0.0248	323.36	451.55	175.61	242.27
1968/69	0.2463	0.2216	0.0237	325.41	479.14	182.48	248.32
1969/70	0.2438	0.2202	0.0236	368.07	518.70	184.89	254.79
1970/71	0.2434	0.2288	0.0247	417.62	571.35	191.89	248.57
1971/72	0.2461	0.2342	0.0249	431.23	635.78	201.43	254.79
1972/73	0.2423	0.2334	0.0249	413.59	691.91	204.93	282.85
1973/74	0.2432	0.2404	0.0238	608.05	830.76	225.15	334.50
1974/75	0.2432	0.2497	0.0285	839.77	1091.67	306.18	360.33
Mean	0.1971	0.2417	0.0256				

Source: Compiled by the author from TFU (1978) data.

Table 5.3. Aggregate Manufacturing Energy Submodel Dataset (1946/47 to 1974/75)

	Value shares			Price indexes			
Fiscal year	Solid fuels	Oil	Electric	Solid fuels	Oil	Electric	Gas
1946/47	0.4795	0.1102	0.3712	100.00	100.00	100.00	100.00
1947/48	0.4704	0.1170	0.3744	102.29	106.29	104.67	106.90
1948/49	0.4599	0.1336	0.3706	125.88	128.52	104.41	119.59
1949/50	0.4716	0.1437	0.3514	147.17	134.69	109.20	133.88
1950/51	0.4978	0.1424	0.3284	178.29	153.08	115.70	158.29
1951/52	0.4491	0.1373	0.3822	234.84	194.99	116.82	226.23
1952/53	0.5069	0.1289	0.3334	264.90	203.84	133.88	239.06
1953/54	0.4952	0.1093	0.3636	265.55	190.45	136.70	245.89
1954/55	0.4746	0.1071	0.3834	263.13	167.75	136.00	221.75
1955/56	0.4443	0.1225	0.3973	269.97	159.95	136.55	232.69
1956/57	0.4169	0.1416	0.4069	270.14	168.80	143.48	236.64
1957/58	0.3885	0.1533	0.4218	261.12	170.08	140.47	245.24
1958/59	0.3705	0.1588	0.4343	259.56	162.05	143.19	245.76
1959/60	0.3511	0.1606	0.4504	253.27	149.83	143.25	249.74
1960/61	0.3641	0.1482	0.4512	259.89	140.28	148.67	242.78
1961/62	0.3442	0.1518	0.4673	253.59	126.31	147.07	232.73
1962/63	0.3302	0.1501	0.4828	263.13	118.39	143.83	237.94
1963/64	0.3184	0.1515	0.4919	260.16	109.55	136.91	245.50
1964/65	0.3129	0.1531	0.4955	262.41	102.56	127.73	246.30
1965/66	0.3043	0.1579	0.5032	267.38	96.74	126.52	218.47
1966/67	0.3076	0.1579	0.5001	281.94	93.25	124.91	227.25
1967/68	0.2980	0.1723	0.4965	280.68	98.37	126.11	217.39
1968/69	0.2898	0.1825	0.5000	292.37	91.97	138.39	167.31
1969/70	0.3015	0.1700	0.5011	329.89	84.13	137.96	137.76
1970/71	0.3223	0.1816	0.4711	369.58	91.24	137.55	90.00
1971/72	0.3221	0.1768	0.4698	406.60	96.48	140.75	80.74
1972/73	0.3357	0.1567	0.4570	408.44	101.59	143.77	75.62
1973/74	0.3167	0.1798	0.4475	454.96	123.97	152.09	75.59
1974/75	0.3613	0.2073	0.3814	672.71	220.20	172.44	93.93
Mean	0.3829	0.1604	0.4305				

Notes: Solid fuel includes black coal, brown coal, coke, and coke oven gas. Gas includes coal gas only.

Sources: Compiled by the author from TFU data.

Table 5.4. Australian Aggregate Manufacturing KLEM Model (1946/47 to 1974/75)

	At the means	Year					
		1946/47	1952/53	1958/59	1964/65	1970/71	1974/75
σ_{KK}	-4.79	-5.94	-6.19	-5.07	-4.67	-4.56	-3.56
η_{KK}	-0.94	-0.99	-1.00	-0.95	-0.93	-1.11	-0.87
σ_{LL}	-2.70	-2.56	-2.55	-2.71	-2.80	-2.89	-2.60
η_{LL}	-0.65	-0.65	-0.65	-0.65	-0.66	-0.66	-0.65
σ_{EE}	-8.73	-6.43	-10.69	-9.70	-8.28	-8.02	-10.45
η_{EE}	-0.22	0.15	-0.31	-0.26	-0.21	-0.20	-0.30
σ_{MM}	-0.58	-0.54	-0.54	-0.57	-0.57	-0.67	-0.74
η_{MM}	-0.31	-0.30	-0.30	-0.31	-0.31	-0.34	-0.35
σ_{KL}	2.00	2.12	2.16	2.05	2.01	1.85	1.78
η_{KL}	0.48	0.54	0.55	0.49	0.47	0.42	0.44
η_{LK}	0.39	0.35	0.35	0.38	0.40	0.45	0.43
σ_{KE}	2.26	2.63	2.34	2.24	2.26	2.05	1.91
η_{KE}	0.06	0.06	0.07	0.06	0.06	0.05	0.05
η_{EK}	0.44	0.43	0.38	0.42	0.45	0.50	0.46
σ_{KM}	0.74	0.71	0.70	0.73	0.75	0.78	0.77
η_{KM}	0.40	0.39	0.39	0.40	0.40	0.39	0.37
η_{MK}	0.15	0.12	0.11	0.14	0.15	0.19	0.19
σ_{LE}	-2.66	-2.85	-2.06	-2.47	-2.86	-3.00	-2.18
η_{LE}	-0.07	-0.07	-0.06	-0.07	-0.07	-0.07	-0.06
η_{EL}	-0.64	-0.72	-0.52	-0.60	-0.67	-0.69	-0.54
σ_{LM}	0.61	0.64	0.64	0.62	0.60	0.56	0.58
η_{LM}	0.33	0.36	0.36	0.34	0.33	0.28	0.28
η_{ML}	0.15	0.16	0.16	0.15	0.14	0.13	0.14
σ_{EM}	0.79	0.78	0.82	0.80	0.78	0.77	0.79
η_{EM}	0.42	0.43	0.46	0.44	0.42	0.39	0.38
η_{ME}	0.02	0.02	0.02	0.02	0.02	0.02	0.02

Source: Calculated by author from TFU (1978) data.

Table 5.5. U.S. Aggregate Manufacturing KLEM Model (1947-71)

	At the means	Year				
		1947	1953	1959	1965	1971
σ_{KK}	-8.82	-8.74	-8.83	-8.75	-8.72	-8.53
η_{KK}	-0.47	-0.49	-0.46	-0.49	-0.50	-0.44
σ_{LL}	-1.66	-1.79	-1.71	-1.66	-1.61	-1.53
η_{LL}	-0.46	-0.45	-0.46	-0.46	-0.46	-0.45
σ_{EE}	-10.66	-10.69	-10.63	-10.70	-10.70	-10.66
η_{EE}	-0.48	-0.47	-0.49	-0.47	-0.45	-0.49
σ_{MM}	-0.36	-0.33	-0.35	-0.36	-0.38	-0.39
η_{MM}	-0.23	-0.21	-0.22	-0.22	-0.23	-0.24
σ_{KL}	1.01	1.01	1.01	1.01	1.01	1.01
η_{KL}	0.28	0.26	0.27	0.28	0.29	0.30
η_{LK}	0.05	0.06	0.05	0.06	0.06	0.05
σ_{KE}	-3.25	-3.09	-3.25	-3.14	-3.22	-3.53
η_{KE}	-0.15	-0.14	-0.15	-0.14	-0.14	-0.16
η_{EK}	-0.17	-0.17	-0.17	-0.18	-0.18	-0.17
σ_{KM}	0.54	0.58	0.53	0.56	0.56	0.49
η_{KM}	0.34	0.37	0.34	0.35	0.35	0.30
η_{MK}	0.03	0.03	0.03	0.03	0.03	0.02
σ_{LE}	0.65	0.61	0.65	0.64	0.64	0.68
η_{LE}	0.03	0.03	0.03	0.03	0.03	0.03
η_{EL}	0.18	0.16	0.17	0.18	0.18	0.20
σ_{LM}	0.60	0.57	0.59	0.59	0.60	0.61
η_{LM}	0.37	0.37	0.37	0.37	0.37	0.37
η_{ML}	0.16	0.15	0.16	0.16	0.17	0.18
σ_{EM}	0.76	0.76	0.77	0.75	0.74	0.75
η_{EM}	0.48	0.49	0.49	0.47	0.46	0.46
η_{ME}	0.03	0.03	0.04	0.03	0.03	0.03

Source: E. R. Berndt and D. O. Wood, "Technology, Prices and the Derived Demand for Energy," *Review of Economics and Statistics,* 57(1975):264–65, Tables 2 to 5.

stable over time than do the Australian results, possibly because the latter dataset includes greater variation in the value shares than is the case for the former. The Pearson coefficient of variation is calculated for those datasets and for the iron and steel data (see Table 5.6).

In the fuels submodel, Turnovsky, Folie, and Ulph report only upon a four-fuel model that includes solid fuels, oil, electricity, and gas, as described at the beginning of this section, and the AES and price elasticities are calculated assuming a constant level of total energy demand, with the results presented in Table 5.7. Therein solid fuels and electricity exhibit an inelastic response to price chanes, the oil demand price elasticity is close to unity, and gas is found to be price elastic. This last elasticity should be considered in the context that over the observation set natural gas accounted for only about 2.6 percent of total energy input; the AES would therefore be sensitive to small changes in its value share and could as a consequence vary dramatically. The total price elasticities for fuels, holding output constant, may be derived from Equations 5.14 and 5.16 (Turnovsky, Folie, and Ulph 1982, p. 64). Equation 5.21 relates the total fuel price elasticity to the elasticity presented in Equation 5.16 plus the proportionate share of that fuel contributing to the total energy own-price elasticity:

$$\eta_{ij}^{T} = \eta_{ij} + \mathbf{S}_j \cdot \eta_{EE}. \tag{5.21}$$

These "total fuel" price elasticities of equation 5.21, calculated at the means of the data, appear in Table 5.8.

Table 5.6. Pearson Coefficient of Variation

	Aggregate manufacturing		Iron and steel
Value share	U.S.A. (1947–71)	Australia (1946/47 to 1974/75)	Australia (1959/60 to 1978/79)
K	0.0897	0.1620	0.1912
L	0.0470	0.0393	0.0961
E	0.0691	0.0718	0.1202
M	0.0230	0.0461	0.1164

Source: M. H. L. Turnovsky and W. A. Donnelly, "Energy Substitution, Separability, and Technical Progress in the Australian Iron and Steel Industry," *Journal of Business and Economic Statistics,* 2(1984):58, Table 4.

Table 5.7. Australian Aggregate Manufacturing Energy Submodel (1946/47 to 1974/75)

	At the	Year					
	means	1946/47	1952/53	1958/59	1964/65	1970/71	1974/75
σ_{ss}	-1.96	-1.31	-1.17	-2.07	-2.71	-2.59	-2.16
η_{ss}	-0.75	-0.63	-0.59	-0.77	-0.85	-0.84	-0.78
σ_{oo}	-6.56	-9.79	-8.01	-6.12	-6.42	-7.63	-4.31
η_{oo}	-0.99	-1.08	-1.03	-0.97	-0.98	-1.17	-0.89
σ_{ee}	-0.72	-0.88	-0.99	-0.71	-1.03	-0.62	-0.85
η_{ee}	-0.31	-0.33	-0.33	-0.31	-0.45	-0.29	-0.23
σ_{gg}	-40.30	-36.09	-50.02	-39.76	-36.85	-67.16	-26.04
η_{gg}	-1.45	-1.41	-1.54	-1.45	-1.42	-1.68	-1.30
σ_{so}	2.90	3.21	2.79	2.99	3.44	3.00	2.56
η_{so}	0.47	0.35	0.36	0.47	0.53	0.54	0.53
η_{os}	1.11	1.54	1.41	1.11	1.08	0.97	0.93
σ_{se}	0.81	0.83	0.82	0.81	0.80	0.80	0.78
η_{se}	0.35	0.31	0.27	0.35	0.40	0.37	0.30
η_{es}	0.31	0.40	0.41	0.30	0.25	0.26	0.28
σ_{sg}	-2.51	-0.88	-1.25	-1.61	-1.92	-3.37	-0.95
η_{sg}	-0.07	-0.03	-0.04	-0.06	-0.07	-0.08	-0.05
η_{gs}	-0.96	-0.42	-0.64	-0.60	-0.60	-1.09	-0.34
σ_{oe}	-0.67	-1.81	-1.68	-0.67	-0.52	-0.35	-0.46
η_{oe}	-0.29	-0.67	-0.56	-0.29	-0.26	-0.16	-0.17
η_{eo}	-0.11	-0.20	-0.22	-0.11	-0.08	-0.06	-0.09
σ_{og}	5.52	5.41	5.79	4.29	4.22	5.19	2.83
η_{og}	0.14	0.21	0.18	0.16	0.16	0.13	0.14
η_{go}	0.89	0.60	0.75	0.68	0.65	0.94	0.59
σ_{eg}	4.00	3.33	4.29	3.14	2.77	3.87	2.77
η_{eg}	0.10	0.13	0.13	0.11	0.11	0.10	0.14
η_{ge}	1.72	1.24	1.43	1.36	1.37	1.82	1.06

Source: Calculated by author from TFU (1978) data.

Table 5.8. Total Fuel Price Elasticities

Fuel type	Fuel type s	o	e	g
s	-0.83	0.43	0.26	-0.08
o	1.03	-1.03	-0.35	0.13
e	0.23	-0.30	-0.40	0.09
g	-1.04	0.10	1.63	-1.46

Source: Compiled by the author.

IRON AND STEEL INDUSTRY

The general question of factor substitutability is most logically applied to individual manufacturing sectors, each of which exhibits more internal homogeneity than does aggregate manufacturing.[12] This is particularly important in regard to the inclusion of energy in the functional specification, since energy intensities differ substantially across industrial sectors. Hogan and Manne (1979) note that the "specific processes for energy substitution are varied and intricate [and that the] morass of detail may be approached gradually by expanding the simple model for improved description of elasticities through the separate analysis of more representative groupings" (p. 15). However, the requisite modeling data become progressively more difficult to obtain the farther dissaggregation proceeds toward what intuitively "best represents" an industry. The Australian iron and steel industry (ASIC 2941) is selected for study for several reasons. It is indicative first because it is a relatively energy-intensive industry, and therefore the results should be of interest in an environment of rapidly changing energy prices. Whereas the value share of energy for aggregate manufacturing in Australia averaged 2.6 percent for the period 1946/47 to 1974/75 in the iron and steel industry, energy share averaged 14.9 percent over the period 1959/60 to 1978/79. In 1980 BHP used 5.8 million tonnes of raw coal (4.8 million tonnes of coke) in the production of 7.4 million tonnes of iron and 7.8 million tonnes of raw steel;[13] 83 percent of the BHP steel production was used within Australia, and over the period of the analysis BHP was the only steel producer in the country. Second, overall the iron and steel industry is relatively homogeneous, since it relies on only three alternative steel-making processes: open hearth furnace (OH), basic oxygen steel furnace (BOS), and electric arc furnace (EA). BHP phased out its final OH in 1982, having converted to the more efficient BOS technology during the 1960s and 1970s. Table 5.9 provides general information on the history of the main BHP plants in operation in 1983.

Table 5.9. Australian Iron and Steel Dataset

Coke ovens		
Port Kembla		1950(1982); 1938(1972); 1960; 1960; 1966; 1972; 1973
Newcastle		1915[1930]; 1919[1930]; 1921[1939]; 1930[1957]; 1930[1963]; 1939(1969); 1954(1979); 1957(1982); 1963[1982]; 1970; 1979
Whyalla		1969[extended, 1979-80]; 1968
Sinter plants		
Port Kembla		1957(1975); 1960(on standby, 1983); 1975
Newcastle		1944(1961); 1961
Blast furnaces		
Port Kembla		1928(1981); 1938(1982); 1952(1982); 1959; 1972
Newcastle		1915[enlarged, 1934 and 1943](1982); 1918; 1921(1982); 1963
Whyalla		1941(1981); 1965
Steel-making shops		
Port Kembla	OH	8: 1928-53(2: 1972; 6:1977) ⟦245, 6@265, 275⟧; 5: 1956-62(1981-82) ⟦4@365, 550⟧
	BOS	2: 1972[1979-80] ⟦2@230⟧, 1983⟦230⟧
	EA	1941⟦20–Heroult⟧, 1961⟦50–ASEA⟧
Newcastle	OH	14: 1915-40(1962-65)
	BOS	1962⟦200⟧; 1963⟦200⟧; 1967(1982) ⟦50⟧
	EA	? ⟦2@25–coreless induction⟧
Whyalla	BOS	1965⟦2@120⟧
	EA	? ⟦18-Heroult⟧

Note: ?, installation date unknown.

Key: installed [replaced], (closed), ⟦tonnes capacity⟧.

Source: Australian Department of Home Affairs and Environment, *The Cost of Pollution Abatement: The Australian Integrated Steel Industry,* Report No. 13 (Canberra: Australian Government Printing Service, 1983); Broken Hill Proprietary Company Annual Reports.

An integrated iron and steel plant involves several distinct stages of production. Figure 5.1 schematically illustrates these various stages from the conversion of metallurgical coal to coke and the preparation of iron ore (sintering) for the blast furnace to the finishing and the rolling of the steel produced.

The energy intensities, as well as the input and output mixes, of the alternative steel furnace processes differ substantially. The amount of finishing of the steel produced relative to pig iron also affects the energy consumption figures. Table 5.10 provides an indication of the relative energy content of various stages of the production process, and there it can be seen that continuous finishing uses considerably less energy than does the conventional method of finishing that is applied to the cooled ingots. Similarly, this relationship holds for hot rolling vis-à-vis cold rolling.

The data in Table 5.11 are an example of the pseudodata mentioned earlier as being derived from the L-P model of the iron and steel industry constructed by Russell and Vaughan (1976). The values presented illustrate the energy requirements of the three steel processes as derived from simulations of the L-P model. It should be noted that the electricity used by the EA process is generated internally in the model by generators that utilize fuel oil.

Table 5.10. Iron Production and Steel-Finishing Energy Use

Activity	Coke (pounds)	Electricity (kilowatt-hours)	Steam (pounds)	Heat (millions Btu)
Coking	–	37.8	8.3	2.0
Sintering	89-94	14	14	0.16-0.23
Blast furnace	1,480-1,819	26	2,190	–
Finishing				
Convential	–	27.3-62.0	–	1.0-1.125
Continuous	–	4.45-5.55	–	0.13
Hot rolling	–	76.5-156.0	–	2.0-2.5
Pickling	–	18.2	–	–
Cold rolling	–	50.8-298	–	0.93

Source: C. S. Russel and W. J. Vaughan, *Steel Production: Processes, Products, and Residuals* (Baltimore: Johns Hopkins University Press, 1976), pp. 39, 75; p. 94, Table 5.2; pp. 98, 99; pp. 136-37, Table 7.1; p. 139, Table 7.2; p. 142, Table 7.3; p. 144, Table 7.4.

Figure 5.1. Schematic of Iron and Steel Production

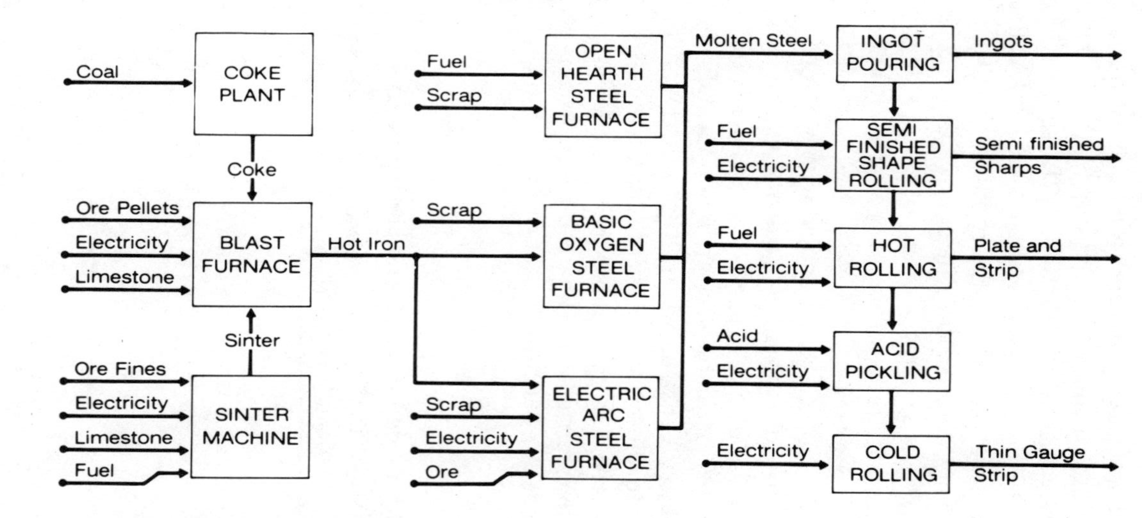

Adapted from Russell and Vaughan (1976, p. 22).

Table 5.11. Steel-Making Energy Use

Process	Coke	Fuel oil	Coke oven gas	Total
Open hearth	11.3	2.8	3.3	17.4
Basic Oxygen	16.3	0.2	4.8	21.3
Electric arc	0.1	8.6	–	8.7

Source: C. S. Russell and W. J. Vaughan, *Steel Production: Processes, Products, and Residuals* (Baltimore: Johns Hopkins University Press, 1976), pp. 39, 75; p. 94, Table 5.2; pp. 98, 99; pp. 136–37, Table 7.1; p. 139, Table 7.2; p. 142, Table 7.3; p. 144, Table 7.4; p. 207, Table 9.7.

Third, it should be noted that a single firm, BHP, is a monopolist in the domestic industry; therefore, technology diffusion among firms is not an issue, whereas the assumption of perfect competition in input markets may well be held suspect. Figures 5.2 through 5.4 illustrate the technology in the BHP sites that are currently in operation. As can be seen from the first of those schematics, Port Kembla specializes in the manufacturing of flat steel products, and it is the largest steel-making facility in Australia with an annual capacity of approximately 6 million tonnes in 1983; this represents two-thirds of the industry capacity. The Newcastle facility has a capacity of around 1.6 million tonnes of mainly bars and rods, or approximately 20 percent of Australian capacity. The final site at Whyalla accounts for the remaining 15 percent, 1.3 million tonnes, and concentrates on the production of structural shapes (see Australian Department of Home Affairs and Environment 1983, p. 7). As well as these differences in output mix, technology among the sites is different in historical terms (see Table 5.9); however, the basic iron and steel-making technology employed is still similar among the locations. Continuous casting accounted for only 14 percent of BHP's installed crude steel capacity in 1978 (Australian Industry Assistance Commission 1980), and so the assumption of a homogeneous steel industry does not seem to be unwarranted.

Finally, the iron and steel-sector modeling data are assembled, with an estimate of the value of capital as $P_K \cdot Q_K$; that is, constructing individual series on the price of capital and on the quantity

Figure 5.2. Schematic of Broken Hill Proprietary Company's Port Kembla Facility

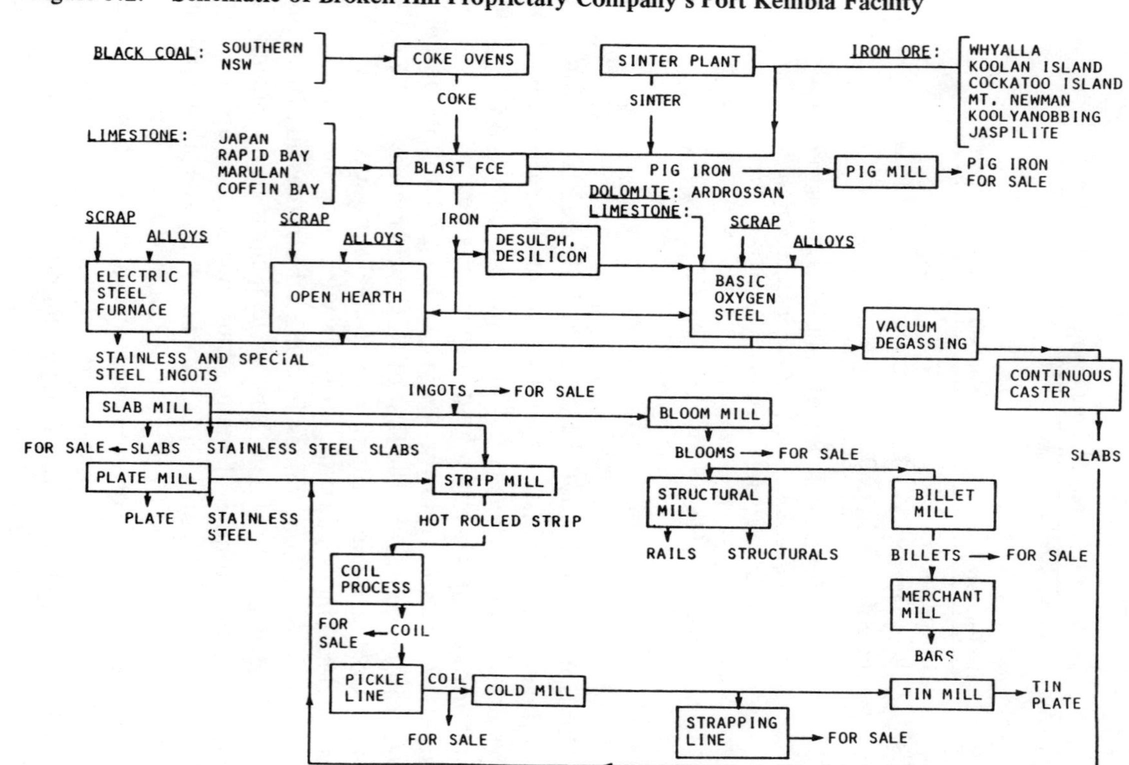

From Australian Department of Home Affairs and Environment (1983, p. 8).

Figure 5.3. Schematic of Broken Hill Proprietary Company's Newcastle Facility

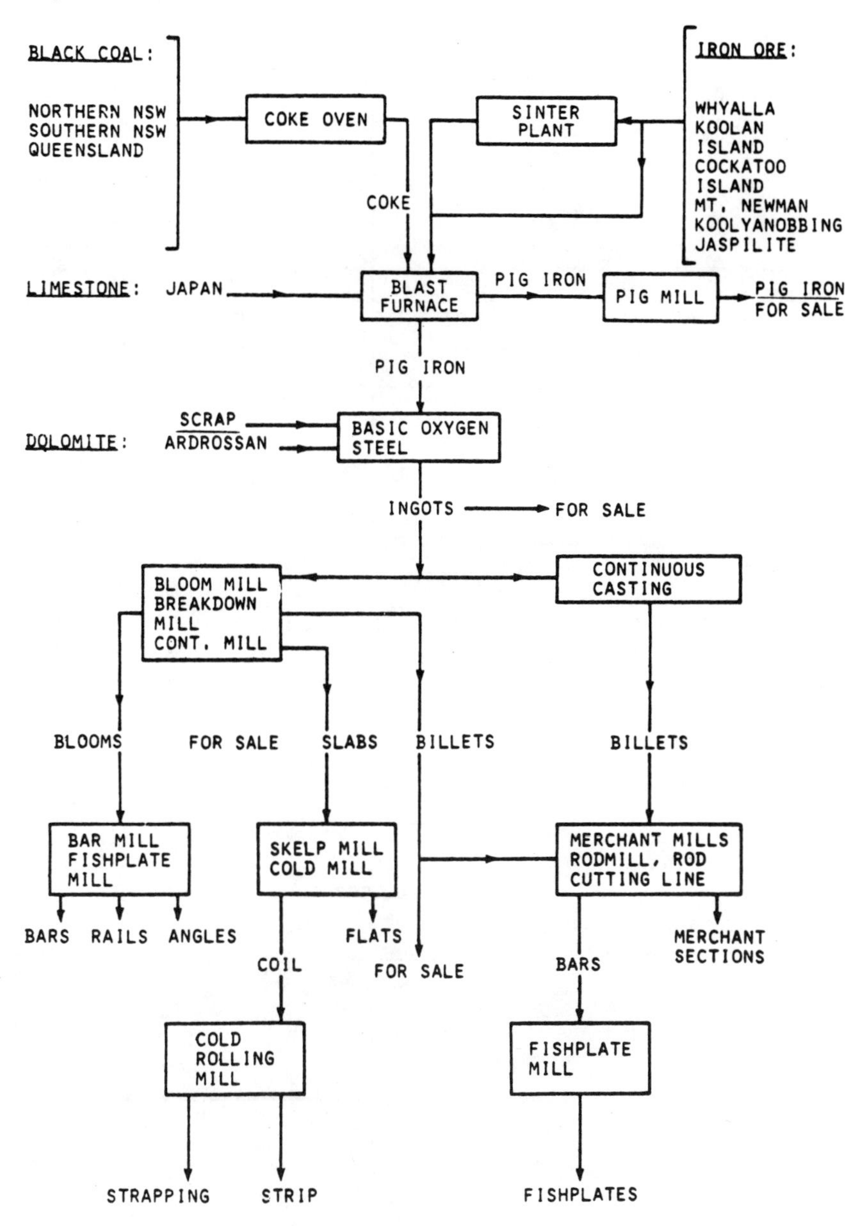

From Australian Department of Home Affairs and Environment (1983, p. 10).

Figure 5.4. Schematic of Broken Hill Proprietary Company's Whyalla Facility

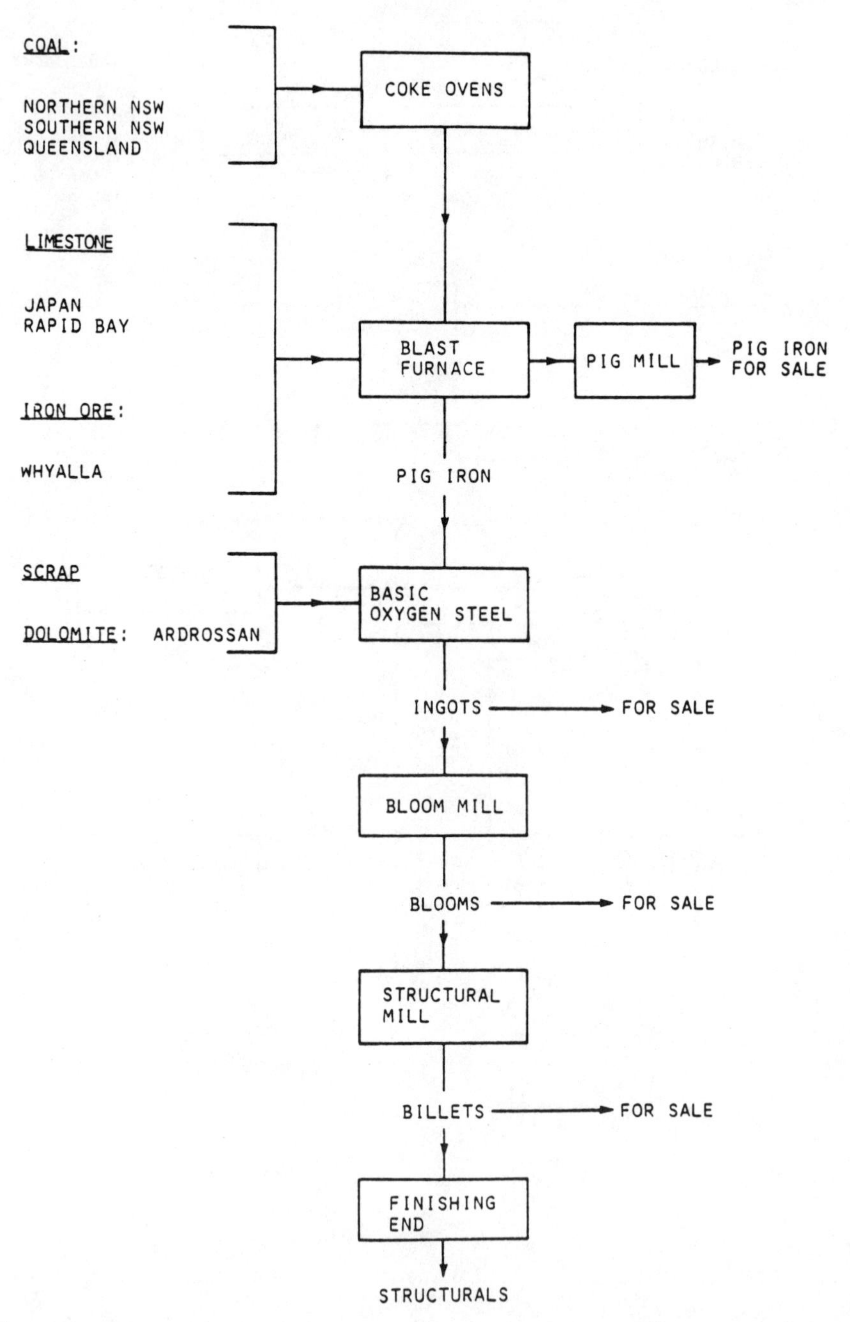

From Australian Department of Home Affairs and Environment (1983, p. 9). From BHP.

of capital services, and following the approach of Christensen and Jorgenson (1969). This methodology contrasts with the residual variable for the value of capital used in the TFU model, which is measured as the difference between aggregate value added in manufacturing and the value of labor services. The data of the current model encompass the period 1946/47 to 1978/79. Various temporal subsamples are defined from the 33 years covered by the total dataset (the model data appear in Tables 5.12 through 5.15), and the overall results of the 20-observation sample from 1959/60 to 1978/79 are reported here.[14] To evaluate the aggregation and separability assumptions, six production functions as defined in Table 5.16 are tested. The **A** and **P** in the table stand for administrative employees and production workers, respectively. Herein, only two four-factor models, namely, $\mathbf{Q}_1$ and $\mathbf{Q}_5$, are analyzed: an aggregate labor model and a model including only production workers.

As was previously stated, a condition for the application of the duality theorem to the translog function is that it should be well behaved in the neighborhood of the estimate; the concavity and positivity conditions are satisfied at all data points in the observation set, while linear homogeneity requires that restrictions 5.11 and 5.12 be met. The symmetry condition of Equation 5.13 derives from the requirement that the production function be twice differentiable, that is, that the bordered Hessian be symmetric.[15] These restrictions are assessed using an *F* test to compare the results of the unconstrained and restricted estimates of the function. The results for the two models are presented in Table 5.17.

Whereas the assumption of linear homogeneity is often rejected in the literature, in the case of the Australian iron and steel industry it is consistently accepted, while the concavity assumption is also satisfied for both of the $\mathbf{Q}_1$ **and** $\mathbf{Q}_5$ models. Models including both categories of labor, $\mathbf{Q}_3$ and $\mathbf{Q}_4$, fail the concavity tests, and the asymptotic standard deviations of the estimated coefficients, as reported in Turnovsky and Donnelly (1982), show that models with fewer inputs have a higher proportion of significant parameters. This result may accrue from the high level of collinearity, tending to be lessened in the models with fewer explanatory variables and generally yielding consistently poor results. The test results presented in Table 5.18 suggest that energy exhibits weak approximate separability in the full **KLEM model.**[16] The value-added specification is not rejected in the **KLEM** model; **K** and **E** form an aggregate in the **KPEM** (capital,

Table 5.12. Iron and Steel Model Cost-Share Dataset (1946/47 to 1978/79)

Fiscal year	Total input cost	Cost shares				
		Capital	Administrative employees	Production workers	Energy	Material
1946/47	67,255	0.04968	0.03851	0.13376	0.12866	0.64939
1947/48	77,908	0.05103	0.03618	0.13855	0.13742	0.63682
1948/49	84,442	0.05648	0.03889	0.14602	0.13621	0.62240
1949/50	94,516	0.05800	0.03639	0.13914	0.14807	0.61840
1950/51	129,333	0.05607	0.03132	0.13871	0.15923	0.61467
1951/52	173,936	0.06192	0.02932	0.13337	0.17499	0.60040
1952/53	237,165	0.06145	0.02399	0.11941	0.16072	0.63443
1953/54	268,854	0.06439	0.02167	0.11401	0.15656	0.64337
1954/55	287,338	0.06252	0.02314	0.12448	0.14885	0.64101
1955/56	308,294	0.06375	0.02273	0.13661	0.15090	0.62601
1956/57	406,387	0.07694	0.02321	0.13245	0.13004	0.63736
1957/58	420,588	0.08648	0.02484	0.13998	0.13293	0.61577
1958/59	467,655	0.08977	0.02485	0.13614	0.12501	0.62423
1959/60	509,060	0.09500	0.02700	0.14153	0.12511	0.61137
1960/61	557,269	0.09972	0.02874	0.14516	0.13068	0.59570
1961/62	547,138	0.12036	0.03238	0.14765	0.13234	0.56727
1962/63	602,017	0.11066	0.03231	0.14326	0.12835	0.58542
1963/64	636,116	0.10907	0.03228	0.14572	0.13010	0.58283
1964/65	701,701	0.11203	0.03411	0.14947	0.12603	0.57836
1965/66	720,003	0.12414	0.03564	0.15311	0.13127	0.55584
1966/67	760,783	0.12701	0.03700	0.14971	0.14090	0.54538
1967/68	812,978	0.13379	0.03851	0.15422	0.13885	0.53463
1968/69	857,692	0.12366	0.04420	0.15799	0.15140	0.52275
1969/70	998,289	0.11802	0.04535	0.14424	0.14762	0.54477
1970/71	1,080,192	0.14011	0.05267	0.14567	0.15449	0.50706
1971/72	1,145,808	0.14949	0.05983	0.14722	0.16205	0.48141
1972/73	1,291,028	0.12800	0.05674	0.15094	0.17002	0.49430
1973/74	1,547,952	0.15856	0.05704	0.15560	0.15841	0.47039
1974/75	1,980,816	0.17568	0.06023	0.15886	0.18440	0.42083
1975/76	2,028,997	0.18780	0.07257	0.14980	0.17355	0.41628
1976/77	2,367,583	0.16573	0.06893	0.15239	0.16367	0.44928
1977/78	2,465,922	0.15713	0.07748	0.15278	0.16175	0.45086
1978/79	2,840,824	0.15521	0.07195	0.15183	0.15971	0.46130
1959/60 to 1978/79 Mean		0.13456	0.04825	0.14986	0.14854	0.51880

Source: Compiled by the author.

Table 5.13. Iron and Steel Energy Submodel Dataset (1946/47 to 1978/79)

Fiscal year	Total fuel cost	Cost shares: Black coal	Coke	Oil	Electricity	Coke oven gas	Natural gas
1946/47	8,348	0.06493	0.53630	0.01090	0.08265	0.30522	NA
1947/48	10,450	0.05359	0.58555	0.01301	0.07962	0.26823	NA
1948/49	11,158	0.06067	0.57000	0.01855	0.08254	0.26824	NA
1949/50	12,631	0.05906	0.61096	0.03879	0.08265	0.20854	NA
1950/51	20,017	0.05895	0.54718	0.02763	0.06939	0.29685	NA
1951/52	29,452	0.05735	0.56743	0.02217	0.06882	0.28423	NA
1952/53	37,087	0.06121	0.56745	0.02071	0.06824	0.28239	NA
1953/54	40,094	0.05572	0.56106	0.01808	0.08350	0.28164	NA
1954/55	40,715	0.05136	0.53442	0.02530	0.10465	0.28427	NA
1955/56	43,460	0.02952	0.54190	0.04841	0.11694	0.26323	NA
1956/57	49,604	0.02439	0.49075	0.07792	0.14761	0.25933	NA
1957/58	52,155	0.02186	0.48055	0.08870	0.16272	0.24617	NA
1958/59	54,473	0.02460	0.46280	0.10673	0.17383	0.23204	NA
1959/60	59,748	0.02214	0.45120	0.11130	0.18747	0.22789	NA
1960/61	67,773	0.01647	0.48192	0.08360	0.17893	0.23908	NA
1961/62	66,719	0.01433	0.48525	0.08020	0.19438	0.22584	NA
1962/63	72,042	0.01160	0.50494	0.07067	0.21846	0.19433	NA
1963/64	77,730	0.00880	0.52539	0.06413	0.21824	0.18344	NA
1964/65	83,062	0.00673	0.51651	0.07820	0.21341	0.18515	NA
1965/66	88,249	0.00573	0.52909	0.07956	0.20866	0.17695	NA
1966/67	101,384	0.00457	0.55125	0.07613	0.19752	0.17053	NA
1967/68	104,761	0.00331	0.54632	0.08660	0.20018	0.16359	NA
1968/69	123,119	0.00243	0.58768	0.06599	0.20562	0.13118	0.00710
1969/70	140,721	0.00223	0.60449	0.06425	0.19743	0.12380	0.00780
1970/71	160,138	0.00135	0.62926	0.07054	0.17697	0.11471	0.00717
1971/72	179,111	0.00112	0.66176	0.06924	0.16555	0.09544	0.00689
1972/73	206,992	0.00091	0.68359	0.05214	0.17164	0.08499	0.00673
1973/74	235,631	0.00087	0.68755	0.05208	0.16593	0.08756	0.00601
1974/75	345,840	0.00061	0.74221	0.04253	0.13410	0.07544	0.00511
1975/76	332,451	0.00076	0.65912	0.05942	0.18663	0.08930	0.00477
1976/77	366,143	0.00078	0.63176	0.06859	0.19838	0.09405	0.00644
1977/78	367,235	0.00126	0.60964	0.07494	0.23347	0.07137	0.00932
1978/79	427,559	0.00157	0.56510	0.09561	0.22993	0.09412	0.01367
1959/60 to 1978/79 Mean		0.00538	0.58270	0.07229	0.19414	0.14144	0.00736

Note: NA, not applicable.
Source: Compiled by the author.

Table 5.14. Iron and Steel Model Factor Price Indexes

Fiscal year	Capital	Total labor	Administrative employees	Production workers	Energy	Material
1946/47	100.00	100.00	100.00	100.00	100.00	100.00
1947/48	113.66	114.34	105.85	116.79	112.79	110.61
1948/49	126.67	132.42	125.03	134.56	138.40	140.15
1949/50	137.57	139.32	130.02	141.99	155.51	162.12
1950/51	165.39	168.63	150.27	173.81	182.92	193.94
1951/52	201.88	199.30	175.80	205.89	245.18	259.85
1952/53	234.88	220.45	194.02	227.85	271.12	296.97
1953/54	257.44	223.51	189.06	232.82	270.90	293.94
1954/55	262.99	249.98	216.03	259.34	263.85	296.21
1955/56	280.50	251.89	220.99	260.63	277.27	306.06
1956/57	282.84	259.15	222.21	269.19	275.82	309.85
1957/58	291.76	265.06	229.08	274.94	267.65	301.52
1958/59	297.34	274.05	232.92	285.12	267.67	296.97
1959/60	299.88	293.52	256.25	303.86	259.62	299.24
1960/61	302.46	317.50	280.67	327.88	272.08	302.27
1961/62	285.04	313.08	289.93	320.20	260.05	296.97
1962/63	262.03	323.12	306.11	328.81	267.78	293.94
1963/64	256.33	334.61	311.57	341.81	263.17	290.15
1964/65	264.06	370.64	350.85	377.21	261.58	296.21
1965/66	275.42	390.08	375.43	395.49	265.47	295.45
1966/67	279.27	399.55	388.09	404.20	280.38	300.00
1967/68	287.13	421.72	408.78	426.85	281.95	300.76
1968/69	286.37	455.93	445.93	460.38	313.40	308.33
1969/70	315.97	487.27	480.30	490.89	346.74	333.00
1970/71	357.09	520.57	486.62	533.90	390.53	307.71
1971/72	370.40	597.06	616.22	589.53	423.60	310.80
1972/73	335.22	661.09	662.08	660.49	428.51	357.35
1973/74	505.10	793.75	777.21	799.69	474.98	433.82
1974/75	707.81	1,049.40	1,036.38	1,053.88	685.08	460.34
1975/76	788.77	1,087.10	1,151.16	1,059.49	660.62	502.58
1976/77	817.16	1,304.83	1,312.11	1,302.85	719.38	590.45
1977/78	831.91	1,478.37	1,613.29	1,419.40	759.43	614.19
1978/79	982.92	1,580.09	1,724.67	1,516.91	796.27	701.14

Source: Compiled by the author.

Table 5.15. Iron and Steel Model Fuels Price Indexes (1946/47 to 1978/79)

Fiscal year	Solid fuel	Solid fuel and gas	Oil	Electricity	Coke oven gas
1946/47	100.00	100.00	100.00	100.00	100.00
1947/48	124.83	123.12	90.95	104.67	93.34
1948/49	157.36	155.05	109.40	104.41	114.53
1949/50	178.46	175.84	111.55	109.20	129.94
1950/51	209.70	206.80	144.07	115.70	155.75
1951/52	297.14	291.74	163.97	116.82	201.32
1952/53	323.36	318.29	156.64	133.88	231.21
1953/54	321.61	316.93	133.92	136.70	234.51
1954/55	309.02	305.56	116.00	136.00	235.55
1955/56	335.15	329.08	117.27	136.55	237.09
1956/57	324.72	319.70	131.85	143.48	235.28
1957/58	317.82	311.78	127.40	140.47	223.50
1958/59	324.46	316.77	115.81	143.19	219.70
1959/60	313.25	306.32	107.03	143.25	214.86
1960/61	343.00	332.20	99.18	148.67	218.00
1961/62	338.13	323.66	84.69	147.07	197.36
1962/63	384.29	356.47	75.20	143.83	177.60
1963/64	397.85	363.28	69.22	136.91	160.43
1964/65	410.23	371.33	70.96	127.72	152.72
1965/66	430.84	387.82	63.18	126.52	152.72
1966/67	472.42	422.20	65.54	124.91	156.41
1967/68	477.19	423.02	73.16	126.11	146.62
1968/69	532.43	467.46	96.97	138.39	149.15
1969/70	652.22	563.75	71.10	137.96	149.69
1970/71	770.30	660.22	84.03	137.55	156.08
1971/72	863.72	735.58	87.53	140.75	156.71
1972/73	867.05	740.34	87.65	143.77	165.45
1973/74	965.48	825.49	100.80	152.09	189.10
1974/75	1,500.70	1,270.12	129.81	172.44	237.45
1975/76	1,327.51	1,139.19	194.80	190.84	274.59
1976/77	1,424.04	1,231.09	212.80	205.92	335.75
1977/78	1,465.68	1,270.48	252.00	222.67	362.52
1978/79	1,465.57	1,275.60	329.51	240.52	387.82

Source: Compiled by the author.

Table 5.16. Alternative Production Function Specifications

$Q_1 = f_1$ (K, L, E, M)	$Q_4 = f_4$ (K, A, P, E)
$Q_2 = f_2$ (K, L, E)	$Q_5 = f_5$ (K, P, E, M)
$Q_3 = f_3$ (K, A, P, E, M)	$Q_6 = f_6$ (K, P, E)

Source: Compiled by the author.

Table 5.17. Translog Function Homogeneity and Significance Test Results (1959/60 to 1978/79)

Function	Homogeneity (*F* statistic)	Cross-product coefficient significance
KLEM	2.6	4 of 6
KPEM	2.8	4 of 6

Source: M. H. L. Turnovsky and W. A. Donnelly, "Energy Substitution, Separability, and Technical Progress in the Australian Iron and Steel Industry," *Journal of Business Economic Statistics,* 2(1984):58, Table 3.

Table 5.18. Translog Function Separability Tests (1959/60 to 1978/79)

Type of separability	Factors	KLEM	KPEM
Global $F(6,48)$		13.598	26.338
Strong $F(3,18)$	Capital	6.782	13.124
	Labor	8.488	5.397
	Energy	22.801	32.151
	Material	11.326	16.743
Weak Approximate $\chi^2(2)$	Capital	10.070	3.216
	Labor	9.467	11.393
	Energy	2.647	6.538
	Material	8.249	13.200
	Capital and labor	7.534	9.394
	Capital and energy	10.241	6.896
	Capital and material	6.835	1.353

Source: M. H. L. Turnovsky and W. A. Donnelly, "Energy Substitution, Separability, and Technical Progress in the Australian Iron and Steel Industry," *Journal of Business Economic Statistics,* 2(1984): 59, Table 5.

production workers, energy, and materials) model. Berndt and Wood (1975) do not reject the existence of a consistent aggregate for **K** and **E**, an "index of utilized capital," in the U.S. aggregate manufacturing model either. Finally, weak approximate separability of capital and material from labor and energy is observed. It has been suggested by Diewert (1975) and Berndt and Wood (1975) that even if the above separability conditions were rejected, aggregation would still be valid, as the conditions for Hicksian aggregation are met. It is argued in regard to instances in which the price of some inputs and the value of total gross output move in fixed proportions that those inputs are separable from the remaining inputs, the latter forming a consistent aggregate. This hypothesis can be tested by regressing the various input prices against the output price (Denny and May 1977); in the iron and steel industry model, Hicksian aggregation is rejected for all cases except for that of energy.[17]

Table 5.19 presents the econometric results for the models $\mathbf{Q}_1$ and $\mathbf{Q}_5$ and for the corresponding fuels submodels. It can be seen that neither the capital-labor nor the capital-energy parameter estimates are statistically different from 0 in the factor models. The only other parameter that is not significant is the labor–technological change parameter γ_L in the **KLEM** model. In the fuels submodels the oil-electricity parameters are insignificant, and the overall explanatory power of the three-fuel system of equations leaves much to be desired; in addition, both submodels are severely affected by the collinearity in the price variables.

The AES and partial demand elasticities are presented in Tables 5.20 and 5.21 for both of the factor models. The issue as to whether capital and energy (**KE**) are substitutes is not resolved in the results of either of the factor models, but both suggest them to be substitutes because the β_{KE} terms are not statistically significant.

The controversial issue of the KE relation can be analyzed from the results of various temporal subsamples presented in Table 5.22. All the samples that include the latest data points show substitutability at the mean values, with the exception of the longest sample. That particular one reflects a long time span of price stability that weighs heavily against the shorter period of major price changes; samples that indicate substantial variations over time, as reflected in the yearly elasticities, also exhibit a definite trend from complementarity to substitutability. But the question in regard to whether or not K and E are complements or substitutes, as measured by the AES, may

Table 5.19. Econometric Model Results (1959/60 to 1978/79)

		Factor models				Fuel models			
			KLEM		KPEM		soe		soeg
Observations			20		20		20		20
System R^2			0.996		0.993		0.614		0.991
χ^2			109.65		99.13		19.02		93.89
DF			9		9		3		6
Equation $\bar{R}^2$	1		0.852		0.878		0.401		0.870
	2		0.973		0.765		0.567		0.550
	3		0.909		0.929				0.008
	4		0.944		0.921				
Number of iterations			4		4		5		6
		β_{KK}	0.0671 (4.22)	β_{KK}	0.0772 (4.70)	β_{ss}	0.0645 (4.18)	β_{ss}	0.1748 (12.72)
		β_{LL}	0.0713 (4.13)	β_{PP}	0.0723 (4.98)	β_{oo}	0.0287 (3.54)	β_{oo}	0.0477 (5.83)
		β_{EE}	0.0830 (7.47)	β_{EE}	0.0928 (8.23)	β_{ee}	0.0175 (1.41)	β_{ee}	-0.0405 (-2.38)
		β_{MM}	0.1856 (5.64)	β_{MM}	0.1721 (4.82)	β_{so}	0.0379 (-5.74)	β_{gg}	0.0853 (5.03)
		β_{KL}	0.0105 (1.87)	β_{KP}	-0.0014 (-0.26)	β_{se}	-0.0266 (2.08)	β_{so}	-0.0370 (-6.34)
		β_{KE}	0.0002 (0.03)	β_{KE}	0.0045 (0.50)	β_{oe}	0.0091 (1.36)	β_{se}	-0.0283 (-2.41)
		β_{KM}	-0.0778 (-3.81)	β_{KM}	-0.0804 (-3.70)	α_s	0.6322 (27.43)	β_{sg}	-0.1094 (-24.47)
		β_{LE}	-0.0286 (-3.22)	β_{PE}	-0.0382 (-4.60)	α_o	0.1344 (11.93)	β_{oe}	0.0170 (1.81)
		β_{LM}	-0.0532 (-4.29)	β_{PM}	-0.0326 (-2.93)	α_e	0.2333 (12.36)	β_{og}	-0.0277 (-2.99)
		β_{EM}	-0.0546 (-3.87)	β_{EM}	-0.0591 (-4.00)			β_{eg}	0.0518 (4.09)
		α_K	0.0880 (6.29)	α_K	0.0802 (5.49)			α_s	0.3441 (15.59)
		α_L	0.1664 (16.84)	α_P	0.1618 (20.84)			α_o	0.1520 (13.16)
		α_E	0.1086 (11.31)	α_E	0.1057 (11.08)			α_e	0.2288 (11.45)
		α_M	0.6375 (28.75)	α_M	0.6523 (27.85)			α_g	-0.2759 (11.45)
		γ_K	0.0016 (2.27)	γ_K	0.0024 (3.30)				
		γ_L	0.0001 (0.09)	γ_P	-0.0014 (2.55)				
		γ_E	0.0020 (3.83)	γ_E	0.0026 (4.99)				
		γ_M	-0.0037 (-3.22)	γ_M	-0.0036 (-2.99)				

Source: M. H. L. Turnovsky and W. A. Donnelly, "Energy Substitution, Separability and Technical Progress in the Australian Iron and Steel Industry," *Journal of Business and Economic Statistics,* 2(1984):57, 58, Tables 2 and 7.

Table 5.20. KLEM Allen Elasticities of Substitution and Demand Price Elasticities (1959/60 to 1978/79)

	At the means	Fiscal year				
		1959/60	1964/65	1968/69	1973/74	1978/79
σ_{KK}	-2.73	-2.09	-2.58	-2.70	-2.64	-2.66
η_{KK}	-0.37	-0.20	-0.29	-0.33	-0.42	-0.41
σ_{LL}	-2.23	-2.42	-2.33	-2.20	-2.13	-2.04
η_{LL}	-0.44	-0.41	-0.43	-0.45	-0.45	-0.46
σ_{EE}	-1.97	-1.69	-1.71	-2.00	-2.00	-1.99
η_{EE}	-0.29	-0.21	-0.22	-0.32	-0.31	-0.30
σ_{MM}	-0.24	-0.14	-0.17	-0.23	-0.29	-0.30
η_{MM}	-0.12	-0.09	0.10	-0.12	-0.14	-0.14
σ_{KL}	1.39	1.66	1.50	1.44	1.34	1.27
η_{KL}	0.28	0.28	0.28	0.28	0.28	0.29
η_{LK}	0.19	0.16	0.17	0.18	0.20	0.22
σ_{KE}	1.01	1.02	1.02	1.01	1.01	1.01
η_{KE}	0.15	0.13	0.13	0.15	0.16	0.16
η_{EK}	0.14	0.10	0.11	0.12	0.15	0.17
σ_{KM}	-0.11	-0.34	-0.20	-0.20	-0.06	-0.09
η_{KM}	-0.06	-0.21	-0.12	-0.11	-0.03	-0.04
η_{MK}	-0.02	-0.03	-0.02	-0.03	-0.01	-0.01
σ_{LE}	0.03	-0.36	-0.18	-0.01	0.12	0.22
η_{LE}	0.00	-0.04	-0.02	-0.00	0.02	0.04
η_{EL}	0.01	-0.06	-0.03	-0.00	0.03	0.05
η_{LM}	0.48	0.48	0.50	0.49	0.48	0.47
η_{LM}	0.25	0.30	0.29	0.26	0.23	0.20
η_{ML}	0.10	0.08	0.09	0.10	0.10	0.11
σ_{EM}	0.29	0.29	0.27	0.30	0.28	0.24
η_{EM}	0.15	0.17	0.15	0.16	0.14	0.11
η_{ME}	0.04	0.04	0.03	0.04	0.04	0.04

Source: Compiled by the author.

Table 5.21. KPEM Allen Elasticities of Substitution and Demand Price Elasticities (1959/60 to 1978/79)

	At the means	Fiscal year				
		1959/60	1964/65	1968/69	1973/74	1978/79
σ_{KK}	-2.19	-1.14	-1.86	-2.09	-2.23	-2.24
η_{KK}	-0.30	-0.11	-0.21	-0.27	-0.36	-0.36
σ_{PP}	-2.44	-2.46	-2.45	-2.42	-2.43	-2.44
η_{PP}	-0.38	-0.36	-0.38	-0.39	-0.39	-0.38
σ_{EE}	-1.57	-1.16	-1.19	-1.59	-1.64	-1.65
η_{EE}	-0.24	-0.15	-0.15	-0.25	-0.27	-0.27
σ_{MM}	-0.27	-0.16	-0.20	-0.27	-0.33	-0.34
η_{MM}	-0.14	-0.10	-0.12	-0.14	-0.16	-0.16
σ_{KP}	0.94	0.92	0.92	0.93	0.94	0.95
η_{KP}	0.15	0.14	0.14	0.15	0.15	0.16
η_{PK}	0.13	0.10	0.11	0.12	0.15	0.18
σ_{KE}	1.20	1.31	1.29	1.23	1.17	1.14
η_{KE}	0.19	0.17	0.17	0.19	0.19	0.20
η_{EK}	0.17	0.15	0.15	0.16	0.19	0.21
σ_{KM}	-0.04	-0.17	-0.17	-0.11	0.01	0.08
η_{KM}	-0.02	-0.10	-0.10	-0.06	0.01	0.04
η_{MK}	-0.01	-0.02	-0.02	-0.01	0.00	0.02
σ_{PE}	-0.55	-1.00	-0.81	-0.60	-0.43	-0.31
η_{PE}	-0.09	-0.13	-0.11	-0.09	-0.07	-0.06
η_{EP}	-0.09	-0.15	-0.13	-0.09	-0.07	-0.05
σ_{PM}	0.62	0.64	0.65	0.63	0.61	0.58
η_{PM}	0.34	0.39	0.39	0.35	0.31	0.28
η_{MP}	0.10	0.09	0.10	0.10	0.10	0.10
σ_{EM}	0.31	0.25	0.26	0.31	0.30	0.29
η_{EM}	0.17	0.15	0.15	0.17	0.16	0.14
η_{ME}	0.05	0.03	0.03	0.05	0.05	0.05

Source: Compiled by the author.

Table 5.22. KE and KM Elasticities in Various Samples

Elasticities at the mean	σ_{KE}	σ_{KM}
1946/47 ⟷ 1967/68	-4.52	0.93
1946/47 ⟷ 1974/75	-2.25	0.31
1959/60 ⟷ 1974/75	-0.28	0.27
1946/47 ⟷ 1978/79	-0.22	0.45
1954/55 ⟷ 1978/79	0.85	0.24
1959/60 ⟷ 1978/79	1.01	-0.11
1964/65 ⟷ 1978/79	0.76	-0.59
1968/69 ⟷ 1978/79	0.62	-0.21

Source: M. H. L. Turnovsky and W. A. Donnelly, "Energy Substitution, Separability and Technical Progress in the Australian Iron and Steel Industry," *Journal of Business Economic Statistics,* 2(1984):60, Table 8.

not be as important as is sometimes stated. Indeed, the elasticities measured at each observation point seem to follow a pattern, but the value of the elasticity at the mean of the data will not evidence such patterns. Table 5.23 presents the AES values for several points in the dataset, and the trend in these is evident. For predictive purposes, the more informative results to report are thus (1) the trends in the elasticity of substitution and (2) the elasticity for the most recent observation.

Table 5.23. KE Elasticities Within the Longest Sample (1946/47 to 1978/79)

1946/47	-1.73	1968/69	-0.04
1954/55	-0.86	1973/74	0.16
1959/60	-0.48	1978/79	0.28
1964/65	-0.32	At the mean	-0.22

Source: M. H. L. Turnovsky and W. A. Donnelly, "Energy Substitution, Separability and Technical Progress in the Australian Iron and Steel Industry," *Journal of Business Economic Statistics,* 2(1984):60, Table 9.

In the models that incorporate the most recent data, K and E are always substitutes at the last data point: Those elasticities vary between 0.28 and 1.01. From these results it may be inferred that the trend has been toward the installation of more energy-efficient capital. This is in fact confirmed by the transition from the OH to the BOS technology, a changeover that is known to have occurred.[18] However, the newer capital may be less efficient in its use of material, and this contingency may possibly result in the KM complementarity and the EM substitutability results evidenced in the 1978/79 data. These speculations could be tested only by estimating a proper vintage capital model.

None of the model results indicates that total labor (the sum of administrative employees and production workers) forms a consistent aggregate—a similar conclusion is reached by Berndt and Christensen (1974). Indeed, in the disaggregated labor models, namely, the capital, administrative employees, production workers, energy, and materials model **(KAPEM)** and capital, administrative employees, production workers, and energy model **(KAPE)** versions, the various elasticities between A and P and the other inputs usually have opposite signs (or are very different), while the elasticity between aggregate labor and these same inputs is always bounded by the former, forming a weighted average of the two (see Table 5.24). These results are surprisingly consistent, although the KAPEM model does not pass all of the regularity tests. But Wales (1977) argues that

Table 5.24. Cross-Elasticities in the Labor Factor Models (1959/60 to 1978/79)

KAPEM	KLEM	KPEM
σ_{KA} = 3.52		
	σ_{KL} = 1.39	
σ_{KP} = 0.79		σ_{KP} = 0.94
σ_{AE} = 0.87		
	σ_{LE} = 0.03	
σ_{PE} = -0.53		σ_{PE} = -0.55
σ_{AM} = -0.57		
	σ_{LM} = 0.48	
σ_{PM} = 0.73		σ_{PM} = 0.62

Source: Compiled by the author.

the rejection of either the monotonicity or the concavity conditions does not necessarily imply that the elasticity estimates are incorrect.

Hicks's neutral technical progress (Hicks 1965, pp. 171–72 and Lau 1978, p. 202) is consistently rejected. The parameter estimates are all positive for capital and energy, which implies that the iron and steel industry is capital and energy using. The KPEM results suggest that technical progress is labor saving,[19] while in the **KLEM** model the technical change parameter estimate is not statistically different from 0. Finally, technical progress is consistently shown to be material saving. All of the results are included in Table 5.19 and are consistent with those provided by the aggregate manufacturing model.

Turning to the energy submodel, several alternative formulations are estimated. These include three- and four-fuel models, that is, solid fuel (including gas), oil, and electricity, with gas as a separate input in the four-fuel model.[20] None of the four-fuel models exhibits global concavity, and since the three-fuel model without any time trend is well behaved in most respects, the elasticities of substitution of that model as presented in Table 5.25 probably should be given most attention, wherein the fuels are all substitutes. In the four-fuel model, solid fuel and gas are found to be complements, as would be expected for the joint products of coke and coke oven gas. Also, oil and gas are found to be complements (see Table 5.26). The three- and four-fuel total price elasticities of demand appear in Tables 5.27 and 5.28 for the KLEM model.

The theory of production in the Australian iron and steel industry has been statistically tested to determine the various substitution relationships, the possible factor bias, and the separability implied by the data. Among the more interesting findings are the following:

1. Capital and energy have become substitutes over the years.[21]
2. Technical progress is capital using, energy using, labor saving, and material saving.[22]
3. Administrative employees and production workers do not form a consistent aggregate.
4. Separability for energy is shown in two different testing procedures.[23]
5. The value-added specification is not rejected.

A comparison of the elasticities of substitution between the aggregate manufacturing model and the iron and steel model is

Table 5.25. Three-Fuel Model (1959/60 to 1978/79)

	At the means	Fiscal year				
		1959/60	1964/65	1968/69	1973/74	1978/79
σ_{ss}	-0.24	-0.29	-0.28	-0.25	-0.17	-0.34
η_{ss}	-0.18	-0.21	-0.20	-0.18	-0.14	-0.23
σ_{oo}	-7.35	-5.67	-7.09	-7.56	-7.62	-6.32
η_{oo}	-0.53	-0.63	-0.55	-0.50	-0.40	-0.60
σ_{ee}	-3.69	-3.84	-3.30	-3.45	-4.39	-3.02
η_{ee}	-0.72	-0.72	-0.70	-0.71	-0.73	-0.69
σ_{so}	0.28	0.51	0.32	0.21	0.07	0.41
η_{so}	0.02	0.06	0.02	0.01	0.00	0.04
η_{os}	0.21	0.36	0.22	0.15	0.05	0.28
σ_{se}	0.81	0.80	0.82	0.82	0.79	0.83
η_{se}	0.16	0.15	0.18	0.17	0.13	0.19
η_{es}	0.60	0.56	0.58	0.60	0.62	0.56
σ_{oe}	1.67	1.45	1.56	1.69	2.09	1.43
η_{oe}	0.32	0.27	0.33	0.35	0.35	0.33
η_{eo}	0.12	0.16	0.12	0.11	0.11	0.14

Source: Compiled by the author.

revealing in that, first, the unusual estimate for the labor-energy elasticity in the former model can perhaps be attributed to the fact that total labor does not form a consistent aggregate. Second, as the iron and steel industry is more energy intensive relative to aggregate manufacturing, the greatest disparities appear in the cross-elasticities involving those inputs.

Extensions of this modeling are suggested. It would be interesting to compare these results with those obtained from a multiproduct model including both finished and semifinished steel products. In the iron and steel industry, this would be most informative since, as we have noted, the energy intensities differ substantially for the different classes of products. Chapter 1 introduced the discussion of the multiproduct translog model:

Table 5.26. Four-Fuel Model (1959/60 to 1978/79)

	At the means	Fiscal year				
		1959/60	1964/65	1968/69	1973/74	1978/79
σ_{ss}	-0.20	-0.33	-0.27	-0.19	-0.08	-0.22
η_{ss}	-0.12	-0.16	-0.14	-0.11	-0.06	-0.12
σ_{oo}	-3.70	-4.13	-3.99	-3.20	-0.61	-4.24
η_{oo}	-0.27	-0.46	-0.31	-0.21	-0.03	-0.41
σ_{ee}	-5.24	-5.49	-4.58	-4.82	-6.50	-4.12
η_{ee}	-1.02	-1.03	-0.98	-0.99	-1.08	-0.95
σ_{gg}	-1.85	-1.75	-1.91	-1.77	-0.06	-0.94
η_{gg}	-0.27	-0.40	-0.35	-0.24	-0.01	-0.10
σ_{so}	0.13	0.30	0.10	0.05	-0.03	0.32
η_{so}	0.01	0.03	0.01	0.00	-0.00	0.03
η_{os}	0.07	0.14	0.05	0.03	-0.02	0.18
σ_{se}	0.75	0.68	0.75	0.77	0.75	0.78
η_{se}	0.15	0.13	0.16	0.16	0.12	0.18
η_{es}	0.44	0.32	0.39	0.45	0.52	0.44
σ_{sg}	-0.27	-0.01	-0.13	-0.34	-0.70	-0.79
η_{sg}	-0.04	-0.00	-0.02	-0.05	-0.07	-0.09
η_{gs}	-0.16	-0.01	-0.07	-0.20	-0.48	-0.45
σ_{oe}	2.22	1.81	2.02	2.25	2.97	1.77
η_{oe}	0.43	0.34	0.43	0.46	0.49	0.41
η_{eo}	0.16	0.20	0.16	0.15	0.15	0.17
σ_{og}	-1.62	-0.09	-0.91	-2.04	-4.68	-1.69
η_{og}	-0.24	-0.02	-0.17	-0.28	-0.44	-0.18
η_{go}	-0.12	-0.01	-0.07	-0.13	-0.24	-0.16
σ_{eg}	2.82	2.21	2.31	2.82	4.34	3.09
η_{eg}	0.41	0.50	0.43	0.39	0.41	0.33
η_{ge}	0.55	0.41	0.49	0.58	0.72	0.71

Source: Compiled by the author.

Table 5.27. Total Fuel Price Elasticity in the Three-Fuel Model (output held constant)

Fuel type	Fuel type s	o	e
s	-0.39	0.00	0.10
o	0.00	-0.55	-0.26
e	0.39	-0.10	-0.78

Source: Compiled by the author.

Table 5.28. Total Fuel Price Elasticity in the Four-Fuel Model (output held constant)

Fuel type	Fuel type s	o	e	g
s	-0.29	-0.06	0.09	-0.08
o	-0.10	-0.34	0.37	-0.28
e	0.27	0.09	-1.08	0.37
g	-0.33	-0.19	0.49	-0.31

Source: Compiled by the author.

$$\mathbf{C} = \prod_{i=1}^{n} \mathbf{P}_i^{(\alpha_i + \frac{1}{2} \sum_{j=1}^{n} \beta_{ij} \cdot \ln \mathbf{P}_j + \theta_i \cdot \ln \mathbf{A})} \cdot$$

$$\prod_{j=1}^{m} \mathbf{Y}_j^{(\gamma_j + \frac{1}{2} \sum_{i=1}^{m} \delta_{ij} \cdot \ln \mathbf{Y}_i + \sum_{i=1}^{n} \rho_{ij} \cdot \ln \mathbf{P}_i + \theta_{j+n} \cdot \ln \mathbf{A})} \cdot$$

$$\prod_{i=1}^{2} e^{[\theta_{i+m+n} \cdot (\ln \mathbf{A})^i]} \cdot e^{\alpha_0}, \tag{5.22}$$

where $\mathbf{C}$ = total cost
$\mathbf{P}$ = input prices
$\mathbf{Y}$ = output prices
$\mathbf{A}$ = technical change
n = number of inputs
m = number of outputs
e = base of the natural logarithm
ln = natural logarithm.

But the additional information is derived at a cost of estimating $(n^2 + 2n + n \cdot m + 2m + m^2 + 2)$ unknowns, which, when reduced by the assumption of symmetry, $\beta_{ij} = \beta_{ji}$ and $\delta_{ij} = \delta_{ji}$, leaves just $[(n^2 + n) / 2 + 2n + 2m + n \cdot m + (m^2 + m) / 2 + 3]$ parameters. The marginal productivity relationships in conjunction with Shephard's lemma yield a set of cost-share equations that include the *m* product output terms in each share equation, as in

$$\begin{aligned} \partial \ln \mathbf{C} / \partial \ln \mathbf{P}_i &= \mathbf{M}_i \\ &\equiv \mathbf{S}_i = \alpha_i + \sum_{j=1}^{n} \beta_{ij} \cdot \ln \mathbf{P}_j \\ &\quad + \sum_{j=1}^{m} \rho_{ij} \cdot \ln \mathbf{Y}_j + \theta_i \cdot \ln \mathbf{A}, \\ &\text{for } i = 1, \ldots, n. \end{aligned} \tag{5.23}$$

To avoid singularity in estimation, one equation is dropped from the system, and so there are $(n - 1)$ share equations with the number of unknowns numbering $[(n^2 + n) / 2 + 2n + m \cdot n - (n + m + 2)]$. Without considering any additional restrictions on the equations other than symmetry, a model that includes four-factor inputs and a single homogeneous output entails estimating 15 coefficients in its cost-share representation as opposed to the 28 coefficients present in the total cost function. To extend this model to include two outputs increases the unknowns to 36, which number, admittedly, may be reduced somewhat by imposing various homogeneity and separability restrictions on the inputs, such as

$$\begin{aligned} &\sum_{i=1}^{n} \alpha_i = 1 \\ &\sum_{i=1}^{n} \beta_{ij} = 0, \text{ for } j = 1, \ldots, n \\ &\sum_{i=1}^{n} \rho_{ij} = 0, \text{ for } j = 1, \ldots, m \end{aligned} \tag{5.24}$$

$$\sum_{i=1}^{n} \theta_i = 0.$$

Requiring output homogeneity necessitates

$$\sum_{j=1}^{m} \delta_{ij} = 0, \text{ for } i = 1, \ldots, m$$

$$\sum_{j=1}^{m} \rho_{ij} = 0, \text{ for } i = 1, \ldots, n \tag{5.25}$$

$$\sum_{j=n+1}^{n+m} \theta_j = 0 .$$

Only $(n - 1)$ of the restrictions on the ρ terms are independent in the context of the earlier restrictions required for factor price homogeneity. Output separability implies further that

$$\rho_{ij} = 0, \text{ for all } i = 1, \ldots, n \quad j = 1, \ldots, m . \tag{5.26}$$

Finally, constant returns to scale would require

$$\sum_{j=1}^{m} \gamma_j = 1 . \tag{5.27}$$

The AES and the Marshallian income-compensated demand elasticities in Equations 5.16 and 5.17 are not changed in this multifactor formulation. However, the cost elasticities that can be used to calculate the marginal rates of transformation of the outputs are derived from information contained only in the total cost function. These cost elasticities are

$$\partial \ln \mathbf{C} / \partial \ln \mathbf{Y}_j = \gamma_j + \sum_{i=1}^{m} \delta_{ij} \cdot \ln \mathbf{Y}_i + \sum_{i=1}^{n} \rho_{ij} \cdot \ln \mathbf{P}_i + \theta_j \cdot \ln \mathbf{A}, \text{ for } j = 1, \ldots, m. \tag{5.28}$$

It can be seen that the cost elasticity depends upon the levels of the outputs, the prices of the inputs, and a technical change factor. These relationships are used to calculate the scale elasticity and the marginal costs of the outputs. The scale elasticity (SCE), as defined by Christensen, Cummings, and Schoech (1983), is

$$\mathrm{SCE} = 1 / \sum_{j=1}^{m} (\partial \ln \mathbf{C} / \partial \ln \mathbf{Y}_j). \tag{5.29}$$

The marginal costs (MC) derived by multiplying the cost elasticity of each output by its average cost is then

$$\mathbf{MC}_j = (\partial \ln \mathbf{C} / \partial \ln \mathbf{Y}_j) \cdot (\hat{\mathbf{C}} / \mathbf{Y}_j). \tag{5.30}$$

Thus, the marginal rate of transformation (**MRT**) of output i for output j is represented by the ratio of the marginal cost of i to that of j times the ratio of output j to output i:

$$\mathbf{MRT}_{ij} = (\mathbf{MC}_i / \mathbf{MC}_j) \cdot (\mathbf{Y}_j / \mathbf{Y}_i). \tag{5.31}$$

The precise values of several of these important derivative measures—in fact, those very measures of interest that necessitate the estimation of the total cost function along with the share equations (the cost elasticities, the marginal costs, the marginal rates of transformation, and the scale elasticities)—may be open to question because of the inherent collinearity introduced into the model. The exemption from the need to estimate the cost function directly is one of the very reasons that the flexible functional forms enjoy the degree of acceptance they do. So although studies using microeconomic data may be implemented, their accuracy is somewhat doubtful.

In addition, estimation of alternative, and potentially more robust, functional forms should be attempted: for example, the generalized BC, which includes the several other flexible functional forms as special cases and thus can be used to test for alternative ones (see Berndt and Khaled 1979). The generalized BC function is

$$c(\mathbf{y}, \mathbf{p}) = [\![1 + [\alpha_o + \Sigma_i\, \alpha_i \cdot \mathbf{p}_i(\lambda) + \tfrac{1}{2}\Sigma_i \Sigma_j \beta_{ij} \cdot \mathbf{p}_i(\lambda) \cdot \mathbf{p}_j(\lambda)]]\!]^{1/2} \cdot \mathbf{y}(\alpha_y + \tfrac{1}{2}\beta_{yy} \cdot \ln \mathbf{y} + \Sigma_i \gamma_i \cdot \ln \mathbf{p}_i), \tag{5.32}$$

$$\text{where } p_i(\lambda) = (p_i^{\lambda/2} - 1) / (\lambda/2), \text{ for } \lambda \neq 0$$
$$p_i(\lambda) = \ln p_i, \qquad \text{for } 1 = 0,$$

with assumptions:

symmetry $\beta_{ij} = \beta_{ji}$

linear homogeneity in prices

$$\Sigma_i \alpha_i = 1 + \lambda \cdot \alpha_0$$
$$\Sigma_j \beta_{ij} = [\lambda/2] \cdot \alpha_i$$
$$\Sigma_i \gamma_i = 0.$$

This leaves the generalized function of

$$c(y, p) = (2/\lambda \cdot \Sigma_i \Sigma_j \beta_{ij} \cdot p_i^{\lambda/2} \cdot p_j^{\lambda/2})^{1/\lambda} \cdot y(\alpha_y + \tfrac{1}{2}\beta_{yy} \cdot \ln y + \Sigma_i \gamma_i \cdot \ln p_i) . \quad (5.33)$$

In estimation, the value derived for λ can be used to identify alternative functional forms, since if $\lambda = 2$, then the generalized BM yields a nonhomothetic generalized square root quadratic model:

$$c(y, p) = (\Sigma_i \Sigma_j \beta_{ij} \cdot p_i \cdot p_j) \cdot y(\alpha_y + \tfrac{1}{2}\beta_{yy} \cdot \ln y + \Sigma_i \gamma_i \cdot \ln p_i) ; \quad (5.34)$$

if $\lambda = 1$, then the generalized BC yields a nonhomothetic generalized Leontief model:

$$c(y, p) = 2 \cdot (\Sigma_i \Sigma_j \beta_{ij} \cdot p_i^{1/2} \cdot p_j^{1/2}) \cdot y(\alpha_y + \tfrac{1}{2}\beta_{yy} \cdot \ln y + \Sigma_i \gamma_i \cdot \ln p_i); \quad (5.35)$$

if $\lambda \rightarrow 0$, then the generalized BC yields a nonhomothetic translog:

$$\ln c(y, p) = \alpha_0 + \Sigma_i \alpha_i \cdot \ln p_i + \tfrac{1}{2}\Sigma_i \Sigma_j \beta_{ij} \cdot \ln p_i \cdot \ln p_j + \alpha_y \cdot \ln y + \tfrac{1}{2}\beta_{yy} \cdot (\ln y)^2 + \Sigma_i \gamma_i \cdot \ln p_i \cdot \ln y . \quad (5.36)$$

Thus, the analyst is provided with a number of interesting paths to pursue.

NOTES

1. Some of the material presented in this chapter is derived from Turnovsky, Folie, and Ulph (1982) and Turnovsky and Donnelly (1984), along with the various supporting working papers cited, and is quoted with the written permission of the director of the Centre for Resource and Environmental Studies and the editor of the *Journal of Business and Economic Statistics.*

2. See Phillips (1955) for a discussion of the contemporary relevance of Quesnay's work.

3. Returns to scale are defined in terms of degree of homogeneity of the particular function, which is determined by the total differential of the production function. Thus,

$$dP' = (\partial P'/\partial L) \cdot dL + (\partial P'/\partial C) \cdot dC.$$

If $dC = dL = 1$, then $dP' = \text{MPPL} + \text{MPPC}$, where MPPL is $\partial P'/\partial L$ and MPPC is the marginal physical product of capital $\partial P'/\partial C$. In this case, if both inputs change by any particular percentage, then output changes by that same percentage.

4. For a discussion of the regularity conditions in general, see Diewert (1974).

5. The industry sectors analyzed in the energy submodel are (1) treatment of nonmetal mine and quarry products; (2) bricks, pottery, and glass; (3) chemicals, dyes, explosives, paints, etc.; (4) industrial metals, machines, and conveyances; (5) precious metals, jewelry, etc.; (6) textiles and textile goods (not dress); (7) skins and leather (not clothing or footwear); (8) joinery, etc.; (11) wood furniture, bedding, etc.; (12) paper stationery, printing, bookbinding, etc.; (13) rubber; (15) miscellaneous products; and (16) heat, light, and power. Industry number 14, "musical instruments," was omitted from analysis because it was a "small energy user" (Duncan and Binswanger 1976, p. 294, Table 1).

6. For some of the earliest CD work done on Australian data, see Handsaker and Douglas (1937, 1938).

7. In the estimation of industry cost functions, Duncan and Binswanger (1974) analyze only 13 sectors, for some reason omitting the precious metals and jewelry industry.

8. This occurs in 10 of 23 estimated own-price elasticities.

9. The sectors are skins and leather, sawmills, wood furniture, rubber, and miscellaneous products.

10. The statistical results are derived from the TFU (1978) Centre for Resources and Environmental Studies CRES working Paper No. R/WP31, and these and the modeling data are reproduced with the written permission of Professor S. Harris, Director of the Centre for Resource and Environmental Studies at that time.

11. M. Turnovsky provided valuable research assistance, and is to be credited with the construction of the data for this model and for the iron and steel model.

12. Portions of this section are reproduced from Turnovsky and Donnelly (1984) with permission of the editor of the *Journal of Business and Economic Statistics* and the managing director of the American Statistical Association.

13. A tonne has a mass equal to 1,000 kilograms (a kilogram equals approximately 2.2046 pounds); thus, it is roughly 10 percent more than a short ton (2,000 pounds) and approximately 0.984 percent of a long ton (2,240 pounds).

14. The results obtained from the other models estimated appear in Turnovsky and Donnelly (1982).

15. Under the accounting identity where the value shares sum to 1, the symmetry and linear homogeneity conditions are equivalent (Christensen, Jorgenson, and Lau 1973).

16. Also, material is weakly separable in that model at the 1 percent significance level. The existence of a weakly separable material input allows the estimation of the KLE model (see Turnovsky and Donnelly 1982).

17. To test the validity of an aggregate labor index, five- and four-input models are estimated, including both administrative employees and production workers. Unfortunately, since most of the functions that include both forms of labor are not well behaved, it thus proves futile to attempt to test the separability restrictions formally.

18. An alternative explanation offered by Martin and Selowsky (1984) and Solow (1984) is that changes in relative energy prices, when combined with distinctly different energy/factor ratios among outputs, may affect compositional changes in those outputs, and thereby the measure of any aggregate elasticity of substitution. Thus, apparent substitution in production may in fact reflect changes in the composition of output. As has been pointed out already, in the case of the iron and steel industry, finished products are more energy intensive than are semifinished products. When the need arises for maintaining continual blast furnace production, changes in relative factor prices and the overall state of the economy will influence the composition of the output of the industry. To test such a hypothesis empirically would require much more detailed microeconomic data on the industry than are currently available. Analysis of this output substitution was attempted using export data, but to no avail.

19. This result agrees with several other studies that tend to confirm the existence of labor-saving technical change [see Sato and Suzawa (1983, p. 53) for several references].

20. Models with linear and logarithmic time trends are also evaluated in Turnovsky and Donnelly (1982).

21. As substitutability is also obtained in the aggregate manufacturing model, the manner in which capital is measured does not seem to have any particular impact on this result.

22. A similar conclusion is reached in the TFU aggregate manufacturing model.

23. This is by the Denny and Fuss form (1977) of weak approximate separability and by the Hicksian aggregation condition.

REFERENCES

Allen, R. G. D. 1938. *Mathematical Analysis for Economists.* London: Macmillan.
Arrow, K. J., H. B. Chenery, B. S. Minhas, and R. M. Solow. 1961. "Capital-

Labor Substitution and Economic Efficiency." *Review of Economics and Statistics,* 43:225–50.

Australian Department of Home Affairs and Environment. 1983. *The Cost of Pollution Abatement: The Australian Integrated Steel Industry.* Prepared for the Australian Environment Council. Report No. 13. Canberra: Australian Government Publishing Service.

Australian Industries Assistance Commission. 1983. *Certain Iron and Steel Products and Certain Alloy Steel Products.* Report No. 321. Canberra: Australian Government Publishing Service.

———. 1980. *Iron and Steel Industry.* Report No. 249. Canberra: Australian Government Publishing Service.

Barnett, W. A., Y. W. Lee, and M. Wolfe. 1984. "The Three Dimensional Global Properties of the Minflex Laurent, Generalized Leontief, and Translog Flexible Functional Forms." Austin: University of Texas. Mimeo.

Berndt, E. R., and L. R. Christensen. 1974. "Testing for the Existence of Consistent Aggregate Index of Labor Inputs." *American Economic Review,* 64:391–404.

———. 1973a. "The Translog Function and the Substitution of Equipment, Structures, and Labor in U.S. Manufacturing 1929–68." *Journal of Econometrics,* 1:81–114.

———. 1973b. "The Internal Structure of Functional Relationships: Separability, Substitution and Aggregations." *Review of Economic Studies,* 50:403–10.

Berndt, E. R., and M. S. Khaled. 1979. "Parametric Productivity Measurement and Choice Among Flexible Functional Forms." *Journal of Political Economy,* 87:1220–45.

Berndt, E. R., and C. M. White. 1979. "Income Redistribution and Employment Effects of Rising Energy Prices." Department of Economics Resources Paper No. 30. Vancouver: University of British Columbia.

Berndt, E. R., and D. O. Wood. 1981. "Engineering and Econometric Interpretations of Energy-Capital Complementarity: Reply and Further Results." *American Economic Review,* 71:1105–12.

———. 1975. "Technology, Prices and the Derived Demand for Energy." *Review of Economics and Statistics,* 57:259–68.

Binswanger, H. P. 1974. "The Measurement of Technical Change Biases with Many Factors of Production." *American Economic Review,* 64:964–76.

Blackorby, C., D. Primont, and R. R. Russell. 1977. "On Testing Separability Restrictions with Flexible Functional Forms." *Journal of Econometrics,* 5:195–209.

Brain, P., and G. S. Schuyers. 1981. *Energy and the Australian Economy.* Melbourne: Longman Cheshire.

Burgess, D. F. 1975. "Duality Theory and Pitfalls in the Specification of Technologies." *Journal of Econometrics,* 3:105–21.

Christensen, L. R., D. Cummings, and P. E. Schoech. 1983. "Econometric Estimation of Scale Economies in Telecommunications." In *Economic Analysis of Telecommunications: Theory and Applications,* edited by L. Courville, A. DeFontenay, and R. Dobell, pp. 27–53. New York: North-Holland.

Christensen, L. R., and D. W. Jorgenson. 1969. "The Measurement of U.S. Real Capital Input, 1929–67." *Review of Income and Wealth,* 15:293–320.

Christensen, L. R., D. W. Jorgenson, and L. J. Lau. 1973. "Transcendental Logarithmic Production Functions." *Review of Economics and Statistics,* 55:28–45.

Cobb, C. W., and P. H. Douglas. 1928. "A Theory of Production." *American Economic Review,* 18(suppl):139–65.

Cramer, J. S. 1969. *Empirical Econometrics,* Amsterdam: North-Holland.

Denny, M., and M. A. Fuss. 1977. "The Use of Approximate Analysis to Test for Separability and the Existence of Consistent Aggregates." *American Economic Review,* 67:404–18.

Denny, M., and D. May. 1977. "The Existence of a Real Value-Added Function in the Canadian Manufacturing Sector." *Journal of Econometrics,* 5:55–69.

Diewert, W. E. 1976. "Exact and Superlative Index Numbers," *Journal of Econometrics,* 4:115–47.

——. 1978. "Superlative Index Numbers and Consistent Aggregation," *Econometrica,* 46:883–900.

Diewert, W. E. 1975. "Hicks Aggregation Theorem and the Existence of a Real Value-Added Function." In *Production Economics: A Dual Approach to Theory and Applications,* edited by M. A. Fuss and D. McFadden, pp. 17–51. Amsterdam: North-Holland.

——. 1974. "Applications of Duality Theory." In *Frontiers of Quantitative Economics, Vol. 2,* edited by M. D. Intriligator and D. A. Kendrick, Amsterdam: North-Holland.

——. 1971. "An Application of the Shephard Duality Theorem: A Generalized Leontief Production Function." *Journal of Political Economy,* 79:481–507.

Diewert, W. E., and T. J. Wales. 1984. "Flexible Functional Forms and Global Curvature Conditions." Technical Working Paper No. 40. Cambridge, MA: National Bureau of Economic Research.

Douglas, P. H. 1948. "Are There Laws of Production?" *American Economic Review.* 38:1–41.

Duncan, R. C., and H. P. Binswanger. 1976. "Energy Sources: Substitutability and Biases in Australia." *Australian Economic Papers,* 15:289–301.

——. 1974a. "Factor Biases and Induced Innovation in Australian Manufacturing Industries." Mimeo draft.

——. 1974b. "Production Parameters in Australian Manufacturing Industries." Mimeo draft.

Field, B. C., and C. Grebenstein. 1980. "Capital-Energy Substitutions in U.S. Manufacturing." *Review of Economics and Statistics,* 62:207–12.

Fishelson, G., and T. V. Long. 1978. "An International Comparison of Energy and Materials Use in the Iron and Steel Industry." In *International Comparisons of Energy Consumption,* edited by T. Dunkerley. Published for Resources for the Future. Baltimore: Johns Hopkins University Press.

Fuss, M. A. 1977. "The Demand for Energy in Canadian Manufacturing: An Example of the Estimation of Production Structures with Many Inputs." *Journal of Econometrics,* 5:89–116.

Fuss, M. A., D. McFadden, and Y. Mundlak. 1978. "A Survey of Functional Forms in the Economic Analysis of Production." In *Production Economics: A Dual Approach to Theory and Applications, Vol. 1,* edited by M. Fuss and D. McFadden, pp. 219–68. Amsterdam: North-Holland.

Gallant, A. R. 1981. "On the Bias in Flexible Functional Forms and an Essentially Unbiased Form: The Fourier Flexible Form." *Journal of Econometrics,* 15:211–45.

Gibbons, J. 1984. "Capital-Energy Substitution in the Long Run." *Energy Journal,* 5:109–18.

Griffin, J. M. 1981a. "The Energy-Capital Reconciliation Attempts." In *Modeling and Measuring Natural Resource Substitution,* edited by E. R. Berndt and B. C. Field, pp. 70–80. Cambridge, MA: MIT Press.

——. 1981b. "Engineering and Econometric Interpretations of Energy-Capital Complementarity: Comment." *American Economic Review,* 71:1100–104.

——. 1977a. "Inter-fuel Substitution Possibilities: A Translog Application to Intercountry Data." *International Economic Review,* 18:755–70.

——. 1977b. "Long-Run Production Modeling with Pseudo-data: Electric Power Generation." *Bell Journal of Economics,* 8:112–27.

Griffin, J. M., and P. R. Gregory. 1976. "An Intercountry Translog Model of Energy Substitution Responses." *American Economic Review,* 66:845–57.

Halvorsen, R., and J. Ford. 1979. "Substitution Among Energy, Capital and Labor Inputs in U.S. Manufacturing." In *Advances in the Economics of Energy and Resources: Structures of Energy Markets, Vol. 1,* edited by R. S. Pindyck, pp. 51–75. Greenwich, CT: JAI Press.

Handsaker, M. L., and P. H. Douglas. 1938. "The Theory of Marginal Productivity Tested by Data Manufacturing in Victoria, II." *Quarterly Journal of Economics,* 52:215–54.

——. 1937. "The Theory of Marginal Productivity Tested by Data Manufacturing in Victoria, I." *Quarterly Journal of Economics,* 52:1–36.

Hawkins, R. G. 1978. "A Vintage Model of the Demand for Energy and Employment in Australian Manufacturing Industry." *Review of Economic Studies,* 45:479–94.

——. 1977. "Factor Demands and the Production Function in Selected Australian Manufacturing Industries." *Australian Economic Papers,* 16:97–111.

Hicks, J. R. 1965. *Capital and Growth,* London: Oxford University Press.

Hogan, W. W. 1979. "Capital Energy Complementarity in Aggregate Energy-Economic Analysis." Energy and Environmental Policy Center. John F. Kennedy School of Government (E-79-03). Cambridge, MA: Harvard University.

Hogan, W. W., and A. S. Manne. 1979. "Energy-Economy Interactions: The Fable of the Elephant and the Rabbit." In *Advances in the Economics of Energy and Resources: Structures of Energy Markets, Vol. 1,* edited by R. S. Pindyck, pp. 7–17. Greenwich, CT: JAI Press.

Hudson, E. A., and D. W. Jorgenson. 1974. "U.S. Energy Policy and Economic Growth, 1975–2000." *Bell Journal of Economics and Management Science,* 5:461–514.

Humphrey, D. B., and J. R. Moroney. 1975. "Substitution Among Capital, labor and Natural Resource Products in American Manufacturing." *Journal of Political Economy,* 83:57–82.

Kang, H., and G. M. Brown. 1981. "Partial and Full Elasticities of Substitution and the Energy-Capital Complementarity Controversy." In *Modelling and Measuring Natural Resource Substitution,* edited by E. R. Berndt and B. C. Field, pp. 81–89. Cambridge, MA: MIT Press.

Kmenta, J., and R. F. Gilbert. 1968. "Small Sample Properties of Alternative Estimators of Seemingly Unrelated Regressions." *Journal of the American Statistical Association,* 63:1180–200.

Kopp, R. J., and V. K. Smith. 1981. "Measuring the Prospects for Resource Substitution Under Input and Technology Aggregations." In *Modelling and Measuring Natural Resource Substitution,* edited by E. R. Berndt and B. C. Field, pp. 145–73. Cambridge, MA: MIT Press.

Kuh, E. 1976. "Some Preliminary observations on the Stability of the Translog Production Function." Mimeo draft.

Lau, L. J. 1978. "Applications of Profit Functions," In *Production Economics: A Dual Approach to Theory and Applications.* Volume 1, edited by M. Fuss and D. McFadden, pp. 133–216. Amsterdam: North-Holland.

Leontief, W. W. 1947. "Introduction to a Theory of the Internal Structure of Functional Relationships." *Econometrica,* 15:361–73.

Magnus, J. R. 1979. "Substitution Between Energy and Non-energy Inputs in the Netherlands 1950–9176." *International Economic Review,* 20:465–84.

Magnus, J. R., and A. Woodland. 1980. "Interfuel Substitution Possibilities in Dutch Manufacturing: A Multivariate Error Component Approach." Department of Economics Discussion Paper No. 8-39. Vancouver: University of British Columbia.

Martin, R., and M. Selowsky. 1984. "Energy Prices, Substitution and Optimal Borrowing in the Short Run: An Analysis of Adjustment in Oil Importing Developing Countries." *Journal of Development Economics,* 14:331–50.

Phillips, A. 1955. "The *Tableau Economique* as a Simple Leontief Model." *Quarterly Journal of Economics,* 69:137–44.

Pindyck, R. S. 1977. "Interfuel Substitution and the Industrial Demand for Energy: An International Comparison." Energy Laboratory Working Paper No. MIT EL 77-026WP. Cambridge, MA: Massachusetts Institute of Technology.

Pollak, R. A., R. C. Sickles, and T. J. Wales. 1984. "The CES-Translog: Specification and Estimation of a New Cost Function." *Review of Economics and Statistics,* 66:602–7.

Russell, C. S., and W. J. Vaughan. 1976. *Steel Production: Processes, Products, and Residuals.* Published for Resources for the Future. Baltimore: Johns Hopkins University Press.

Sato, R., and G. S. Suzawa. 1983. *Research and Productivity: Endogenous Technical Change.* Boston: Auburn House Publishing Company.

Simmons, P., and D. Weiserbs. 1979. "Translog Flexible Functional Forms and Associated Demand System." *American Economic Review,* 69:892–901.

Solow, J. L. 1984. "The Composition of Output and Aggregate Capital-Energy Complementarity." Department of Economics, University of Iowa, Working Paper Series No. 84-1, January. Paper presented in the Economics Department Seminar Series. Canberra: Australian National University. Mimeo.

Stapleton, D. C. 1981. "Inferring Long-Term Substitution Possibilities from Cross-Section and Time-Series Data." In *Modelling and Measuring Natural Resource Substitution,* edited by E. R. Berndt and B. C. Field, pp. 93–118. Cambridge, MA: MIT Press.

Theil, H. 1980. *The System-wide Approach to Micro-economics.* Chicago: University of Chicago Press.

Turnovsky, M. H. L., and W. A. Donnelly. 1984. "Energy Substitution, Separability and Technical Progress in the Australian Iron and Steel Industry." *Journal of Business and Economic Statistics,* 2:54–63.

——. 1982. "Energy Substitution and Separability in the Australian Iron and Steel Industry." Centre for Resource and Environmental Studies Paper No. 2. Canberra: Australian National University.

Turnovsky, M. H. L., M. Folie, and A. Ulph. 1982. "Factor Substitutability in Australian Manufacturing with Emphasis on Energy Inputs." *Economic Record,* 58:61–72.

——. 1978. "Factor Substitutability in Australian Manufacturing Emphasising Energy Inputs." Centre for Resource and Environmental Studies Working Paper No. R/WP31. Canberra: Australian National University.

Vinals, J. M. 1984. "Energy-Capital Substitution, Wage Flexibility and Aggregate Output Supply." *European Economic Review,* 26:229–45.

Wales, T. J. 1977. "On the Flexibility of Flexible Functional Forms—An Empirical Approach." *Journal of Econometrics,* 5:183–93.

Wicksell, K. 1916. "Den 'kritiska pukten' i lagen fur jordbrukets aftagande produktivitet." *Ekonomisk Tidskrift,* 10:285–92.

Williams, M., and P. Laumas. 1981. "The Relation Between Energy and Nonenergy Inputs in India's Manufacturing Industries." *Journal of Industrial Economics,* 30:113–22.

Wills, J. 1979. "Technical Change in the U.S. Primary Metals Industry." *Journal of Econometrics,* 10:85–98.

Woodland, A. D. 1978. "On Testing Weak Separability." *Journal of Econometrics,* 8:383–98.

Six

Modeling Directions

OVERVIEW

Modeling as a discipline is in a transitional period with ever-increasing detail seeming to be the direction for the future, but the road has several branches. One branch proposes that emphasis be placed upon understanding the effects of discontinuities of the system rather than assuming continuous functions. Another branch moves away from classical statistics and advocates the use of Bayesian statistical concepts, so as to account explicitly for the preconceived notions of the modeler. Elsewhere, greater engineering detail is being incorporated into the behavioral models, those which have already been discussed. Such process modeling is designed to provide the analyst with more information than is available from the reduced-form models. While these diverse approaches have a similar objective, a greater understanding of the system relationships, each poses rather unique problems for empirical implementation. The first branch requires major theoretical advances, the second algorithmic development, and the last greatly expanded databases. It is evident, however, that the use of modeling is becoming increasingly widespread as computer technology advances, and thus it is a most exciting time for the theoretician, for the empiricist, and for the analyst.

INTRODUCTION

As the previous chapters have illustrated, a modeling strategy should be selected for the particular problem under analysis, and due consideration must be given to the availability of data. It follows from the basic precepts of economic theory that a modeling strategy should be sufficient for the analysis and need not (indeed, should not) by the very definition of a model, exceed the requirements of the particular problem under study. Questions arising from the very specific problems that forecasting entails often may be answered with simple single-equation procedures, but these cannot take into account the implications of general equilibrium analysis and may not be of import to the policy issue under consideration with any exactitude. For example, the forecasting of the demand for electricity must make possible the provision of information on the capital and expenditure streams, which will be needed to augment generating capacity, and on the potential substitution that consumers of electricity might choose; as well, it must take into account alternative sources of power. Thus, a single-equation modeling strategy may not apply. These questions also involve long lead times, but this implies, for example, the administrative lags, environmental impact studies, and other delays that might accrue in the bringing of any eventual capacity "on line." These delays can be decades in order of magnitude. In the instance of a regulated monopoly such as electricity, there must necessarily be caution in regard to reliance solely upon individuals who are within the industry and who might supply biased forecasts of its demand; this is because of the inevitable potential of a conflict of interest. Examples of the problems and inefficiencies resulting from such conflicts are numerous. Either independent analyses should be required, or, as a workable compromise, decision makers should seek to encourage adoption of structured forecasting techniques that utilize explicit assumptions, with the intent always in mind that the results should be suitable for replication by other analysts. Such approaches do not ensure the accuracy of forecasts so developed, but they do produce findings that can be evaluated objectively by allowing for the identifying and highlighting of different relationships and subjectively by providing for the specification of alternative scenarios. Indeed, the forecasts presented for consideration, those conditional upon the scenario assumptions, can be criticized in an analytic framework that provides for a more rational debate of the issues. Then, what is required is an appreciation of the implications of the assumptions underlying the modeling strategy; these issues have been discussed in Chapter 2 in particular, and also in each of the subsequent chapters in regard to individual models. Selection of an appropriate modeling strategy can facilitate policy analysis and thereby assist decision making.

It was widely accepted before the Arab oil embargo (EC1) that that industry would be unable to adapt quickly to any major change in the price of energy, its having become acclimated to decades of declining relative prices. In fact, it was discovered quickly that industries can and do modify their energy usage to an unexpected extent. This was seen to occur when the price of fuels jumped during the late 1970s and as spot shortages of fuels were experienced; these substitutions continued throughout that decade and into the middle 1980s.[1] As long as energy expenditures represent only a small proportion of total operating expenditure, business has little incentive to ensure that it is most efficiently used, since the marginal cost of monitoring energy usage then can easily exceed the marginal return; thus, the "waste" of energy is to be anticipated. The rapidly escalating energy prices immediately after EC1 and at the time of the overthrow of the Shah of Iran (EC2) caused much closer scrutiny of energy usage patterns. It is hoped that the disintegration of the Arab cartel within OPEC in early 1986 and the ensuing "price wars" during the first and second quarters of 1986 (EC3) will not induce a return by the industrialized nations to the pre-1972 relatively profligate use of energy, since improvements in management practices represent only the initial response to changing energy prices, while the longer-term response involves the "retrofitting" of plants (and homes) and the replacement of obsolete with energy-efficient equipment. Changing expectations concerning future price path trajectories can negate the energy efficiency gains already achieved. In energy-intensive manufacturing industries where the in situ capital stock is specifically engineered for low energy prices and is combined with the expectation of continuing decline in those prices, the adjustment process is anticipated to be prolonged and more difficult; but as the production function work emphasizing energy substitution generally demonstrates, this conversion is undertaken. Energy elasticities estimated by the production function model reinforce such an assertion, both in the aggregate manufacturing and in the iron and steel–sector results, and indicate that energy demand has become more elastic over time, albeit still quite inelastic, relatively speaking.[2] The somewhat greater responsiveness indicated in the iron and steel energy elasticities signifies that a disaggregate modeling strategy is essential in that it does not obscure the interfactor and interfuel substitution, these potentials differing among industry sectors. Unfortunately, the paucity of sectoral data precludes many such undertakings. Bohi and Zimmerman (1984) argue in regard to the commercial and industrial sectors that "until observations at a firm level become available, there is little hope for a better understanding of [energy] demand behavior" (p. 148).

This continuing criticism of modelers should be attended to more acutely by all organizations that gather and disseminate statistics. Since most of such data are collected by governments, innovative thinking would be worthwhile among statisticians in any efforts that might be made to provide data of greater utility to analysts and decision makers. But until such a state of affairs pertains, data-related issues serve to constrain to a large extent what can be proposed in terms of future research, because many of the most interesting problems can be tackled only with significantly improved data. The question of energy substitution in industrial use is but one instance; while the translog and other flexible functional forms have provided insights, these research strategies have been successful largely because the models specified to date have imposed minimal data demands upon the modeler. Extensions to the basic "flexible functional form," such as the attempt to construct cost functions, greatly increase the data requirements, in that there are additional parameters to be defined and estimated, and such research necessitates the augmentation of the time-series data with cross-sectional data. The efficacy of these extensions has been demonstrated in several isolated studies, but they have little chance of broad implementation in the absence of a more disaggregate database. The attempts to replace the reduced form models with structural models that provide further insights will continue, since, as Bohi and Zimmerman (1984) in their review of energy models observe, "As a group, the structural models perform better than the reduced form models; they provide more information, produce statistically significant results, and exhibit greater consistency across the separate studies" (p. 150).

The emphasis in regard to directions for future research remains clear: Models that incorporate greater definition in the sense of their structural elaboration and precision are required. Whether this transition necessitates a development of comprehensive engineering/economic end-use models or not, or whether this objective can be accomplished instead by an acceptance of the importance of the influences of the stock of energy-using appliances and the incorporation of that consideration into the model structure or not, remains open to debate. These issues will be addressed below, but other generic modeling strategies will be considered first.

OTHER MODELING STRATEGIES

Alternative modeling strategies that have yet to be used extensively in policy analysis include Markov processes, Bayesian statistics, and catastrophe theory. The potential of these methodologies needs must be examined in regard to the socioeconomic import of each. The preceding models have assumed continuous functions for which first and second derivatives exist; and while this assumption is convenient and facilitates the derivation of solutions, it may not be applicable to a number of contemporary questions. For instance, do relationships exhibit discontinuities that are important when attempting to make policy? Berlinski (1976) argues, for example:

> The properties that mark [Rene] Thom's [catastrophe] theory and that are notably absent from classical physics are *discontinuity* and *divergence*. By discontinuity, [E.C.] Zeeman means that in catastrophe theory a slow passage through a space of causes is liable to produce an abrupt change in a corresponding space of effects. . . . A theory is divergent, in contrast, if small changes in its initial conditions are not preserved, with subsequent states spreading apart like a perpetually expanding accordion. (p. 105)

The prospect that catastrophe theory can be used to model phenomena exhibiting such system behavior is an exciting one for social scientists as well as philosophers, since the traditional assumption of continuity does not accommodate invention, innovation, or any drastic changes in the parameters of the system. But philosophical problems are posed by such an idea in regard to the social sciences in particular for if such occurrences can be foreseen, such foresight may, in and by itself, inevitably alter outcomes and consequences and thereby vitiate the potential for the theory to be adequately tested.

Bayesian statistical theory is based upon the premise that prior knowledge about a system's parameters is adjusted by the data in order to provide posterior estimates of those parameters. This approach is explicitly subjective, assumes learning, is difficult to implement, and therefore has not enjoyed widespread application.

An alternative modeling strategy that does not presume learning is a Markov process, of which we have seen special cases—the first-order AR model and random walk are two examples—of conditional transition probabilities.[3] The Markov process can be combined with Bayesian decision theory to yield an optimal control modeling strategy wherein the decision maker may sample, at some cost, in order to update the prior distribution before selecting a policy alternative. Once the next state of the process has been achieved, the sampling can resume, presumably continuing only until the prior distribution becomes sufficiently acceptable within some given probability boundaries, and such that a given policy may be adopted and then followed indefinitely without further change. Incorporating principles of Bayesian statistics into the Markov process provides for the systematic inclusion of the analyst's prior beliefs concerning parameter values into the latter procedure and allows for the modification of these, based upon the sample data.

The traditional statistical approach implicitly assumes a diffuse prior or "ignorance prior"; that is to say, classic statistics does not pose a prior probability density function for the unknown parameters and merely derives point estimates for these unknowns as if these were developed from repeated sampling from some population. Conversely, Bayesian methods do not presuppose a sampling process, but rather derive the posterior probability density function that incorporates extraneous information, the prior probability density function, into the analysis in a consistent fashion. The formalization of the prior beliefs in conjunction with the mechanical difficulty of determining the posterior probability density function poses formidable tasks, and therefore Bayesian procedures are not widely used in policy analysis applications.

In catastrophe theory the movement of the system from one state to another is determined by the behavioral convention adopted. The extremes on the continuum of possible conventions are the "delay" and "Maxwell" conventions. The former assumes that the system's state remains unchanged until the state disappears, whereas the latter requires the achievement of a stable minimum. The attention of the analyst is directed toward the "set of points in the space of control parameters at which transition occurs from one local minimum to another . . . the *bifurcation set*" (Gilmore 1981, p. 144). Unless a specific convention for transition is defined, then the solution may not be deterministic but may necessitate a resolution through fuzzy set theory in order to identify the critical values for the control parameters in the neighborhood of a jump from one state to another.[4] A problem suitable for the application of catastrophe theory can be identified by considering the following characteristics of the system: (1) the existence of more than two possible states (modality); (2) an unstable equilibrium (inaccessibility); (3) a large change in the value of the state variable, which may unexpectedly occur (sudden jumps); (4) small perturbations in control parameters occasionally precipitating large changes in state values (divergence); and (5) systems not being reversible (hysteresis). Fortunately, most topics suitable for inquiry by economists and other social scientists appear to behave in a smooth and regular manner; however, whenever important policy issues arise, these inevitably are in the context of some crisis and the attendant questions. The crisis being defined by some unexpected occurrence, the question as to whether the response to this unanticipated change remains similar to responses to the more ordinary changes experienced in the past may be more amenable to manipulation through the techniques of catastrophe theory. Therefore, this area of research deserves more of the attention of individuals engaged in social policy analysis than it has received to date.

END-USE AND MICROSIMULATION MODELS

In considering the derived demand nature of gasoline consumption in Chapter 4, the prospect of constructing an end-use model in terms of the characteristics and usage of the motor vehicle stock is discussed. Such modeling strategy requires information on the vintages of vehicles, their efficiencies, and how drivers respond to changing economic conditions–all factors important in their effects on the level of gasoline demand. Studies such as those of Train (1986), Lin, Botsas, and Moore (1985), Wheaton (1982), Acton, Mitchell, and Mowill (1976), Acton, Mitchell, and Sohlberg (1980), Archibald and Gillingham (1980), and Sweeney (1978) attempt to quantify these influences, but the constraints imposed owing to scanty data may be detected in this line of research.[5] Because of data considerations, most gasoline modeling is typified by reduced form specifications. Similar issues arise with respect to household nontransport demand for energy, wherein the types of appliances in use, their capacities, their efficiencies, and their "market penetration ratios"[6] serve to establish upper limits on energy consumption [Garbacz (1983, 1984) utilizes the National Interim Energy Consumption Survey (NIECS) (DOE 1980a, 1980b, 1980c) to incorporate an appliance index equation along with equations for demand and price in order to capture some of the effects of appliances without attempting an actual "vintage capital" model]. Engineering information is valuable in providing an understanding of the technical aspects of energy-using appliances and their conversion efficiencies, and it can provide details about the thermal characteristics of various types of housing construction. Surveys may be utilized to ascertain the quantities and distributions of appliances, and the combination of the engineering and survey data may be incorporated into an overall modeling strategy.

This microapproach to energy analysis, one that begins from specific end-use applications or from the individual consumer or household and that then aggregates upward to the total, the "bottom-up" approach, contrasts with the models discussed so far, since all of the former rely upon aggregate reduced form structures that ignore the microeconomic detail. Whereas consistency is easily ensured with the "top-down" models—and this is a feature that the bottom-up approach does not exhibit—their structure prevents the analysis of many policy-oriented questions, such as the impact of mandated vehicle mileage standards and required appliance efficiency labeling. The impetus for the development of the Sweeney model (1978) at DOE relates to impacts of the establishment of mileage standards, Train's scenario analyses of gasoline demand for the California Energy Commission (CEC) (1986), the work on residential energy consumption at Oak Ridge National Laboratory (ORNL), and the Electric Power Research Institute (EPRI). The last analyzes questions relating to the energy substitution potential of the household. Abstracting temporarily from the inherent data problems that accrue from such bottom-up models, there are questions as to how engineering and economic models are to be integrated to provide a policy model that responds to those variables over which some degree of direct or indirect control can be exerted. The direct controls that could be considered would include excise taxes, usage fees, mileage standards, thermal efficiency standards, and moratoriums on hookups, this last being intended to encourage conservation of a particular source of energy. The indirect procedures might include labeling requirements and the use of advertising as a form of "moral suasion" directed toward an overall conservation "ethos." The efficacy of such alternatives cannot always be evaluated by the top-down models, which, by their very nature, disregard engineering characteristics; but unless such details as these, for example, can be incorporated usefully into economic models, such aforementioned questions remain intractable in respect to attempts at clarification by the bottom-up models also.

The ORNL end-use and EPRI microsimulation models are of interest because they provide insight into how some of these issues may be eliminated in practice. While these models have similarities, their differences are nevertheless important in determining the appropriate applications of each. The discussion here will not be comprehensive, since both the ORNL Residential Energy Consumption (REC) model and the EPRI Residential End-Use Energy Planning System (REEPS) are well documented elsewhere.[7] The theoretical structure of a microsimulation model of household energy use exploits the concept of the production function as applied to the household. It begins with the premise that the household provides utility to the inhabitants in the form of shelter, succor, convenience, and comfort, and with the recognition that household production can be achieved with differing mixtures of the factor inputs: land, labor, capital (the land factor includes both materials and energy). Technological constraints define the production function, its position, and its shape; the isoquants, contours of equal output levels, provide the marginal rates of substitution between the inputs that must be equated to their relative prices in order to achieve an optimal mix of the inputs. If such equality is not satisfied, then a reallocation among the inputs can increase total output with no corresponding increase in expenditures, so the household might possibly improve its situation at no additional cost by changing its mix of inputs. The simulation model then provides for selection among the various alternative combinations of inputs that will provide the specified level of utility to the household. Considering the question of residential energy usage in this context provides a methodology by which engineering information may be combined with economic relationships, the former helping to define the production function in terms of the technical transformation possibilities and the latter incorporating behavioral response factors. For example, space-heating energy consumption is determined by technological constraints (appliance and building efficiencies), behavioral responses (thermostat settings), and external factors (climatic conditions). The top-down models can incorporate only implicitly those technical aspects of space-heating energy demand, whereas the bottom-up energy model takes the form of the accounting identity

$$\mathbf{E}_t \equiv \mathbf{S}_t \cdot \mathbf{F}_t \cdot \mathbf{U}_t \cdot \mathbf{X}_t \,, \tag{6.1}$$

where **E** = energy consumption
S = stock of appliances
F = efficiency factor
U = utilization rate
X = exogenous factors.

Each of these components of energy demand can be modeled separately. Increments in the stock of appliances depend upon the "new" investment decision that may be posited to be determined by the life cycle cost of the appliance (purchase price plus expected operating costs), socioeconomic conditions (neighborhood and income), and exogenous factors (natural gas hookup moratoriums). "Replacement" decisions may also depend upon the existing portfolio of appliances; for example, use of oil space heating may bias decisions toward oil or electric water heating and away from natural gas. The efficiencies of the appliances can be specified in technological terms, but these too can respond to changing economic conditions; for example, when changing consumer demand increases the "economic rent" to a producer of an efficient small car that relies upon a newly developed technology, these decisions will spur other manufacturers to emulate the competition in order to maintain their market shares.[8] Such "rent-seeking" behavior will occasion improvements in efficiency.[9] The utilization of existing appliances may be expected to respond to changing operating costs, but such response is muted by the (usually) long life of these devices, and little actual information is available on the consequence of such responses. Finally, the exogenous factors, for example, climate, will exert both a direct effect on energy consumption and an indirect one. Its influence on the evolution of the characteristics of appliances and their efficiencies, and a colder than normal winter, will increase space-heating fuel consumption but will not necessarily induce increased thermal efficiency in buildings, whereas houses in northern climates will generally have better insulation than those in more moderate ones. Therefore, some of these exogenous factors can be characterized by way of technological parameters and others by way of behavioral parameters. How, then, are these aspects actually modeled?

The ORNL REC models end-use energy consumption for all consumers in a given region, while the EPRI REEPS simulates individual household behavior, providing regional totals by tabulation. Because of its structure, the REC model contains less detail on the capacity, efficiency, and vintage of energy-using appliances than is contained in REEPS; the advantage gained in reduced data requirements of the former serves to limit its effectiveness in accurately reflecting the heterogeneity present among categories of households. Even so, the REC model is a large one: "A rough estimate [of the REC model] is that 500 behavioral and technological parameters, plus approximately 450 base-year data points plus inputs . . . determined by engineering or econometric study or by substantive judgmental assumption [are required]" (Cowing and McFadden 1984, p. 46).

As can be appreciated, the end-use model strategy depends for its success upon very much more information than do the reduced form models. While the REC model, originally constructed to operate at the national level and subsequently regionalized, provides forecasts for "5 fuel types, 3 dwelling types, and 8 end uses, for up to 30 years" (Cowing and McFadden 1984, p. 48), the microsimulation approach of REEPS requires even more data detail, because it models the life cycle appliance decisions for various representative categories of households on a regional basis. Initially, REEPS modeled the 10 DOE regions, and currently it models the 48 contiguous states.

Information on the household stock of appliances comes from a 1975 DOE-commissioned survey by the Washington Center for Metropolitan Studies, which questioned 3,149 households on dwelling characteristics and energy use patterns (Cowing and McFadden 1984, p. 110). The NIECS (DOE 1980a, 1980b, 1980c) clustered random samples of 4,081 households over the monthly period extending from April 1978 to March 1979 and the Pacific Northwest survey prepared by the Bonneville Power Administration (Dubin 1985, pp. 195–215, xviii) provides additional information; but these two studies, along with the one formerly mentioned, remain the principal sources, and are not the best in regard to their currency and comprehensiveness. These surveys are utilized by REEPS to construct contingency tables on housing stock and appliance characteristics that relate to space heating, water heating, air conditioning, cooking, and dishwashing energy uses, which represent about 80 percent of household nontransport consumption. The tables provide the basis for simulating the individual households in "essentially [a] complex accounting system" tabulation framework (Cowing and McFadden 1984, p. 17) that is balanced to observed consumption levels and then used to prepare forecasts. Whereas REEPS considers the determinants of the elements in Equation 6.1 at the level of the individual household and aggregates up, REC assumes that those determinants can be consistently modeled at the aggregate level. Inevitably, aggregation bias will ensue from the latter assumption, unless either the efficiency factor of the utilization rates are constants in their aggregate representations; this is an unlikely possibility in that the socioeconomic and geographic factors are diverse in regard to households. Conversely, REEPS allows the entire portfolio of energy-using appliances in the household to adjust, within certain limits, to a changing economic environment, but at the considerable expense imposed by the level of detail required; add to that the potential for biases introduced because of sampling errors and also because of the assumptions in regard to behavioral response that follow from the lack of sufficient survey data with which to develop statistical estimates. When considering either of these approaches as an alternative to reduced form modeling, the analyst must weigh the increased complexity of the model against the increased costs of development, maintenance, and operation of the model, and must do so while attempting to assess whether or not the detail to be furnished represents useful information. Because of all the assumptions and judgments included in both the ORNL REC model and the EPRI REEPS, this detail may be spurious.[10] Both models, perhaps, should be evaluated not on their abilities to provide accurate baseline forecasts outside the observation set, being something difficult to assess because of the limited data, but rather on the insights with regard to the relative impacts of alternative policy strategies, since, in validating any model, calibration to replicate as accurately as possible the history is often of concern.

Train (1986) defines the set of "qualitative choice" models as ones addressing questions of selecting among a finite set of mutually exclusive, exhaustive alternatives such as a "worker's choice of mode for travel to work (with the alternatives being auto, transit, walk, etc.); a household's choice of make and model of automobile (VW Rabbit, Olds, Omega, etc.)" (p. 4). Such situations yield a "discrete choice" versus a "continuous choice" wherein the question is "how much" or "how many"; for instance, How many miles should one drive a vehicle? The qualitative choice models are amenable to the logit, probit, or generalized extreme value probability distributions. In the instance of the demand for gasoline, it is argued that the first choice made by the consumer is a discrete one, that is, the option in regard to the mode of travel, and the second choice is a continuous one, that is, the number of vehicle miles traveled. Dubin and McFadden (1984) and Heckman (1978, 1981) provide methods for specifying and estimating these types of models involving a discrete/continuous choice situation.

Approaching the question of gasoline demand from this perspective, Train defines and simulates 25 classes of vehicles covering 10 vintages, which means, in the case of two-vehicle households, over 30,000 pairs of class/vintage to be modeled; this feat is accomplished for CEC using a national dataset with many simplifying and scaling assumptions, so as to confine the problem within tractable limits. The class/vintage selection is determined by the price of the vehicle, its operating cost per mile, and some size, vintage, and type characteristics using a multinomial logit function fitted by a maximum likelihood estimation procedure. The solution of this first module, the discrete choice component, specifies the efficiency of the vehicle stock, because classes of vehicles are essentially defined on this criterion; such classes would be gas minicar, diesel minicar, gas subcompact, diesel subcompact, small gas vans, small diesel vans, large gas vans, large diesel vans, etc.[11] The second module involves determining the vehicle miles traveled by vehicle class, a function of various economic (gasoline price, household income, vehicle operating expenses) and demographic (number of workers in the household, household size, urban or rural location, broad geographic region) factors. A further distinction is made between intracity/nonintracity work/nonwork trips.

The simulation of these components for base case and alternative scenarios provides the analyst with the effects that might be anticipated as a result of a particular policy option as opposed to other choices that might be suitable. But as with the REC and REEPS models, this one suffers from lack of data along with the computational burdens imposed, and these limitations must be kept in mind. But the results of the scenarios developed for CEC suggest that while gasoline consumption is expected to decline between now and the turn of the century, this is *not* attributable to either an overall change in consumer behavior or a reduction in the use of private vehicles or a switch to alternative fuels (diesel, methanol, liquid petroleum gas, or electric), but rather is "nearly entirely due to the increased fuel efficiency that the Commission projects for gasoline vehicles" (Train 1986, p. 190).

The caveat that the author mentions as relating to these results is that they depend upon the scenario specifications as to the vehicle characteristics, household demographics, and fuel price trajectories provided by CEC; that is, these are conditional forecasts. It can be seen that these types of microsimulation models provide interesting insights for the policymaker, ones that may well serve to modify preconceived notions such as the conventional wisdom that alternatively fueled vehicles will displace the gasoline-powered automobile. Whether their usefulness in terms of predictive accuracy over the simpler reduced form model (which suggests a long-term inelastic response to changing regimes of energy prices and, by implication, provides similar conclusions about market potential for alternative fuel types) remains to be demonstrated.

MODELING THEORY

The precept of parsimony known as "Occam's razor"[12] is a fundamental rule for modelers in all of the sciences, social or physical. The description of Adam Smith, quoted in Chapter 1, accurately presents the essential elements of the conflict between the Ptolemaic and Copernican theories of cosmology, the former involving a complex geocentric system of eccentric spheres and epicycles and the latter a simpler, more elegant heliocentric representation. In addition, the Copernican model provides better predictions. The absence of the need for greater accuracy as well as the deficiency in the data are two reasons, among others, that the geocentric model took so long to be displaced, given that the basic description is attributed to Aristarchus by Johnannes Kepler himself. In his notes to *Mysterium Cosmographicum,* he advises

> not [to] conduct yourselves as enemies of the most excellent display of divine works. . . . as Aristarchus says that the Sun is the center both of the five planetary orbits and also of the sixth, which carries the Earth, so that with the Sun at rest the Earth is carried round the Sun . . . and the motion both of the Earth and of the other five is controlled from the single source of the solar body. (Kepler 1981 reprint, pp. 207, 209)

But Kepler could never have completed his theory without the meticulously collected planetary observations of Tycho Brahe, since it is argued that many of the observations of Ptolemy in actuality are predictions from the geocentric model rather than inputs to it. The divergence of results over the centuries became increasingly evident when the Brahe data were analyzed by Kepler, and thus after 2,000 years this system was replaced by the more powerful model. Similarly, Newtonian physics was supplanted by Einsteinian relativity theory, "elements of which broach upon metaphysics," as Koestler (1982) recognized and foresaw: "Eminent physicists, [are] aware of the revolutionary implications of quantum theory and the new cosmology—which are bound to transform man's image of the universe even more radically than the Copernican revolution had done" (p. 661).

Some evolutionary development might possibly be expected in public policy models as the requirements of decision makers change and the concomitant data are developed, but it is advisable that these not be viewed from such a great height and length as stars and millennia. Several of the modeling strategies discussed are more fully developed conceptually than empirically for both of these reasons, but the questions of society are, by necessity, becoming progressively more sophisticated as the level of technology advances. For example, it is not uncommon today for the media to consult experts on the results of models in evaluating items of contemporary interest. Daily, the newspapers and television news reports carry stories about the state of the economy and utilize macroeconomic forecasts from some of the recognized national forecasting groups mentioned in the first chapter. The disaster at the nuclear facility at Chernobyl in the Soviet Union and the uncertainties concerning the short-term and potential long-term consequences of the radiation release led the news media to consult scientists for the results of their models of such mishaps. Now, the interest of the media in such a circumstance is, to a great extent, pecuniary, since any information is better than none when deadlines are to be met; but the overview is indicative of the curiosity of the public in regard to the implications of complicated phenomena. The generally increasing acceptance of "electronic spreadsheet" programs on personal computers, which are touted as providing the capability of addressing "what if" questions, is another such indication. These are the same types of questions as the larger ones that have been mentioned on a slightly different scale, but with the rapidly changing microcomputer technology, this dichotomy becomes smaller and smaller each day. Most of the models discussed in this book can be accommodated on the current generation of personal computers; thus, it is to be expected that the evolution of these types of policy tools will accelerate rapidly as more and more analysts place reliance upon formalized models to ponder the questions of alternatives. Questions that previously could be handled only by in-house mainframe computers or through the use of a "service bureau" now can be inexpensively and easily addressed with a microcomputer sitting on the analyst's desk. In consequence, city and county planning authorities no longer have to rely upon high-priced consultants for developing I-O model scenarios, modeling property assessment, reviewing budgets, or projecting revenues, but can undertake these projects themselves.

Among the benefits to be derived from these changes are increased understanding and confidence in the model results on the part of local decision makers. A magazine dedicated to computer matters devoted an entire issue to computer-based simulation, including articles on modeling strategies, simulations, and policy analysis.[13] In this issue Miller and Kelso (1985) present a model using an electronic spreadsheet to analyze the proposed funding of the federal government's environmental Superfund. Previously, Johansson (1985) illustrated how the Klein I model could be implemented in a spreadsheet and iteratively solved for equilibrium levels. The magazine has also reviewed microcomputer-based econometric modeling packages [see Davenport (1986) on microTSP]. Thus, a formerly esoteric subject is losing its mystique and is increasingly used. The prospects for the future are even brighter, given that undergraduate courses in modeling now reinforce the theoretical foundations with applications that the students can replicate and extend.

THE FUTURE

Policy modeling involves (1) identifying the problem, (2) selecting a strategy, (3) specifying relationships, (4) obtaining data, (5) verifying the data, (6) estimating parameters, (7) calibrating the model, (8) validating the results, (9) evaluating sensitivity analyses, (10) preparing scenarios, (11) documenting, (12) maintaining, and (13) extending. Although each of these aspects is important, some are given more attention than others in actual practice. Statisticians may consider collection of the data and verifying these; theoretical econometricians will concentrate on deriving the parameters and extending the tools available to modelers; while applied econometricians may direct more attention to obtaining data and calibrating and validating the model. However, it is necessary for policy analysts to be involved actively in all of these components of modeling, although their major interests may be in the identification of the problem and the preparation of scenarios. Developing the appropriate balance among all of the components of modeling is mandatory in order that models be of use in solving questions of policy. Failure to recognize this can vitiate the formalized model. For example, the course of debate in regard to an issue often leads analysts to argue that the data available are inappropriate for the analysis and therefore the modeling should not proceed. This outcome does not mean that decisions will not be made, but that these will be based on an analysis accomplished on an ad hoc basis instead. The advantages of formalized modeling are numerous, yet perhaps the greatest one is that the structuring of the problem that modeling requires often forces the analyst to consider explicitly more of the elements at issue than otherwise. Aspects of problems previously overlooked may thus be identified, and the nature of the relationships accurately established.

NOTES

1. Surveys conducted during the initial months after EC1 at the plant level in the United States revealed that energy consumption could be substan-

tially reduced through the modification of standard operating procedures in recognition of the changing economic realities. For example, at one of the General Motors regional assembly plants in Baltimore, it was found that by keeping the doors closed at the end of the production line except when vehicles were actually leaving the building, sufficient energy savings would be achieved to more than offset the cost of opening and closing the doors. Management practice can, under the proper circumstances, adapt quickly.

2. The manufacturing energy elasticity increased from -0.26 in 1958/59 to -0.30 in 1974/75, while the iron and steel industry exhibited a change from -0.21 to -0.31 between the years 1959/60 and 1973/74 (see Tables 5.4 and 5.20).

3. The concept of a Markov process is that the future of the system depends only upon its present state and not upon its past states. Thus, such processes do not have any heredity, only an evolution from the present—a drunk's next lurch is totally independent of his other previous staggering path, depending solely upon his current position.

4. Fuzzy sets do not provide deterministic monotonic solutions, but define a set of potential answers among which the solution obtains.

5. See Green (1985) for a good description of the theory employed in the Sweeney model and how it may be used in policy analyses.

6. A market penetration ratio measures the proportion of the potential market for a product that is currently being served.

7. See Lin, Hirst, and Cohn (1976), Hirst and Carney (1978), and Hirst, Goeltz, and Carney (1981) for descriptions of the REC model and Dubin (1985) and Goett and McFadden (1984, 1982) for the REEPS. Cowing and McFadden (1984) provide a definitive comparison of REC and REEPS.

8. An economic "rent" accrues to a factor input that is in fixed supply and, correspondingly, when that return to the factor exceeds the factor's opportunity cost (which is 0 if supply is perfectly inelastic); but such rent cannot be sustained in the long run if markets are competitive with low barriers to entry.

9. This is the ultimate concern of a cartel, perhaps such as OPEC, whose joint profit maximization strategy in restricting production and thereby increasing prices would be constrained by the threat of a "backstop" technology (solar energy or nuclear fusion) supplanting oil. Thus, the cost of the backstop fuel will establish the upper limit on oil price. Recall that during the decade of OPEC's ascendancy, none of the potential substitute energy sources ever achieved economic viability (see Griffin and Steele 1980, pp. 81-82). This is probably not merely a happenstance, since many of the OPEC ministers and advisors received their training in economics (and some even taught it) in the United States, the United Kingdom, and Western Europe.

10. Again, the reader is referred to Morgenstern (1963), which should provide the beginning modeler with a healthy skepticism as to the import to assign to socioeconomic data.

11. See Train (1986, p. 174, Table 9.1) for a complete listing of the 25 vehicle types modeled.

12. This "principle" is named after one William of Ockham of c.1300–349, who rejected scholasticism, teaching that general ideas have no reality. From these tenets there follow the constraints in regard to simplicity with which

the widely used phrase is now associated.

13. *Byte,* "The Small Systems Journal," devoted the October 1985 issue to "Simulating Society."

REFERENCES

Acton, J., B. Mitchell, and R. Mowill. 1976. "Residential Demand of Electricity in Los Angeles: An Econometric Study of Disaggregated Data." Report prepared for the National Science Foundation, R-1899-NSF.

Acton, J., B. Mitchell, and R. Sohlberg. 1980. "Estimating Residential Electricity Demand Under Declining-Block Tariffs: An Econometric Study Using Micro-data." *Applied Economics,* 12:145–61.

Archibald, R., and R. Gillingham. 1980. "An Analysis of the Short-Run Consumer Demand for Gasoline Using Household Survey Data." *Review of Economics and Statistics,* 62:622–28.

Berlinski, D. 1976. *On Systems Analysis: An Essay Concerning the Limitations of Some Mathematical Methods in the Social, Political and Biological Sciences.* Cambridge, MA: MIT Press.

Bohi, D. R., and M. B. Zimmerman. 1984. "An Update of Econometric Studies of Energy Demand Behavior." *Annual Review of Energy,* 9:105-54.

Byte. 1985. "Simulating Society." 10(10) (October issue).

Chung, K. L. 1967. *Markov Chains with Stationary Transition Probabilities.* 2nd ed. New York: Springer-Verlag.

Cowing, T. G., and D. L. McFadden. 1984. *Microeconomic Modeling and Policy Analysis: Studies in Residential Energy Demand.* Orlando, FL: Academic Press.

Davenport, P. 1986. "microTSP." *Byte,* 11:257–63.

De la Barriere, R. P. 1967. *Optimal Control Theory.* New York: Dover.

Dubin, J. A. 1985. *Consumer Durable Choice and the Demand for Electricity.* Amsterdam: North-Holland.

____. 1983. "The National Interim Energy Consumption Survey (NIECS) and the Pacific Northwest Data Base (PNW)–A Summary and Collected Programs." Working Paper No. 469. Los Angeles: California Institute of Technology.

Dubin, J. A., and D. McFadden. 1984. "An Econometric Analysis of Residential Electric Appliance Holdings and Consumption." *Econometrica,* 52:345–62.

Garbacz, C. 1984. "Residential Electricity Demand: A Suggested Appliance Stock Equation." *Energy Journal,* 5:151–54.

____. 1983. "A Model of Residential Demand for Electricity Using a National Household Sample." *Energy Economics,* 5:124–28.

Gilmore, R. 1981. *Catastrophe Theory for Scientists and Engineers.* New York: John Wiley & Sons.

Goett, A., and D. McFadden. 1984. "The Residential End-Use Energy Planning System: Simulation Model Structure and Empirical Analysis." In *Advances in the Economics of Energy and Resources,* edited by J. R. Moroney,

Greenwich, CT: JAI Press.

Goett, A., and D. McFadden. 1982. "Residential End-Use Energy Planning System (REEPS)," Final report, EPRI Report EA-2512. Palo Alto: Electric Power Research Institute.

Green, R. D. 1985. *Forecasting with Computer Models: Econometrics, Population and Energy Forecasting.* New York: Praeger.

Griffin, J. M., and H. B. Steele. 1980. *Energy Economics and Policy.* New York: Academic Press.

Heckman, J. 1981. "Statistical Models for Discrete Panel Data." In *Structural Analysis of Discrete Data with Econometric Applications,* edited by C. Manski and D. McFadden. Cambridge, MA: MIT Press.

——. 1978. "Dummy Endogenous Variables in a Simultaneous Equation System." *Econometrica,* 46:931–59.

Hey, J. D. 1983. *Data in Doubt: An Introduction to Bayesian Statistical Inference for Economists.* London: Basil Blackwell.

Hirst, E., and J. Carney. 1978. "The ORNL Engineering-Economic Model of Residential Energy Use." ORNL Report No. CON-24. Oak Ridge, TN: Oak Ridge National Laboratory.

Hirst, E., R. Goeltz, and J. Carney. 1981. "Residential Energy Use and Conservation Actions: Analysis of Disaggregate Household Data." ORNL Report No. CON-68. Oak Ridge, TN: Oak Ridge National Laboratory.

Johansson, J-H. 1985, "Simultaneous Equations with Lotus 1-2-3," *Byte.* 10: 399–405.

Kennedy, P. 1985. *A Guide to Econometrics.* 2nd ed. Cambridge, NA: MIT Press.

Kepler, J. 1981. Reprint of 1596 edition. *Mysterium Cosmographicum,* trans. by A. M. Duncan. New York: Abaris Books.

Koestler, A. 1982. *Bricks to Babel: Selected Writings with Author's Comments.* Bungay, Suffolk: Pan Books.

Lin, A., E. N. Botsas, and S. A. Moore. 1985. "State Gasoline Consumption in the USA: An Econometric Analysis." *Energy Economics,* 7:29–36.

Lin, W., E. Hirst, and S. Cohn. 1976. "Fuel Choice in the Household Sector." ORNL Report No. CON-3. Oak Ridge, TN: Oak Ridge National Laboratory.

Martin, J. J. 1967. *Bayesian Decision Problems and Markov Chains.* New York: John Wiley & Sons.

Miller, R. M., and A. S. Kelso. 1985. "Analyzing Government Policies: Economic Modeling with Lotus 1-2-3." *Byte,* 10:199–210.

Morgenstern, O. 1963. *On the Accuracy of Economic Observations.* 2nd ed. Princeton: Princeton University Press.

Sweeney, J. L. 1978. "The Demand for Gasoline in the United States: A Vintage Capital Model." In *Workshops on Energy Supply and Demand,* pp. 240–77. International Energy Agency. Paris: Organization for Economic Cooperation and Development.

Train, K. 1986. *Qualitative Choice Analysis: Theory, Econometrics, and an Application to Automobile Demand.* Cambridge, MA: MIT Press.

U.S. Department of Energy. 1979. *State Energy Fuel Prices by Major Economic Sectors from 1960–1977.* DOE/EIA-0190. Washington, D.C. U.S. Department of Energy.

U.S. Department of Energy, Energy Information Administration. 1981a. *Residential Energy Consumption Survey: 1979–1980 Consumption and Expenditures, Part II: Regional Data.* DOE/EIA-0262/2. Washington, D.C. U.S. Department of Energy.

———. 1981b. *Residential Energy Consumption Survey: 1979–1980 Consumption and Expenditures, Part I: National Data (Including Conservation).* DOE/EIA-0262/1. Washington, D.C.: U.S. Department of Energy.

———. 1980a. *Residential Energy Consumption Survey: Consumption and Expenditures, April 1978 Through March 1979.* DOE/EIA-027/5. Washington, D.C.: U.S. Department of Energy.

———. 1980c. *Residential Energy Consumption Survey: Characteristics of the Housing Stocks and Households.* DOE/EIA-0207/2. Washington, D.C.: U.S. Department of Energy.

———. 1979. *Single-Family Households: Fuel Inventories and Expenditures: National Interim Energy Consumption Survey.* DOE/EIA-0207/1. Washington, D.C.: U.S. Department of Energy.

U.S. Department of Energy, Energy Information Administration, Office of Energy Markets and End Use. 1981. "Technical Documentation for the Residential Energy Consumption Survey: National Interim Energy Consumption Survey 1978–1979, Household Monthly Energy Consumption and Expenditures—Public Use Data Tapes—User's Guide." Washington, D.C.: U.S. Department of Energy.

Wheaton, W. C. 1982. "The Long-Run Structure of Transportation and Gasoline Demand." *Bell Journal of Economics,* 13:439–54.

Bibliography

Acton, J., B. Mitchell, and R. Mowill. 1976. "Residential Demand for Electricity in Los Angeles: An Econometric Study of Disaggregated Data." Report prepared for the National Science Foundation, R-1899-NSF.

Acton, J., B. Mitchell, and R. Sohlberg. 1980. "Estimating Residential Electricity Demand Under Declining-Block Tariffs: An Econometric Study Using Micro-data." *Applied Economics,* 12:145-61.

Adams, F. G., and J. Marquez. 1984. "Petroleum Price Elasticity, Income Effects, and OPEC Pricing Policy." *Energy Journal,* 5:115-28.

Adelman, M. A. 1980. "Energy-Income Coefficients and Ratios: A Reply." *Energy Economics,* 2:184-85.

——. 1980. "Energy-Income Coefficients and Ratios: Their Use and Abuse." *Energy Economics,* 2:2-4.

Aigner, D. J. 1979. "Bayesian Analysis of Optimal Sample Size and a Best Decision Rule for Experiments in Direct Load Control." *Journal of Econometrics,* 9:209-21.

Aigner, D. J., and J. A. Hausman. 1980. "Correcting for Truncation Bias in the Analysis of Experiments in Time-of-Day Pricing of Electricity." *Bell Journal of Economics,* 11:131-42.

Aigner, D. J., C. A. Lovell, and P. Schmidt. 1977. "Formulation and Estimation of Stochastic Frontier Production Function Models." *Journal of Econometrics,* 6:21-37.

Allen, R. G. D. 1938. *Mathematical Analysis for Economists.* London: Macmillan.

Almon, C. 1967. *Matrix Methods in Economics.* Reading, MA: Addison-Wesley.

——. 1966. *The American Economy to 1975.* New York: Harper & Row.

Almon, C., M. R. Buckler, L. M. Horwitz, and T. C. Reimbold. 1974. *1985: Industry Forecasts of the American Economy.* Lexington, MA: Lexington Books.

Alt, C., A. Bopp, and G. Lady. 1979. "Econometric Analysis of the 1974/75 Decline in Petroleum Consumption in the USA: Some Policy Implications." *Energy Economics,* 1:27-32.

Anderson, K. P. 1973. "Residential Demand for Electricity: Econometric Estimates for California and the United States." *Journal of Business,* 46:526-53.

Applebaum, E. 1979. "On the Choice of Functional Forms." *International Economic Review,* 20:449-58.

Archibald, R., and R. Gillingham. 1983. "An Analysis of the Short-Run Consumer Demand for Gasoline Using Household Survey Data: A Reply."

Review of Economics and Statistics. 65:533-34.

———. 1981. "A Decomposition of the Price and Income Elasticities of the Consumer Demand for Gasoline." *Southern Economic Journal,* 47:1021-31.

———. 1980. "An Analysis of the Short-Run Consumer Demand for Gasoline Using Household Survey Data." *Review of Economics and Statistics,* 62: 622-28.

Arrow, K. J., H. B. Chenery, B. S. Minhas, and R. M. Solow. 1961. "Capital-Labor Substitution and Economic Efficiency." *Review of Economics and Statistics,* 43:225-50.

Ascher, W. 1978. *Forecasting: An Appraisal for Policy-makers and Planners.* Baltimore: Johns Hopkins University Press.

Askin, A. B. 1976. "The Macroeconomic Implications of Alternative Energy Scenarios." In *Econometric Dimensions of Energy Supply and Demand,* edited by A. B. Askin and J. Kraft, pp. 99-109. Lexington, MA: Lexington Books.

Askin, A. B., and J. Kraft, eds. 1976. *Econometric Dimensions of Energy Supply and Demand.* Lexington, MA: Lexington Books.

Association of University Business and Economic Research. 1978. *Socio-economic Impact of Electrical Energy Construction: Proceedings of the first AUBER Energy Workshop.* Washington, D.C.: Association of University Business and Economic Research.

Atkinson, S. E. 1981. "Reshaping Residential Electricity Load Curves Through Time-of-Use Pricing." *Resources and Energy,* 3:175-94.

———. 1979. "Responsiveness to Time-of-Day Electricity Pricing: First Empirical Results." *Journal of Econometrics,* 9:79-95.

Atkinson, S. E., and R. Halvorsen. 1981. "Automatic Price Adjustment Clauses and Input Choice in Regulated Utilities." *Journal of Business Administration,* 12:185-96.

Australian Bureau of Statistics. 1983. "Projections of Population of the States and Territories of Australia, 1981 to 2021." Catalogue No. 3214.0. Canberra: Government Publishing Service.

———. Quarterly. "Time-Series Data on Magnetic Tape and Microfiche." Catalogue number 1311.0. Canberra: Government Publishing Service.

Australian Department of Home Affairs and Environment. 1983. *The Cost of Pollution Abatement: The Australian Integrated Steel Industry.* Prepared for the Australian Environment Council. Report No. 13. Canberra: Australian Government Publishing Service.

Australian Department of National Development. 1978. *Demand for Primary Fuels: Australia 1976-77 to 1986-87.* Canberra: Australian Government Publishing Service.

———. 1978. *End-Use Analysis of Primary Fuels Demand: Australia 1973-74 to 1986-87.* Canberra: Australian Government Publishing Service.

Australian Department of National Development and Energy. 1982. *Energy*

Forecasts for the 1980's. Canberra: Australian Government Publishing Service.

——. 1981. *Forecasts of Energy Demand and Supply: Primary and Secondary Fuels, Australia: 1980–81 to 1991–92.* Canberra: Australian Government Publishing Service.

Australian Department of Resources and Energy. 1983. *Forecasts of Energy Demand and Supply, Australia: 1982–83 to 1991–92.* Canberra: Australian Government Publishing Service.

——. Monthly. "Major Energy Statistics" and "Sales of Petroleum Products by State Marketing Area," Canberra.

Australian Industries Assistance Commission. 1983. *Certain Iron and Steel Products and Certain Alloy Steel Products.* Report No. 321. Canberra: Australian Government Publishing Service.

——. 1980. *Iron and Steel Industry.* Report No. 249. Canberra: Australian Government Publishing Service.

Australian Institute of Petroleum. 1973. "List Prices of Petrol, Australian Capital Cities, 1945–1973." Melbourne: Petroleum Information Bureau. Mimeo.

——. Annual. *Oil and Australia: The Figures Behind the Facts.* Melbourne: Petroleum Information Bureau.

Australian National Energy Advisory Committee. 1978. *Some Aspects of Energy Modeling in Australia.* Canberra: Australian Government Publishing Service.

Australian Senate Select Committee on South-West Tasmania. 1982. *Hearings, Thursday, 4 February 1982.* Parliament of the Commonwealth of Australia, Official Hansard Report. Canberra: Australian Government Publishing Service.

——. 1982. *Report on Demand and Supply of Electricity for Tasmania and Other Matters.* Parliament of the Commonwealth of Australia. Canberra: Australian Government Publishing Service.

Bacharach, M. 1970. *Biproportional Matrices and Input-Output Change.* Cambridge: Cambridge University Press.

Balestra, P., and M. Nerlove. 1966. "Pooling Cross Section and Time Series Data in the Estimation of a Dynamic Model: The Demand for Natural Gas." *Econometrica,* 34:585–612.

Baltagi, B. H., and J. M. Griffin. 1984. "U.S. Gasoline Demand: What Next?" *Energy Journal,* 5:129–40.

——. 1983. "Gasoline Demand in the OECD: An Application of Pooling and Testing Procedures." *European Economic Review,* 22:117–37.

Banks, F. E. 1983. *Resources and Energy: An Economic Analysis.* Lexington, MA: Lexington Books.

——. 1980. *The Political Economy of Oil.* Lexington, MA: Lexington Books.

Barnes, R., R. Gillingham, and R. Hagemann. 1981. "The Short-Run Residential Demand for Electricity." *Review of Economics and Statistics,* 63:541–51.

Barnett, W. A. 1984. "The Minflex Laurent Translog Flexible Functional Form." Center for Economic Research Discussion Paper No. 84-7. Austin: University of Texas.

——. 1983. "Definitions of Second Order Approximation and of Flexible Functional Form." *Economics Letters,* 12:31-35.

Barnett, W. A., Y. W. Lee, and M. Wolfe. 1984. "The Three Dimensional Global Properties of the Minflex Laurent, Generalized Leontief, and Translog Flexible Functional Forms." Austin: University of Texas. Mimeo.

Bartels, R. 1984. "Models for the Shares of Fuel Expenditures in NSW Industries: Report Prepared for the Energy Authority of New South Wales." Department of Econometrics. Sydney: University of Sydney.

——. 1984. "Econometric Models of the Demand for Electricity in the Commercial and Industrial Sectors: A Literature Survey and Suggestions for Modelling the Demand in Victoria." Report prepared for the State Electricity Commission of Victoria. Department of Econometrics. Sydney: University of Sydney. Mimeo.

——. 1983. "Modelling Household Demand for Electricity in N.S.W.: Literature Review and Suggestions." Department of Economic Statistics. Sydney: University of Sydney. Mimeo.

Bartels, R., P. Lopert, and S. Williamson. 1985. "The Residential Demand for Electricity in N.S.W." Sydney: Energy Authority of New South Wales.

Barten, A. P. 1964. "Consumer Demand Functions Under Conditions of Almost Additive Preferences." *Econometrica,* 32:1-38.

Barten, A. P., and S. J. Turnovsky. 1966. "Some Aspects of the Aggregation Problem for Composite Demand Equations." *International Economic Review,* 7:231-59.

Battalio, R. C., J. H. Kagel, R. C. Winkler, and R. A. Winett. 1979. "Residential Electricity Demand: An Experimental Study." *Review of Economics and Statistics,* 61:180-89.

Baughman, M. L., P. L. Joskow, and D. P. Kamat. 1979. *Electric Power in the United States: Models and Policy Analysis.* Cambridge, MA: MIT Press.

Baumol, W. J. 1977. *Economic Theory and Operations Analysis.* 4th ed. Englewood Cliffs, NJ: Prentice-Hall.

Baxter, R., and R. Rees. 1968. "The Analysis of the Industrial Demand for Electricity." *Economic Journal,* 78:277-98.

Beattie, B. R., and C. R. Taylor. 1985. *The Economics of Production.* New York: John Wiley & Sons.

Behling, D. J. 1976. "A Dynamic Systems Analysis of the Relation Between Energy and the Economy." Report No. BNL 21281. Upton, NY: Brookhaven National Laboratory.

Beierlein, J. G., J. W. Dunn, and J. C. McConnon. 1981. "The Demand for Electricity and Natural Gas in the Northeastern United States." *Review of Economics and Statistics,* 63:403-8.

Belsley, D. A., E. Kuh, and R. E. Welsh. 1980. *Regression Diagnostics: Identifying Differential Data and Sources of Collinearity.* New York: John Wiley & Sons.

Bera, A. K. 1982. "A New Test for Normality." *Economics Letters,* 9:263-68.

Bera, A. K., and R. P. Byron. 1983. "A Note on the Effects of Linear Approximation on Hypothesis Testing." *Economics Letters,* 12:251-54.

Bera, A. K., R. P. Byron, and C. M. Jarque. 1981. "Further Evidence on Asymptotic Tests for Homogeneity and Symmetry in Large Demand Systems." *Economics Letters,* 8:101-5.

Bera, A. K., and C. M. Jarque. 1981. "Efficient Tests for Normality, Homoscedasticity and Serial Independence of Regression Residuals: Monte Carlo Evidence." *Economics Letters,* 7:313-18.

——. 1981. "Tests for Specification Errors: A Simultaneous Approach." Paper presented at the Tenth Conference of Economists of the Economic Society of Australia and New Zealand, Canberra, August 24-28. Mimeo.

Berg, S. V., and J. P. Herden. 1976. "Electricity Price Structures: Efficiency, Equity and the Composition of Demand." *Land Economics,* 52:169-78.

Berg, S. V., and W. E. Roth. 1976. "Some Remarks on Residential Electricity Consumption and Social Rate Restructuring." *Bell Journal of Economics,* 7:690-98.

Berg, S. V., and A. Savvides. 1983. "The Theory of Maximum kW Demand Charges for Electricity." *Energy Economics,* 5:258-66.

Bergendahl, P., and C. Bergstrom. 1981. "Long-Term Oil Substitution: The IEA-MARKAL Model and Some Simulation Results for Sweden." *Scandinavian Journal of Economics,* 83:237-52.

Berlinski, D. 1976. *On Systems Analysis: An Essay Concerning the Limitations of Some Mathematical Methods in the Social, Political and Biological Sciences.* Cambridge, MA: MIT Press.

Berndt, E. R. 1983. "Modelling the Aggregate Demand for Electricity: Simplicity vs. Virtuosity." Alfred P. Sloan School of Management Working Paper No. 1415-83. Cambridge, MA: Massachusetts Institute of Technology.

——. 1982. "From Technocracy to Net Energy Analysis: Engineers, Economists and Recurring Energy Theories of Value." Alfred P. Sloan School of Management Working Paper No. 1353-82. Cambridge, MA: Massachusetts Institute of Technology.

——. 1978. "Aggregate Energy Efficiency and Productivity Measurement." *Annual Review of Energy,* 3:225-73.

——. 1976. "Reconciling Alternative Estimates of the Elasticity of Substitution." *Review of Economics and Statistics,* 58:59-68.

Berndt, E. R., and L. R. Christensen. 1974. "Testing for the Existence of Consistent Aggregate Index of Labor Inputs." *American Economic Review,* 64:391-404.

——. 1973. "The Translog Function and the Substitution of Equipment, Structures, and Labor in U.S. Manufacturing 1929-68." *Journal of Economet-*

rics, 1:81-114.

——. 1973. "The Internal Structure of Functional Relationships: Separability, Substitution and Aggregations." *Review of Economic Studies,* 50:403-10.

Berndt, E. R., and B. C. Field, eds. 1981. *Modeling and Measuring Natural Resource Substitution.* Cambridge, MA: MIT Press.

Berndt, E. R., M. A. Fuss, and L. Waverman. 1978. "Factor Markets in General Disequilibrium: Dynamic Models of the Industrial Demand for Energy." In *Workshop on Energy Supply and Demand,* pp. 222-39. International Energy Agency. Paris: Organization for Economic Co-operation and Development.

Berndt, E. R., and M. S. Khaled. 1979. "Parametric Productivity Measurement and Choice Among Flexible Functional Forms." *Journal of Political Economy,* 87:1220-45.

Berndt, E. R., G. May, and G. C. Watkins. 1980. "An Econometric Model of Alberta Electricity Demand." In *Energy Policy Modeling: United States and Canadian Experiences: Specialized Energy Policy Models,* edited by W. T. Ziemba, S. L. Schwartz, and E. Koenigsberg, pp. 103-16. Boston: Martinus Nijhoff.

Berndt, E. R., and C. M. White. 1979. "Income Redistribution and Employment Effects of Rising Energy Prices." Department of Economics Resources Paper No. 30. Vancouver: University of British Columbia.

Berndt, E. R., and D. O. Wood. 1984. "Energy Price Changes in the Induced Revaluation of Durable Capital in U.S. Manufacturing During the OPEC Decade." MIT Energy Laboratory Report No. 84-003. Cambridge, MA: Massachusetts Institute of Technology. Mimeo.

——. 1983. "Data Development for U.S. Manufacturing Energy Price of Quantity Series, 1906-1947." MIT Energy Laboratory. Mimeo.

——. 1981. "Engineering and Econometric Interpretations of Energy-Capital Complementarity: Reply and Further Results." *American Economic Review,* 71:1105-12.

——. 1979. "Engineering and Econometric Interpretations of Energy-Capital Complementarity." *American Economic Review,* 69:342-54.

——. 1978. "Engineering and Econometric Approaches to Industrial Energy Conservation and Capital Formation: A Reconciliation." In *Workshop on Energy Supply and Demand,* pp. 278-314. International Energy Agency. Paris: Organization for Economic Co-operation and Development.

——. 1976. "Some Preliminary Observations on the Stability of the Translog Production Function." Draft.

——. 1975. "Technology, Prices and the Derived Demand for Energy." *Review of Economics and Statistics,* 57:259-68.

Berzeg, K. 1982. "Demand for Motor Gasoline: A Generalized Error Components Model." *Southern Economic Journal,* 49:462-71.

Betancourt, R. R. 1981. "An Econometric Analysis of Peak Electricity Demand in the Short Run." *Energy Economics,* 3:14-29.

Betancourt, R., and C. Clague. 1981. *Capital Utilization: A Theoretical and Empirical Analysis.* New York: Cambridge University Press.

Bewley, R. A. 1982. "The Generalized Addilog Demand System Applied to Australian Time Series and Cross Section Data." *Australian Economic Papers,* 21:177-92.

Bickel, P. J., and K. A. Doksum. 1981. "An Analysis of Transformations Revisited." *Journal of the American Statistical Association,* 76:296-311.

Bigman, D. 1983. "Technical Change, Distributive Shares, and Aggregation." *Oxford Economic Papers,* 30:434-46.

Billings, R. B. 1983. "Specification of Block Rage Price Variables in Demand Models: Reply." *Land Economics,* 59:370-71.

——. 1983. "Specification of Block Rate Price Variables in Demand Models." *Land Economics,* 58:386-94.

Binswanger, H. P. 1974. "The Measurement of Technical Change Biases with Many Factors of Production." *American Economic Review,* 64:964-76.

Black, J., ed. 1982. *Liquid Fuels in Australia: A Social Science Perspective.* Sydney: Pergamon Press.

Blackorby, C., D. Primont, and R. R. Russell. 1978. *Duality, Separability and Functional Structure: Theory and Economic Applications.* New York: North-Holland.

——. 1977. "On Testing Separability Restrictions with Flexible Functional Forms." *Journal of Econometrics,* 5:195-209.

Blair, J. M. 1978. *The Control of Oil.* New York: Vintage Books.

Blair, P., T. Cassel, and R. Edelstein. 1981. "Optimal Investments in Geothermal Electricity Facilities: A Theoretic Note." *Journal of Business Administration,* 12:197-212.

Blair, R. D., D. L. Kaserman, and R. C. Tepel. 1984. "The Impact of Improved Mileage on Gasoline Consumption." *Economic Inquiry,* 22:209-17.

Blattenberger, G. R., L. D. Taylor, and R. K. Rennhack. 1983. "Natural Gas Availability and the Residential Demand for Energy." *Energy Journal,* 4:23-45.

Blomquist, G. 1984. "The 55 m.p.h. Speed Limit and Gasoline Consumption." *Resources and Energy,* 6:21-39.

Bohi, D. R. 1981. *Analyzing Demand Behavior: A Study of Energy Elasticities.* Published for Resources for the Future. Baltimore: Johns Hopkins University Press.

Bohi, D. R., and M. B. Zimmerman. 1984. "An Update of Econometric Studies of Energy Demand Behavior and Review." *Annual Review of Energy,* 9:105-54.

Bopp, A. E. 1983. "The Demand for Kerosene: A Modern Giffen Good." *Applied Economics,* 15:459-67.

——. 1980. "A Combined Decline-Curve and Price Analysis of U.S. Crude Oil Production, 1968-76." *Energy Economics,* 2:111-14.

Bopp, A. E., and G. M. Lady. 1982. "On Measuring the Effects of Higher Energy

Prices." *Energy Economics,* 4:218-24.

Bopp, A. E., and J. A. Neri. 1978. "The Price of Gasoline: Forecasting Comparisons." *Quarterly Review of Economics and Business,* 18:23-31.

Borg, S., W. A. Donnelly, T. Eagan, D. Knapp, and D. H. Nissen. 1975. "The Econometric Regional Demand Model–ERDM." Energy Information and Analysis Technical Report No. EIATR 75-18. Washington, D.C.: U.S. Federal Energy Administration.

Box, G. E. P., and D. R. Cox. 1982. "An Analysis of Transformations Revisited, Rebutted." *Journal of the American Statistical Society,* 77:209-10.

——. 1964. "An Analysis of Transformations." *Journal of the Royal Statistical Society,* B26:211-43.

Box, G. E. P., and B. M. Jenkins. 1970. *Time Series Analysis, Forecasting and Control.* San Francisco: Holden-Day.

Brain, P., and G. S. Schuyers. 1981. *Energy and the Australian Economy.* Melbourne: Longman Cheshire.

Breusch, T. S., and A. R. Pagan. 1980. "The Lagrangian Multiplier Test and Its Application to Model Specifications in Econometrics." *Review of Economic Studies,* 47:239-53.

——. 1979. "A Simple Test for Heteroscedasticity and Random Coefficient Variation." *Econometrica,* 47:1287-94.

Bronfenbrenner, M., and P. H. Douglas. 1939. "Cross-Section Studies in the Cobb-Douglas Function." *Journal of Political Economy,* 47:761-85.

Brookes, L. 1980. "Energy-Income Coefficients and Ratios: A Comment." *Energy Economics,* 2:184.

Brown, M., ed. 1967. *The Theory and Empirical Analysis of Production. Studies in Income and Wealth, Vol. 31.* New York: National Bureau of Economic Research. Distributed by Columbia University Press.

Brown, R. L., J. Durbin, and J. M. Evans. 1975. "Techniques for Testing the Consistency of Regression Relationships." *Journal of the Royal Statistical Society,* B37:149-63.

Brown, R. S., D. W. Caves, and L. R. Christensen. 1979. "Modeling the Structure of Cost and Production for Multiproduct Firms." *Southern Economic Journal,* 46:256-73.

Buchanan, J. M. 1966. "Peak Loads and Efficient Pricing: Comment." *Quarterly Journal of Economics,* 80:463-71.

Bullard, C. W., and R. A. Herendeen. 1977. "The Energy Cost of Goods and Services: An Input-Output Analysis for the U.S.A., 1963 and 1967." In *Energy Analysis,* edited by J. A. G. Thomas, pp. 71-81. Surrey: IPC Science and Technology Press.

Burgess, D. F. 1976. "Tariffs and Income Distribution: Some Empirical Evidence for the United States." *Journal of Political Economy,* 84:17-45.

——. 1975. "Duality Theory and Pitfalls in the Specification of Technologies." *Journal of Econometrics,* 3:105-21.

——. 1974. "A Cost Minimization Approach to Import Demand Equations."

Review of Economics and Statistics, 56:225-34.

Byron, R. P., and A. K. Bera. 1983. "Least Square Approximations to Unknown Regression Functions: A Comment." *International Economic Review,* 24: 255-60.

Byte, 1985. "Simulating Society." 10.

Caldwell, S., W. Greene, T. Mount, S. Saltzman, and R. Broyd. 1981. "A Comprehensive Approach to Forecasting Long Term Regional Electricity Demand." In *Mathematical Modeling of Energy Systems,* edited by I. Kavrakoglu, pp. 277-99. Alphen van der Rijn, the Netherlands: Sijthoff and Noordhoff.

Calloway, T. A., and R. G. Thompson. 1976. "An Integrated Industry Model of Petroleum Refining, Electric Power and Chemicals Industries for Costing Pollution Control and Estimating." *Engineering and Process Economics,* 1:199-216.

Cameron, T. A., and S. L. Schwartz. 1979. "Sectoral Energy Demand in Canadian Manufacturing Industries." *Energy Economics,* 1:112-18.

Cargill, T. F., and R. A. Meyer. 1971. "Estimating the Demand for Electricity by Time of Day." *Applied Economics,* 3:233-46.

Cato, D., M. Rodekohr, and J. L. Sweeney. 1976. "The Capital Stock Adjustment Process and the Demand for Gasoline: A Market Share Approach." In *Econometric Dimensions of Energy Demand and Supply,* edited by B. Askin and J. Kraft, pp. 29-52. Lexington, MA: Lexington Books.

——. 1975. "Demand for Gasoline: Application of Commodity Hierarchy Theory." *Proceedings of the American Statistical Association,* :260-64.

Caves, D. W., and L. R. Christensen. 1980. "Global Properties of Flexible Functional Forms." *American Economic Review,* 70:422-32.

Caves, D. W., L. R. Christensen, and J. A. Herriges. 1984. "Modelling Alternative Residential Peak-Load Electricity Rate Structures." *Journal of Econometrics,* 24:249-68.

Caves, D. W., L. R. Christensen, and T. A. Swanson. 1981. "Productivity Growth, Scale Economies, and Capacity Utilization in U.S. Railroads, 1955-74." *American Economic Review,* 71:994-1002.

——. 1980. "Productivity in U.S. Railroads, 1951-1974." *Bell Journal of Economics,* 11:166-81.

Caves, D. W., L. R. Christensen, and M. W. Tretheway. 1980. "Flexible Cost Functions for Multiproduct Firms." *Review of Economics and Statistics,* 62:477-81.

Chao, H. 1983. "Peak Load Pricing and Capacity Planning with Demand and Supply Uncertainty." *Bell Journal of Economics,* 14:179-90.

Chatterji, M., and P. Van Rompuy, eds. 1976. *Energy, Regional Science and Public Policy.* Berlin: Springer-Verlag.

Chern, W. S., and R. E. Just. 1982. "Assessing the Need for Power: A Regional Econometric Model." *Energy Economics,* 4:232-39.

——. 1980. "A Generalized Model for Fuel Choices with Application to the

Paper Industry." *Energy Systems and Policy,* 4:273-94.

Chetty, V. K. 1969. "Econometrics of Joint Production: A Comment." *Econometrica,* 37:731.

Chow, G. C. 1983. *Econometrics.* New York: McGraw-Hill.

Christ, C. F. 1966. *Econometric Models and Methods.* New York: John Wiley & Sons.

Christensen, L. R., D. Cummings, and P. E. Schoech. 1983. "Econometric Estimation of Scale Economies in Telecommunications." In *Economic Analysis of Telecommunications: Theory and Applications,* edited by L. Courville, A. deFontenay, and R. Dobell, pp. 27-53. New York: North-Holland.

Christensen, L. R., and W. M. Greene. 1976. "Economies of Scale in U.S. Electric Power Generation." *Journal of Political Economy,* 84:655-76.

Christensen, L. R., and D. W. Jorgenson. 1969. "The Measurement of U.S. Real Capital Input, 1929-1967." *Review of Income and Wealth,* 15:293-320.

Christensen, L. R., D. W. Jorgenson, and L. J. Lau. 1975. "Transcendental Logarithmic Utility Functions." *American Economic Review,* 65:367-83.

——. 1973. "Transcendental Logarithmic Production Functions." *Review of Economics and Statistics,* 55:28-45.

——. 1971. "Conjugate Duality and the Transcendental Logarithmic Utility Functions." *Econometrica,* 39:255-56.

Chung, J. W. 1981. "The Price of Gasoline, the Oil Crisis, and the Choice of Transportation Mode." *Quarterly Review of Economics and Business,* 21:77-86.

Chung, K. L. 1967. *Markov Chains with Stationary Transition Probabilities.* 2nd ed. New York: Springer-Verlag.

Clements, K. W. 1983. "The Demand for Energy Used in Secondary Industry." Department of Economics. Nedlands: University of Western Australia. Mimeo.

——. 1982. "The Demand for Energy Used in Transport." Department of Economics Discussion Paper No. 82.29. Nedlands: University of Western Australia.

Cobb, C. W., and P. H. Douglas. 1928. "A Theory of Production." *American Economic Review,* 18(suppl.):139-65.

Cochrane, D., and G. H. Orcutt. 1949. "Application of Least Squares Regressions to Relationships Containing Autocorrelated Error Terms." *Journal of the American Statistical Association,* 44:32-61.

Cockerill, A. 1974. "Economies of Scale and the Structure of the Steel Industry." *Business Economist,* 6:43-51.

Cooley, T. T., and E. C. Prescott. 1973. "Systematic (Non-random) Variation Models Varying Parameter Regression: A Theory and Some Applications." *Annals of Economic and Social Measurement,* 2:463-73.

Corum, K. R., and T. H. Morlan. 1984. "Improving the Efficiency of Electricity Use in the Pacific Northwest: Programs Versus Pricing." Paper presented at the Fifty-Ninth Annual Western Economic Association, Las Vegas, NV,

June 26. Mimeo.

Cowing, T. G., and D. L. McFadden. 1984. *Microeconomic Modeling and Policy Analysis: Studies in Residential Energy Demand.* Orlando, FL: Academic Press.

Crew, M. A., and P. R. Kleindorfer. 1976. "Peak Load Pricing with a Diverse Technology." *Bell Journal of Economics,* 7:207-31.

Crow, R. T., and B. T. Ratchford. 1979. "An Econometric Approach to Forecasting the Market Potential of Electric Automobiles." In *International Studies of the Demand for Energy,* edited by W. D. Nordhaus, pp. 45-64. Amsterdam: North-Holland.

Cumberland, J. H., W. A. Donnelly, C. S. Gibson, and C. E. Olson. 1976. "Forecasting Alternative Regional Energy Requirements and Environmental Impacts." In *Energy, Regional Science and Public Policy,* edited by M. Chatterji and P. Van Rompuy, pp. 32-57. Amsterdam: North-Holland.

Currie, D. 1981. "Some Long Run Features of Dynamic Time Series Models." *Economic Journal,* 91:704-15.

Dahl, C. A. 1984. "Gasoline Demand Survey." Department of Economics. Baton Rouge: Louisiana State University. Mimeo draft.

———. 1984. "Vertical Equity Effects and Total Consumer Losses for Emergency Allocation Schemes for the Gasoline Market." *Applied Economics,* 16:25-32.

———. 1983. "An Analysis of the Short-Run Consumer Demand for Gasoline Using Household Survey Data: A Comment." *Review of Economics and Statistics,* 65:532-33.

———. 1982. "Do Gasoline Demand Elasticities Vary?" *Land Economics,* 58: 373-82.

———. 1979. "Consumer Adjustment to a Gasoline Tax." *Review of Economics and Statistics,* 61:427-32.

———. 1978. "American Energy Consumption: Extravagant or Economical? A Study of Gasoline Demand." *Resources and Energy,* 1:359-73.

Dahl, C. A., and G. S. Laumas. 1981. "Stability of U.S. Petroleum Refinery Response to Relative Product Prices." *Energy Economics,* 3:30-35.

Daly, G. G., and T. H. Mayor. 1983. "Reason and Rationality During the Energy Crisis." *Journal of Political Economy,* 91:168-81.

Dansby, R. E. 1979. "Multi-period Pricing with Stochastic Demand." *Journal of Econometrics,* 9:223-37.

Dantzig, G. B. 1963. *Linear Programming and Extensions.* Princeton: Princeton University Press.

Dargay, J. M. 1983. "The Demand for Energy in Swedish Manufacturing Industries." *Scandinavian Journal of Economics,* 85:37-51.

Darmstadter, J. with P. D. Teitelbaum and J. G. Polach, 1971. *Energy in the World Economy: A Statistical Review of Trends in Output, Trade, and Consumption since 1925.* Baltimore: Johns Hopkins University Press.

Data Resources Inc. 1973. "A Study of the Quarterly Demand for Gasoline and

Impacts of Alternative Gasoline Taxes." Interim Report to the U.S. Environmental Protection Agency and the President's Council of Environmental Quality.

Davenport, P. 1986. "microTSP." *Byte,* 11:257-63.

Davidson, R., and J. G. MacKinnon. 1983. "Testing the Specification of Multivariate Models in the Presence of Alternative Hypotheses." *Journal of Econometrics,* 23:301-13.

De la Barriere, R. P. 1967. *Optimal Control Theory.* New York: Dover.

Denny, M., and M. A. Fuss. 1977. "The Use of Approximation Analysis to Test for Separability and the Existence of Consistent Aggregates." *American Economic Review,* 67:404-18.

Denny, M., M. A. Fuss, and L. Waverman. 1981. "Substitution Possibilities for Energy: Evidence from U.S. and Canadian Manufacturing Industries." In *Modeling and Measuring Natural Resource Substitution,* edited by E. R. Berndt and B. C. Field, pp. 230-58. Cambridge, MA: MIT Press.

Denny, M., and D. May. 1977. "The Existence of a Real Value-Added Function in the Canadian Manufacturing Sector." *Journal of Econometrics,* 5:55-69.

Denny, M., and C. L. Pinto. 1978. "An Aggregate Model with Multi-product Technologies." In *Production Economics: A Dual Approach to Theory and Applications, Vol. 2,* edited by M. A. Fuss and D. McFadden, pp. 249-67. Amsterdam: North-Holland.

Derrick, S., D. McDonald, and P. Rosendale. 1981. "The Development of Energy Resources in Australia: 1981 to 1990." *Australian Economic Review,* 54:13-55.

DeSouza, G. R. 1981. *Energy Policy and Forecasting: Economic, Financial and Technical Dimensions.* Lexington, MA: Lexington Books.

Dewees, D. N., R. M. Hyndman, and L. Waverman. 1975. "Gasoline Demand in Canada: 1956-1972." *Energy Policy,* 3:116-23.

De Wolff, R. 1938. "The Demand for Passenger Cars in the United States." *Econometrica,* 6:113-29.

Dhrymes, P. J. 1971. *Distributed Lags: Problems of Estimation and Formulation.* San Francisco: Holden-Day.

——. 1970. *Econometrics: Statistical Foundations and Applications.* New York: Harper & Row.

Dhrymes, P. J., and M. Kurz. 1964. "Technology and Scale in Electricity Generation." *Econometrica,* 32:287-315.

Dhrymes, P. J., and B. M. Mitchell. 1969. "Estimation of Joint Production Functions." *Econometrica,* 37:732-40.

Dias-Bandaranaike, R., and M. Munasinghe. 1983. "The Demand for Electricity Services and the Quality of Supply." *Energy Journal,* 4:49-71.

Diewert, W. E. 1978. "Hicks Aggregation Theorem and the Existence of a Real Value-Added Function." In *Production Economics: A Dual Approach to Theory and Applications,* edited by M. A. Fuss and D. McFadden, pp. 17-51. Amsterdam: North-Holland.

———. 1978. "Superlative Index Numbers and Consistent Aggregation." *Econometrica,* 46:883-900.

———. 1976. "Exact and Superlative Index Numbers." *Journal of Econometrics,* 4:115-47.

———. 1974. "Applications of Duality Theory." In *Frontiers of Quantitative Economics, Vol. 2,* edited by M. D. Intriligator and D. A. Kendrick, pp. 176-99. Amsterdam: North-Holland.

———. 1971. "An Application of the Shephard Duality Theorem: A Generalized Leontief Production Function." *Journal of Political Economy,* 79:481-507.

Diewert, W. E., and T. J. Wales. 1984. "Flexible Functional Forms and Global Curvature Conditions." Technical Working Paper No. 40. Cambridge, MA: National Bureau of Economic Research.

Dimopoulos, D. 1981. "Pricing Schemes for Regulated Enterprises and Their Welfare Implications in the Case of Electricity." *Bell Journal of Economics,* 12:185-200.

Dixon, P. B., B. R. Parmenter, J. Sutton, and D. Vincent. 1982. *ORANI, A General Equilibrium Model of the Australian Economy.* Amsterdam: North-Holland.

Donnelly, W. A. 1986. "The Demand for Gasoline and State Tax Revenues." *Kentucky Journal of Economics and Business,* 6:11-17.

———. 1985. "Selecting Appropriate Policy Modeling Strategies." In *Energy Modeling in Australia.* Centre for Applied Research. Kensington: University of New South Wales.

———. 1985. "Electricity Demand Modeling." In *New Mathematical Advances in Economic Dynamics,* edited by D. F. Batten and P. F. Lesse, pp. 179-95. Sydney: Croom-Helm.

———. 1985. "Note on 'The Residential Demand for Electricity: A Variant Parameters Approach'." *Applied Economics,* 17:241-42.

———. 1985. "A State-Level, Variable Elasticity of Demand for Gasoline Model." *International Journal of Transport Economics,* 12:193-202.

———. 1984. "The Australian Demand for Petrol." *International Journal of Transport Economics,* 11:189-205.

———. 1984. "Residential Electricity Demand Modeling in the Australian Capital Territory: Preliminary Results." *Energy Journal,* 5:119-31.

———. 1984. "Gasoline Demand Modeling." Centre for Resource and Environmental Studies Working Paper No. 1984/8. Canberra: Australian National University.

———. 1984. "Energy Modeling." Centre for Resource and Environmental Studies Working Paper No. 1984/6. Canberra: Australian National University.

———. 1984. "Energy Model for Australia: Final Report on the National Energy Research Development and Demonstration Grant: Project Number 78/2447." Centre for Resource and Environmental Studies Working Paper No. 1984/4. Canberra: Australian National University.

——. 1983. "Electricity Demand Modeling." Paper presented at the Fifty-Third ANZAAS Congress, Section 24:Economics, University of Western Australia, Perth, May 16-20. Mimeo.

——. 1982. "The Regional Demand for Petrol in Australia." *Economic Record,* 58:317-27.

——. 1982. "Petrol and Diesel Fuel Demand Elasticities for Germany and Belgium." Report to the Australian Department of Transport, Canberra. Mimeo.

——. 1981. "The Demand for Petrol in Australia." Centre for Resource and Environmental Studies Working Paper No. R/WP-61. Canberra: Australian National University.

——. 1979. "A State-Level Monthly Demand for Gasoline Specification: Kentucky." Unpublished.

——. 1977. "The Regional Energy, Activity and Demographic Model." Paper presented at the Western Regional Science Association Meeting, Tucson, AZ, February. Mimeo.

Donnelly, W. A., and M. Diesendorf. 1985. "Variable Elasticity of Demand Models for Electricity." *Energy Economics,* 7:159-62.

Donnelly, W. A., and A. K. Dragun. 1984. "Production Functions for Australian Coal." Centre for Resource and Environmental Studies Working Paper No. 1984/28. Canberra: Australian National University.

Donnelly, W. A., and W. Gaynor. 1977. "The FEA Short-Term Motor Gasoline Forecasting Equations." FEA 77-WPIA-11. Washington, D.C.: U.S. Federal Energy Administration.

Donnelly, W. A., E. S. Gooneratne, and M. H. L. Turnovsky. 1982. "The Residential Demand for Electricity in the ACT." Centre for Resource and Environmental Studies Working Paper No. R/WP68. Canberra: Australian National University.

Donnelly, W. A., and F. Hopkins. 1976. "The Regional Energy, Activity and Demographic (READ) Model." *Northeast Regional Science Review,* 6:107-17.

Donnelly, W. A., F. Hopkins, A. Havenner, B. Hong, and T. Morlan. 1977. "Estimating a Comprehensive County-Level Forecasting Model for the U.S." In *Socio-economic Impact of Electrical Energy Construction,* pp. 193-255. Washington, D.C.: Association of University Bureaus of Economic Research.

Donnelly, W. A., and E. S. Leung. 1983. "Residential Electricity Demand." *Search,* 14:206-11.

Donnelly, W. A., and A. M. Parhizgari. 1976. "Estimating the Regional Impacts of Energy Shortages." In *Energy, Regional Science and Public Policy,* edited by M. Chatterji and P. Van Rompuy, pp. 178-93. Amsterdam: North-Holland.

Donnelly, W. A., and H. D. W. Saddler. 1984. "The Retail Demand for Electricity in Tasmania." *Australian Economic Papers,* 23:53-59.

———. 1982. "The Retail Demand for Electricity in Tasmania." Centre for Resource and Environmental Studies Working Paper No. R/WP-63. Canberra: Australian National University.

Dorfman, N. S. 1981. "Gasoline Distribution Policies in a Shortage: Welfare Impacts on Rich and Poor." *Public Policy,* 29:473-505.

Dorfman, R., P. A. Samuelson, and R. M. Solow. 1958. *Linear Programming and Economic Analysis.* New York: McGraw-Hill.

Douglas, P. H. 1976. "The Cobb-Douglas Production Function Once Again: Its History, Its Testing, and Some New Empirical Values." *Journal of Political Economy,* 84:903-15.

———. 1948. "Are There Laws of Production?" *American Economic Review,* 38:1-41.

Drollas, L. P. 1984. "The Demand for Gasoline: Further Evidence." *Energy Economics,* 6:71-82.

Dubin, J. A. 1985 *Consumer Durable Choice and the Demand for Electricity.* Amsterdam: North-Holland.

———. 1983. "The National Interim Energy Consumption Survey (NIECS) and the Pacific Northwest Data Base (PNW)–A Summary and Collected Programs." Working Paper No. 469. Los Angeles: California Institute of Technology.

Dubin, J. A., and D. McFadden. 1984. "An Econometric Analysis of Residential Electric Appliance Holdings and Consumption." *Econometrica,* 52:345-62.

Duffins, L. M., and W. S. Chern. 1984. "An Energy and Generalized Fuel Choice Model for the Primary Metals Industry." *Energy Journal,* 5:35-53.

Duncan, R. C., and H. P. Binswanger. 1976. "Energy Sources: Substitutability and Biases in Australia." *Australian Economic Papers,* 15:289-301.

———. 1974. "Factor Biases and Induced Innovation in Australian Manufacturing Industries." Mimeo draft.

———. 1974. "Production Parameters in Australian Manufacturing Industries." Mimeo draft.

Durand, D. 1937. "Some Thoughts on Marginal Productivity with Special Reference to Prof. Douglas' Analysis." *Journal of Political Economy,* 45:740-58.

Durbin, J. 1970. "Testing for Serial Correlation in Least-Squares Regression When Some of the Regressors Are Lagged Dependent Variables." *Econometrica,* 38:410-21.

Durbin, J., and G. S. Watson. 1950. "Testing for Serial Correlation in Least Squares Regressions." *Biometrika,* 37:409-28.

Durnkerley, J., ed. 1978. *International Comparisons of Energy Consumption.* Published for Resources for the Future. Baltimore: Johns Hopkins University Press.

Eden, R., M. Posner, R. Bending, E. Crouch, and J. Stanislaw. 1981. *Energy Economics: Growth, Resources and Policies.* Cambridge: Cambridge University Press.

El Badawi, I., A. R. Galland, and G. Souza. 1982. "An Elasticity Can Be Estimated Consistently Without A Priori Knowledge of Functional Form." Center for Mathematical Studies in Economics and Management Science Discussion Paper No. 542. Evanston, IL: Northwestern University.

Electric Power Research Institute. 1982. "Price Elasticities of Demand for Energy: Evaluating the Estimates: Project 1220-1 Final Report." EPRI EA-2612. Prepared by Resources for the Future. Palo Alto: EPRI.

——. 1977. "Stanford-EPRI Workshop for Considering a Forum for the Analysis of Energy Options Through the Use of Models: Project 875-1 Special Report." EPRI EA-414-SR. Stanford: Stanford University.

——. 1976. "Long-Range Forecasting Properties of State-of-the-Art Models for Demand for Electric Energy." EPRI EA-221, Vol. 1. Prepared by Charles River Associates. Palo Alto: EPRI.

——. 1976. "Proceedings on Forecasting Methodology for Time-of-Day and Seasonal Electric Utility Loads," edited by J. W. Boyd. EPRI SR-31. Palo Alto: EPRI.

Electricity Supply Association of Australia. Annual. *The Electricity Supply Industry in Australia.* Melbourne: ESAA.

Energy Modeling Forum. 1980. "Aggregate Elasticity of Energy Demand." EMF Report No. 4, Vol. 1. Stanford: Stanford University.

Epstein, L. G. 1981. "Quality Theory and Functional Forms for Dynamic Factor Demands." *Review of Economic Studies,* 48:81-95.

Ericsson, N. R. 1982. "Testing Linear Versus Logarithmic Regression Models: A Comment." *Review of Economics and Statistics,* 69:477-81.

Evans, L., and G. Wells. 1983. "Pierce and Haugh on Characterizations of Causality: A Re-examination." *Journal of Econometrics,* 23:331-35.

Farrar, D. E., and R. R. Glauber. 1967. "Multicollinearity in Regression Analysis: The Problem Revisited." *Review of Economics and Statistics,* 49:92-107.

Faruqui, A. 1983. "A Vintage Adjustment Model of Industrial Energy Substitution." *Resources and Energy,* 5:285-301.

Fauchett, J., and Associates. 1973. "Data Development for the I-O Energy Model." Final Report to the Energy Policy Project. Washington, D.C.

Field, B. C., and C. Grebenstein. 1980. "Capital-Energy Substitution in U.S. Manufacturing." *Review of Economics and Statistics,* 62:207-12.

Filmer, R. J., and R. Mannion. 1979. "Petrol Prices and Passenger Motor Vehicle Travel." Paper presented at the Regional Science Association Conference, Albury-Wodonga, New South Wales-Victoria, December 2-5. Mimeo.

Filmer, R. J., and S. Talbot. 1974. "Demand for Passenger Motor Vehicles." Paper presented at the Australian National University, Canberra, July 19. Mimeo.

Fishelson, G. 1982. "Demand for Gasoline Usage by Passenger Cars." *Resources and Energy,* 4:163-72.

——. 1980. "The Effects of Restricted Energy Imports: Implications for the

Economy of a Small Country." *Energy Economics,* 2:166-79.

Fishelson, G., and T. V. Long. 1978. "An International Comparison of Energy and Materials Use in the Iron and Steel Industry." In *International Comparisons of Energy Consumption,* edited by T. Dunkerley. Published for Resources for the Future. Baltimore: Johns Hopkins University Press.

Fisher, F. M. 1965. "The Choice of Instrumental Variables in the Estimation of Economy-wide Econometric Models." *International Economic Review,* 6:245-74.

Fisher, F. M., and C. Kaysen. 1962. *The Demand for Electricity in the United States.* Amsterdam: North-Holland.

Foell, W. K., J. W. Mitchell, and J. L. Pappas. 1975. "The Wisconsin Regional Energy Model: A Systems Approach to Regional Energy Analysis." Energy Systems and Policy Research Group Report No. 56. Madison: University of Wisconsin.

Folie, M. 1977. "Competition in the Australian Retail Market for Petrol." Paper presented at the Sixth Conference of Economists, Economic Society of Australia and New Zealand, University of Tasmania, Hobart, May. Mimeo.

Forrester, J. W. 1971. *World Dynamics.* Cambridge, MA: Wright-Allen Press.

Fraumeni, B. M., and D. W. Jorgenson. 1981. "Capital Formation and U.S. Productivity Growth, 1948-1976." In *Production Analysis: A Range of Perspectives,* edited by A. Dogramaci, pp. 49-70. Boston: Martinus Nijhoff.

Freeman, A. M. 1974. "Wilson 'Electricity Consumption: Supply Requirements, Demand Elasticity and Rate Design': Discussion." *American Journal of Agricultural Economics,* 56:429-31.

Frisch, R. 1965. *Theory of Production.* Dordrecht, Holland: Reidel.

——. 1959. "A Complete Scheme for Computing All Direct and Cross Demand Elasticities in a Model with Many Sectors." *Econometrica,* 27:177-96.

——. 1935. "The Principle of Substitution: An Example of Its Appreciation in the Chocolate Industry." *Nordisk Tideskrift for Teknisk Okonomi,* 1:12-27.

Fuss, M. A. 1977. "The Demand for Energy in Canadian Manufacturing: An Example of the Estimation of Production Structures with Many Inputs." *Journal of Econometrics,* 5:89-116.

Fuss, M. A., R. Hyndman, and L. Waverman. 1977. "Residential, Commercial and Industrial Demand for Energy in Canada." In *International Studies of the Demand for Energy,* edited by W. Nordhaus, pp. 151-79. Amsterdam; North-Holland.

Fuss, M. A., and D. McFadden, eds. 1978. *Production Economics: A Dual Approach to Theory and Applications, Vols. 1 and 2.* Amsterdam: North-Holland.

Fuss, M. A., D. McFadden, and Y. Mundlak. 1978. "A Survey of Functional Forms in the Economic Analysis of Production." In *Production Economics: A Dual Approach to Theory and Applications, Vol. 1,* edited by M. Fuss and D. McFadden, pp. 219-68. Amsterdam: North-Holland.

Gabor, A. 1966. "Peak Loads and Efficient Pricing: Further Comment." *Quarterly Journal of Economics,* 80:472-80.

——. 1955-56. "A Note on Block Tariffs." *Review of Economic Studies,* 23: 32-41.

Galatin, M. 1968. *Economies of Scale and Technological Change in Thermal Power Generation.* Amsterdam: North-Holland.

Gallant, A. R. 1981. "On the Bias in Flexible Functional Forms and an Essentially Unbiased Form: The Fourier Flexible Form." *Journal of Econometrics,* 15:211-45.

Gallini, N. T. 1983. "Demand for Gasoline in Canada." *Canadian Journal of Economics,* 16:299-324.

Gamponia, V., and G. Brown. 1982. "Steel and Energy Substitution in U.S. Manufacturing." *Southern Economic Journal,* 48:785-91.

Garbacz, C. 1984. "Residential Electricity Demand: A Suggested Appliance Stock Equation." *Energy Journal,* 5:151-54.

——. 1983. "A Model of Residential Demand for Electricity Using a National Household Sample." *Energy Economics,* 5:124-28.

——. 1983. "Electricity Demand and the Electricity of Intra-marginal Price." *Applied Economics,* 15:699-701.

——. 1983. "A Model of Residential Demand for Electricity Using a National Household Sample." *Energy Economics,* 5:124-28.

Garvey, G. 1972. *Energy, Ecology, Economy: A Framework for Environmental Policy.* New York: Norton.

Gaudry, M., and M. Dagenais. 1979. "Heteroscedasticity and the Use of Box-Cox Transformations." *Economics Letters,* 2:225-29.

Georgescu-Roegen, N. 1979. "Energy Analysis and Economic Valuation." *Southern Economic Journal,* 45:1023-58.

Gibbons, J. 1984. "Capital-Energy Substitution in the Long Run." *Energy Journal,* 5:109-18.

Gill, G. S., and G. S. Maddala. 1976. "Residential Demand for Electricity in the TVA Area: An Analysis of Structural Change." *Proceedings of the American Statistical Association,* pp. 315-19.

Gilmore, R. 1981. *Catastrophe Theory for Scientists and Engineers.* New York: John Wiley & Sons.

Goett, A., and D. McFadden. 1984. "The Residential End-Use Energy Planning System: Simulation Model Structure and Empirical Analysis." In *Advances in the Economics of Energy and Resources,* edited by J. R. Moroney. Greenwich, CT: JAI Press.

Goldberger, A. S. 1968. *Topics in Regression Analysis.* New York: Macmillan.

——. 1968. "The Interpretation and Estimation of Cobb-Douglas Functions." *Econometrica,* 36:464-72.

——. 1964. *Econometric Theory.* New York: John Wiley & Sons.

Goodwin, P. B., and M. J. H. Mogridge. 1981. "Hypotheses for a Fully Dynamic Model of Car Ownership." *International Journal of Transport Economics,*

8:313-26.

Gordon, R. L. 1974. "Wilson's 'Electricity Consumption: Supply Requirements, Demand Elasticity and Rate Design': Discussion." *American Journal of Agricultural Economics,* 56:431-33.

Gottwald, D., and W. Guth. 1980. "Allocation of Exhaustible Resources in Oligopolistic Markets: A Dynamic Game Approach." *Energy Economics,* 2:208-22.

Granger, C. W. J. 1969. "Investigating Causal Relationships by Econometric Models and Cross-Spectral Methods." *Econometrica,* 37:428-38.

Granger, C. W. J., R. Engle, R. Ramanathan, and A. Andersen. 1979. "Residential Load Curves and Time-of-Day Pricing." *Journal of Econometrics,* 9:13-32.

Granger, C. W. J., and P. Newbold. 1977. *Forecasting Economic Time Series.* New York: Academic Press.

——. 1974. "Spurious Regressions in Econometrics." *Journal of Econometrics,* 2:111-20.

Green, R. D. 1985. *Forecasting with Computer Models: Econometric, Population and Energy Forecasting.* New York: Praeger.

Greene, D. L. 1981. "The Aggregate Demand for Gasoline and Highway Passenger Vehicles in the United States: A Review of the Literature 1938-1978." Energy Division. Oak Ridge, TN: Oak Ridge National Laboratories.

——. 1981. "State-Level Stock System Model of Gasoline Demand." In *Energy: Forecasting, Data, and Conservation,* pp. 44-50. Transport Research Record No. 801. Transport Research Board. Washington, D.C.: National Academy of Sciences.

——. 1980."Regional Demand for Gasoline: Comment." *Journal of Regional Science,* 20:103-9.

——. 1980. "The Spatial Dimension of Gasoline Demand." *Geographical Survey,* 9:19-28.

——. 1979."State Differences in the Demand for Gasoline: An Econometric Analysis." *Energy Systems and Policy,* 3:191-212.

Greene, D. L., and G. Kulp. 1982. "An Analysis of the 1978-80 Decline in Gasoline Consumption in the United States." *Energy,* 7:367-75.

Greene, W. H. 1980. "On Estimation of a Flexible Frontier Production Model." *Journal of Econometrics,* 13:101-15.

Greenlees, J. S. 1980. "Gasoline Prices and Purchases of New Automobiles." *Southern Economic Journal,* 47:167-78.

Griffin, J. M. 1981. "The Energy-Capital Reconciliation Attempts." In *Modeling and Measuring Natural Resource Substitution,* edited by E. R. Berndt and B. C. Field, pp. 70-80. Cambridge, MA: MIT Press.

——. 1981. "Engineering and Econometric Interpretations of Energy-Capital Complementarity: Comment." *American Economic Review,* 71:1100-4.

——. 1981. "Statistical Cost Analysis Re-revisited: Reply." *Quarterly Journal of Economics,* 96:177-81.

——. 1979. "Statistical Cost Analysis Revisited." *Quarterly Journal of Economics,* 93:107-29.

——. 1978. "Joint Production Technology: The Case of Petrochemicals." *Econometrica,* 46:379-96.

——. 1978. "The Econometrics of Joint Production: Another Approach." *Review of Economics and Statistics,* 59:389-97.

——. 1977. "Inter-fuel Substitution Possibilities: A Translog Application to Intercountry Data." *International Economic Review,* 18:755-70.

——. 1977. "Long-Run Production Modeling with Pseudo-data: Electric Power Generation." *Bell Journal of Economics,* 8:112-27.

——. 1974. "The Effects of Higher Prices on Electricity Consumption." *Bell Journal of Economics and Management Science,* 5:515-39.

Griffin, J. M., and P. R. Gregory. 1976. "An Intercountry Translog Model of Energy Substitution Responses." *American Economic Review,* 66:845-57.

Griffin, J. M., and H. B. Steele. 1980. *Energy Economics and Policy.* New York: Academic Press.

Gruen, F. H., and A. L. Hillman. 1981. "A Review of Issues Pertinent to Liquid Fuel Policy." *Economic Record,* 57:111-27.

Guilkey, D. K. 1975. "A Test for the Presence of First-Order Vector Autoregressive Errors When Lagged Endogenous Variables Are Present." *Econometrica,* 43:711-17.

Guilkey, D. K., C. A. K. Lovell, and R. C. Sickles. 1983. "A Comparison of the Performance of Three Flexible Functional Forms." *International Economic Review,* 24:591-616.

Gunn, G., and P. H. Douglas. 1941. "The Production Function for Australian Manufacturing." *Quarterly Journal of Economics,* 56:108-29.

——. 1940. "Further Measurements of Marginal Productivity." *Quarterly Journal of Economics,* 54:399-428.

Hackl, P. 1979. "Moving Sums of Residuals: A Tool for Testing the Constancy of Regression Relationships over Time." In *Models and Decision Making in National Economies,* edited by J. M. L. Janseen, L. F. Pau, and D. Straszak, pp. 219-25. Amsterdam: North-Holland.

Hadley, G. 1964. *Nonlinear and Dynamic Programming.* Reading, MA: Addison-Wesley.

——. 1962. *Linear Programming.* Reading, MA: Addison-Wesley.

Haitovsky, Y. 1969. "Multicollinearity in Regression Analysis: Comment." *Review of Economics and Statistics,* 51:486-89.

Hall, P. H. 1981. "Patterns of Energy Use in Australian Manufacturing Industry: A Disaggregated Study of the Manufacturing Sector, 1950-1968." Department of Economics. Armidale, NSW: University of New England. Mimeo.

Hall, R. E. 1973. "The Specification of Technology with Several Kinds of Output." *Journal of Political Economy,* 81:878-92.

Hall, V. B. 1984. "Some Thoughts on Energy Modeling and Policy in Australia." *Economic Papers,* 3:21-36.

———. 1983. "Industrial Sector Interfuel Substitution Following the First Major Oil Shock." *Economics Letters,* 12:377-82.

———. 1983."Industrial Sector Interfuel Substitution Following the First Major Oil Shock." Department of Economics Working Paper No. 67. Sydney: University of Sydney.

Hall, V. B., and P. Saunders. 1984. "Further Evidence from Survey Data on Australian Manufacturing Price Changes." *Economic Record,* 60:68-84.

Halvorsen, R. 1978. *Econometric Models of U.S. Energy Demand.* Lexington, MA: Lexington Books.

———. 1976. "Demand for Electric Energy in the United States." *Southern Economic Journal,* 42:610-25.

———. 1975. "Residential Demand for Electric Energy." *Review of Economics and Statistics,* 57:12-18.

Halvorsen, R., and J. Ford. 1979. "Substitution Among Energy, Capital and Labor Inputs in U.S. Manufacturing." In *Advances in the Economics of Energy and Resources: Structures of Energy Markets, Vol. 1,* edited by R. S. Pindyck, pp. 51-75. Greenwich, CT: JAI Press.

Handsaker, M. L., and P. H. Douglas. 1937. "The Theory of Marginal Productivity Tested by Data Manufacturing in Victoria, I." *Quarterly Journal of Economics,* 52:1-36.

———. 1938. "The Theory of Marginal Productivity Tested by Data Manufacturing in Victoria, II." *Quarterly Journal of Economics,* 52:215-54.

Hannan, E. J. 1960. *Multiple Time Series.* New York: John Wiley & Sons.

Hannon, B. 1973. "An Energy Standard of Value." *Annals of the American Academy of Political and Social Science,* 410:139-53.

Harris, C. C., V. D. McConnell, and J. H. Cumberland. 1984. "A Model for Forecasting the Economic and Environmental Impact of Energy Policy." *Energy Economics,* 6:167-76.

Harris, S. F. 1982. "Social Aspects of Energy in Australia: A Social Science Literature and Research Review." In *Liquid Fuels in Australia: A Social Science Research Perspective,* edited by J. Black, pp. 7-81. Sydney: Pergamon Press.

Hartman, R. S. 1983. "The Estimation of Short-Run Household Electricity Demand Using Pooled Aggregate Data." *Journal of Business and Economic Statistics,* 1:127-35.

Hartman, R. S., and A. Werth. 1981. "Short-Run Residential Demand for Fuels: A Disaggregated Approach." *Land Economics,* 57:197-212.

Hartmann, J. W., F. E. Hopkins, and D. B. Cato. 1981. "Short-Term Forecasting of Gasoline Demand." In *Energy: Forecasting, Data, and Conservation,* pp. 22-28. Transportation Research Record No. 801. Transportation Research Board. Washington, D.C.: National Academy of Sciences.

Hausman, J. A., M. Kinnucan, and D. McFadden. 1979. "A Two-Level Electricity Demand Model: Evaluation of the Connecticut Time-of-Day Pricing Test." *Journal of Econometrics,* 10:263-89.

Havenner, A. M., and W. A. Donnelly. 1977. "Estimating from a Pooled Time-Series of Cross Sections of Simultaneous Equations." *Proceedings Summer Computer Simulations Conference IEEE.*

Hawkins, R. G. 1978. "A Vintage Model of the Demand for Energy and Employment in Australian Manufacturing Industry." *Review of Economic Studies,* 45:479-94.

——. 1977. "Factor Demands and the Production Function in Selected Australian Manufacturing Industries." *Australian Economic Papers,* 16:97-111.

——. 1975. "The Demand for Electricity: A Cross-Section Study of New South Wales and the Australian Capital Territory." *Economic Record,* 51:1-18.

Heady, E. O., and J. L. Dillon. 1961. *Agricultural Production Functions.* Ames: Iowa State University Press.

Heath, T. 1981. Reprint of 1913 edition. *Aristarchus of Samos: The Ancient Copernicus.* New York: Dover.

Heckman, J. 1981. "Statistical Models for Discrete Panel Data." In *Structural Analysis of Discrete Data with Econometric Applications,* edited by C. Manski and D. McFadden. Cambridge, MA: MIT Press.

——. 1978. "Dummy Endogenous Variables in a Simultaneous Equation System." *Econometrica,* 46:931-59.

——. 1978. "An Analysis of the Changing Location of Iron and Steel Production in the Twentieth Century." *American Economic Review,* 68:123-33.

Heien, D. M. 1969. "Income and Price Lags in Consumer-Demand Analysis." *Journal of the Royal Statistical Society,* 132:265-71.

Helliwell, J. F., R. N. McRae, P. Boothe, A. Hansson, M. Margolick, T. Padmore, A. Plourde, and R. Plummer. 1982. "Energy and the National Economy: An Overview of the MACE Model." Presented at the Canadian Economics Association Meeting, University of Ottawa, Ottawa, June 4. Mimeo.

Hendricks, W., R. Koenker, and D. J. Poirier. 1979. "Residential Demand for Electricity: An Econometric Approach." *Journal of Econometrics,* 9:33-57.

Hermann, G. F. K. 1974. "Electricity Tariffs: The Principles Underlying Their Design and Structure." Based upon doctoral dissertation, University of Stellenbosch, Rome.

Hewitt, D. R., and P. N. James. 1975. "Forecasting Motor Spirit Demand." Petroleum Branch. Canberra: Australian Department of Minerals and Energy. Mimeo.

Hewlett, H. C. 1977. "Changing Patterns of Households' Consumption of Energy Commodities." *Proceedings of the American Statistical Association,* Part I:99-108.

Hey, J. D. 1983. *Data in Doubt: An Introduction to Bayesian Statistical Inference for Economists.* London: Basil Blackwell.

Hildebrand, G. H., and T. Liu. 1965. *Manufacturing Production Functions in the United States, 1957. An Interindustry and Interstate Comparison of Productivity.* New York State School of Industrial Labor Relations.

Ithaca: Cornell University Press.

Hill, D. H., D. A. Ott, L. D. Taylor, and J. M. Walker. 1983. "Incentive Payments in Time-of-Day Electricity Pricing Experiments: The Arizona Experience." *Review of Economics and Statistics,* 65:59-65.

Hillier, F. S., and G. J. Lieberman. 1974. *Operations Research.* 2nd ed. San Francisco: Holden-Day.

Hillinger, C. 1970. "Comment on 'Invariance Axioms and Economic Indexes'." *Econometrica,* 38:773-74.

Hillman, A. L., and E. Katz. 1984. "Oil Price Instability and Domestic Energy Substitution for Imported Oil." *Economic Record,* 60:85-89.

Hipel, K. W., A. I. McLeod, and W. C. Lennox. 1977. "Advances in Box-Jenkins Modelling 1: Model Construction." *Water Resources Research,* 13:567-75.

Hirst, E., and J. Carney. 1978. "The ORNL Engineering-Economic Model of Residential Energy Use." ORNL Report No. CON-24. Oak Ridge, TN: Oak Ridge National Laboratory.

Hirst, E., R. Goeltz, and J. Carney. 1981. "Residential Energy Use and Conservation Actions: Analysis of Disaggregate Household Data." ORNL Report No. CON-68. Oak Ridge, TN: Oak Ridge National Laboratory.

Hoffman, K. C. 1973. "A Linear Programming Model of the Nation's Energy System." Report No. BNL ESAG-4. Upton, NY: Brookhaven National Laboratory.

Hogan, W. T. 1983. *World Steel in the 1980's.* Lexington, MA: Lexington Books.

Hogan, W. W. 1979. "Capital Energy Complementarity in Aggregate Energy-Economic Analysis." Energy and Environmental Policy Center. John F. Kennedy School of Government (E-79-03). Cambridge, MA: Harvard University.

Hogan, W. W., and A. S. Manne. 1979. "Energy-Economy Interactions: The Fable of the Elephant and the Rabbit." In *Advances in the Economics of Energy and Resources: Structures of Energy Markets, Vol. 1,* edited by R. S. Pindyck, pp. 7-17. Greenwich, CT; JAI Press.

Hollander, G., P. Threlfall, and K. A. Tucker. 1982. "Energy and the Australian Tourism Industry." Bureau of Industry Economics Working Paper No. 25. Canberra: Bureau of Industry Economics.

Holloway, M. L., ed. 1980. *Texas National Energy Modeling Project: An Experience in Large-Scale Model Transfer and Evaluation.* London: Academic Press.

Hotard, G. D., C. C. Liu, and J. H. Tistroph. 1983. "Logistic Function Modeling of Natural Resource Production." In *Energy Models and Studies,* edited by B. Lev, pp. 353-64. Amsterdam: North-Holland.

House, P. W., and T. Williams. 1981. "Using Models for Policy Analysis: A Case Study of an Energy-Environment-Economic Issue." *Energy Systems and Policy,* 5:1-24.

Houston, D. A. 1982. "Residential Electricity Demand Revisited." *Energy*

Journal, 1:29-41.

Houthakker, H. S. 1980. "Residential Electricity Demand Revisited." *Energy Journal,* 1:29-41.

———. 1955-56. "The Pareto Distribution and the Cobb-Douglas Production Function in Activity Analysis." *Review of Economic Studies,* 23:27-31.

———. 1951. "Electricity Tariffs in Theory and Practice." *Economic Journal,* 61:1-25.

———. 1951. "Some Calculations of Electricity Consumption in Great Britain." *Journal of the Royal Statistical Society,* A114:359-71.

Houthakker, H. S., and L. D. Taylor. 1970. *Consumer Demand in the United States.* 2nd ed. Cambridge, MA: Harvard University Press.

Houthakker, H. S., P. K. Verleger, and D. P. Sheehan. 1974. "Dynamic Demand Analyses for Gasoline and Residential Electricity." *American Journal of Agricultural Economics,* 56:412-18.

Hudson, E. A. 1981. "Modeling Production and Pricing Within an Interindustry Framework." In *Energy Policy Planning,* edited by B. A. Bayraktar, E. A. Cherniavsky, M. A. Laughton, and L. E. Ruff, pp. 201-14. NATO Advanced Research Institute on the Application of Systems Science to Energy Policy Planning, New York, 1979. New York: Plenum Press.

Hudson, E. A., and D. W. Jorgenson. 1974. "U.S. Energy Policy and Economic Growth, 1975-2000." *Bell Journal of Economics and Management Science,* 5:461-514.

Hudson, E. A., D. W. Jorgenson, and D. C. O'Conner. 1980. "The Impact of Restrictions on the Expansion of Electric Generating Capacity." Mimeo.

Huettner, D., J. Kasulis, and N. Dikeman. 1983. "The Significance of the Insignificant Results from the Oklahoma Time-of-Day Rate Experiment." *Resources and Energy,* 5:181-99.

Hughes, H. 1964. *The Australian Iron and Steel Industry.* Melbourne: Melbourne University Press.

Hughes, W. R. 1981. "Petrol Consumption in New Zealand." *New Zealand Institute of Economic Research Quarterly Predictions,* 65:32-37.

———. 1980. "Petrol Consumption in New Zealand 1969-79: Fixed and Time Varying Parameter Results." *New Zealand Economic Papers,* 14:28-42.

Hull, C. H., ed. 1964. *The Economic Writings of Sir William Petty, Vol. II.* Reprint of *Five Essays in Political Arithmetick.* 1687. New York: August M. Kelley.

Hulten, C. R. 1973. "Divisia Index Numbers." *Econometrica,* 41:1017-25.

Humphrey, D. B., and J. R. Moroney. 1975. "Substitution Among Capital, Labor and Natural Resource Products in American Manufacturing." *Journal of Political Economy,* 83:57-82.

Hutton, S. 1984. "Domestic Fuel Expenditure: An Analysis of Three National Surveys." *Energy Economics,* 6:52-58.

Intriligator, M. A. 1971. *Mathematical Optimization and Economic Theory.*

Englewood Cliffs, NJ: Prentice-Hall.

Intriligator, M. A., and D. A. Kendrick, eds. 1975. *Frontiers of Quantitative Economics, Vols. 1 and 2.* Amsterdam: North-Holland.

Ironmonger, D., I. Manning, and T. Van Hoa. 1984. "Longitudinal Working Models: Estimates of Household Energy Consumption in Australia." *Energy Economics,* 6:41-46.

Isard, W. 1960. *Methods of Regional Analysis: An Introduction to Regional Science.* Cambridge, MA: MIT Press.

——. 1948. "Some Locational Factors in the Iron and Steel Industry Since the Early Nineteenth Century." *Journal of Political Economy,* 56:203-17.

Isard, W., and W. Capron. 1949. "The Future Locational Pattern of Iron and Steel Production in the United States." *Journal of Political Economy,* 57:118-33.

Iulo, W. 1961. *Electric Utilities: Costs and Performance.* Pullman: Washington State University Press.

James, D. E. 1983. *Integrated Energy-Economic-Environmental Modeling with Reference to Australia.* Department of Home Affairs and Environment. Canberra: Australian Government Publishing Service.

——. 1980. "A System of Energy Accounts for Australia." *Economic Record,* 56:171-81.

Jarque, C. M., and A. K. Bera. 1980. "Efficient Tests for Normality, Homoscedasticity and Serial Independence of Regression Residuals." *Economics Letters* 6:255-59.

Johansen, L. 1972. *Production Functions: An Integration of Micro and Macro, Short Run and Long Run Aspects.* Amsterdam: North-Holland.

——. 1960. *A Multi-sectoral Study of Economic Growth.* Amsterdam: North-Holland.

Johnson, L. W. 1980. "Regional Demand for Gasoline: Comment." *Journal of Regional Science,* 20:99-101.

Johnston, J. 1984. *Econometric Methods.* 3rd ed. New York: McGraw-Hill.

Jordan, W. J. 1983. "Heterogeneous Users and the Peak Load Pricing Model." *Quarterly Journal of Economics,* 98:127-38.

Jorgenson, D. W. 1984. "The Role of Energy in Productivity Growth." *Energy Journal,* 5:11-26.

——. 1981. "Energy Prices and Productivity Growth." *Scandinavian Journal of Economics,* 83:165-79.

——. 1974. "The Economic Theory of Replacement and Depreciation." In *Econometrics and Economic Theory: Essays in Honour of Jan Tinbergen,* edited by W. Sellekaerts, pp. 189-221. London: Macmillan.

Jorgenson, D. W., and B. M. Fraumeni. 1981. "Relative Prices and Technical Change." In *Modeling and Measuring Natural Resource Substitution,* edited by E. R. Berndt and B. C. Field, pp. 17-47. Cambridge, MA: MIT Press.

Joskow, P. L. 1976. "Contributions to the Theory of Marginal Cost Pricing." *Bell Journal of Economics,* 7:197-206.

Judge, G. G., W. E. Griffiths, R. C. Hill, and T. Lee. 1980. *The Theory and Practice of Econometrics.* New York: John Wiley & Sons.

Kanetkar, V., P. N. Nemety, S. Schwartz, I. B. Vertinsky, P. Vertinsky, and W. Ziemba. 1983. "Canada and the Changing Economy of the Pacific Basin. Australia and New Zealand: Part I: Energy Prospects, Threat in Opportunity and Opportunity in Threat." Institute of Asian Research Working Paper No. 6. The Asian Center. Vancouver: University of British Columbia.

____. 1983. "Canada and the Changing Economy of the Pacific Basin. Australia and New Zealand: Part II: A First Round of Modeling." Institute of Asian Research Working Paper No. 7. The Asian Center. Vancouver: University of British Columbia.

Kang, H., and G. M. Brown. 1981. "Partial and Full Elasticities of Substitution and the Energy-Capital Complementarity Controversy." In *Modelling and Measuring Natural Resource Substitution,* edited by E. R. Berndt and B. C. Field, pp. 81-89. Cambridge, MA: MIT Press.

Karlson, S. H. 1983. "Modeling Location and Production: An Application to U.S. Fully Integrated Steel Plants." *Review of Economics and Statistics,* 65:41-50.

Karunaratne, N. D. 1981. "An Input-Output Analysis of Australian Energy Planning Issues." *Energy Economics,* 3:159-68.

Kennedy, M. 1974. "An Economic Model of the World Oil Market." *Bell Journal of Economics and Management Science,* 5:540-77.

Kennedy, P. 1985. *A Guide to Econometrics.* 2nd ed. Cambridge, MA: MIT Press.

Kepler, J. 1981. Reprint of 1596 edition. *Mysterium Cosmographicum,* trans. by A. M. Duncan. New York: Abaris Books.

Klein, L. R. 1947. "The Use of Cross-Section Data in Econometrics with Application to a Study of Production of Railroad Services in the United States." New York: National Bureau of Economic Research. Mimeo.

Kmenta, J. 1971. *Elements of Econometrics.* New York: Macmillan.

Kmenta, J., and R. F. Gilbert. 1968. "Small Sample Properties of Alternative Estimators of Seemingly Unrelated Regressions." *Journal of the American Statistical Association,* 63:1180-200.

Koenker, R. 1979. "Optimal Peak Load Pricing with Time-Additive Consumer Preferences." *Journal of Econometrics,* 9:175-92.

Koestler, A. 1982. *Bricks to Bable: Selected Writings with Author's Comments.* Bungay, Suffolk: Pan Books.

Kohler, D. 1983. "The Bias in Price Elasticity Estimates Under Homothetic Separability: Implications for the Analysis of Peak-Load Electricity Pricing." *Journal of Business and Economic Statistics,* 1:202-10.

Kopp, R. J., and V. K. Smith. 1983. "Neoclassical Modeling of Nonneutral Technological Change: An Experimental Appraisal." *Scandinavian Journal of*

Economics, 85:127-46.

——. 1981. "Measuring the Prospects for Resource Substitution Under Input and Technology Aggregations." In *Modeling and Measuring Natural Resource Substitution,* edited by E. R. Berndt and B. C. Field, pp. 143-73. Cambridge, MA: MIT Press.

Koshal, R. K., and J. Bradfield. 1977. "World Demand for Gasoline: Some Empirical Findings." *Keio Economic Studies,* 14:41-48.

Kouris, G. 1983. "Fuel Consumption for Road Transport in the USA." *Energy Economics,* 4:89-99.

——. 1981. "Elasticities: Science or Fiction?" *Energy Economics,* 3:66-70.

Kraft, J., and A. Kraft. 1978. "On the Relationship Between Energy and GNP." *Journal of Energy and Development,* 3:401-3.

Kraft, J., and M. Rodekohr. 1980. "Regional Demand for Gasoline: A Reply to Some Reconsiderations." *Journal of Regional Science,* 20:111-14.

——. 1980. "A Temporal Cross Section Specification of the Demand for Gasoline Using a Random Coefficient Regression Model." *Energy,* 5:1193-202.

——. 1978. "Regional Demand for Gasoline: A Temporal Cross-Section Specification." *Journal of Regional Science,* 18:45-55.

Kuczynski, M. and P. L. Meek, editors. 1972. *Quesney's Tableau Economique.* London: Macmillan.

Kuenne, R. E. 1979. "A Short-Run Demand Analysis of the OPEC Cartel." *Journal of Business Administration,* 129-64.

Kuh, E. 1976. "Some Preliminary Observations on the Stability of the Translog Production Function." Mimeo draft.

——. 1974. "An Essay on Aggregation Theory and Practice." In *Econometrics and Economic Theory: Essays in Honour of Jan Tinbergen,* edited by W. Sellekaerts, pp. 357-99. London: Macmillan.

Kuh, E., and R. E. Welsh. 1980. "Econometric Models and Their Assessment for Policy: Some New Diagnostics Applied to Translog Energy Demand in Manufacturing." In *Validation and Assessment Issues of Energy Models,* edited by S. I. Gass, pp. 445-75. National Bureau of Standards. Washington, D.C.: U.S. Department of Commerce.

Kumar, T. K. 1975. "Multicollinearity in Regression Analysis." *Review of Economics and Statistics,* 57:365-66.

Kwast, M. L. 1980. "A Note on the Structural Stability of Gasoline Demand and the Welfare Economics of Gasoline Taxation." *Southern Economic Journal,* 46:1212-20.

Kydes, A. S., and J. Rabinowitz. 1979. "The Time-Stepped Energy System Optimization Model (TESOM): Overview and Special Features." Upton, NY: Brookhaven National Laboratory.

Labys, W. C. 1982. "Measuring the Validity and Performance of Energy Models." *Energy Economics,* 4:159-68.

Labys, W. C., and P. K. Pollak. 1984. *Commodity Models for Forecasting and Policy Analysis.* Sydney: Croom Helm.

Lakhani, H. A. G. 1982. "Impact of Technological Change on Labor Productivity in U.S. Coal Mines: Evidence from Time Series and Cross Sectional Data." *Energy,* 7:773-82.

Lakskmanan, T. R. 1983. "A Multiregional Model of the Economy, Environment, and Energy Demand in the United States." *Economic Geography,* 59:296-320.

Lau, L. J. 1976. "A Characterization of the Normalized Restricted Profit Function." *Journal of Economic Theory,* 12:131-63.

———. 1972. "Profit Functions of Technologies with Multiple Inputs and Outputs." *Review of Economics and Statistics,* 54:281-89.

Lau, L. J., and P. A. Yotopoulos. 1971. "A Test for Relative Efficiency and Application to Indian Agriculture." *American Economic Review,* 61:94-109.

Lawrence, A., and D. Aigner, eds. 1979. "Modeling and Forecasting Time-of-Day and Seasonal Electricity Demands." *Journal of Econometrics,* 9(suppl).

Lawrence, A., and S. Braithwait. 1979. "The Residential Demand for Electricity with Time-of-Day Pricing." *Journal of Econometrics,* 9:59-77.

Leamer, E. E. 1978. *Specification Searches: Ad Hoc Inference with Nonexperimental Data.* New York: John Wiley & Sons.

Leontief, W. W. 1982. "Academic Economics" [Letter]. *Science,* 217:104-5.

———. 1951. *The Structure of the American Economy 1919-1939.* New York: Oxford University Press.

———. 1947. "Introduction to a Theory of the Internal Structure of Functional Relationships." *Econometrica,* 15:361-73.

Lesuis, P. J., F. Muller, and P. Nijkamp. 1978. "Analytical Methods for Environmental and Energy Policies." Institute for Economic Research Discussion Paper Series No. 7822/G. Rotterdam: Erasmus University.

Lillard, L. A., and J. P. Acton. 1981. "Seasonal Electricity Demand and Pricing Analysis with a Variable Response Model." *Bell Journal of Economics,* 12:71-92.

Lin, A., E. N. Botsas, and S. A. Moore. 1985. "State Gasoline Consumption in the USA: An Econometric Analysis." *Energy Economics,* 7:29-36.

Lin, W., E. Hirst, and S. Cohn. 1976. "Fuel Choice in the Household Sector." ORNL Report No. CON-3. Oak Ridge, TN: Oak Ridge National Laboratory.

Ling, S. 1964. *Economies of Scale in Steam-Electric Power Generating Industry.* Amsterdam: North-Holland.

Liscom, W. L. 1982. *The Energy Decade 1970-1980: A Statistical and Graphic Chronicle.* Cambridge, MA: Ballinger.

Lomax, K. S. 1952. "Cost Curves for Electricity Generation." *Econometrica,* 19:193-97.

Long, T. V., and L. Schipper. 1978. "Resource and Energy Substitution." *Energy,* 3:63-82.

Longva, S., and O. Olsen. 1983. "Price Sensitivity of Energy Demand in Norwegian Industries." *Scandinavian Journal of Economics,* 85:17-36.

Lovell, M. C. 1983. "Data Mining." *Review of Economics and Statistics,* 65:1-12.

——. 1963. "Seasonal Adjustment of Economic Time Series and Multiple Regression Analysis." *Journal of the American Statistical Association,* 58: 993-1000.

Lutton, T. J., and M. R. LeBlanc. 1984. "A Comparison of Mutlivariate Logit and Translog Models for Energy and Nonenergy Input Cost Share Analysis." *Energy Journal,* 5:35-44.

Madan, D. B. 1984. "The Measurement of Capital Utilization Rates." Department of Econometrics. Sydney: University of Sydney. Mimeo.

Maddala, G. S. 1977. *Econometrics.* New York: McGraw-Hill.

Maddala, G. S., and R. B. Roberts. 1981. "Statistical Cost Analysis Re-revisited." *Quarterly Journal of Economics,* 96:177-81.

Maddigan, R. J., W. S. Chern, and C. G. Rizy. 1983. "Rural Residential Demand for Electricity." *Land Economics,* 59:150-61.

Magnus, J. R. 1979. "Substitution Between Energy and Non-energy Inputs in the Netherlands 1950-1976." *International Economic Review,* 20:465-84.

Magnus, J. R., and A. Woodland. 1980. "Interfuel Substitution Possibilities in Dutch Manufacturing: A Mutlivariate Error Component Approach." Department of Economics Discussion Paper No. 8-39. Vancouver: University of British Columbia.

Malabre, A. L. 1986. "Kondratieff Rolls on, as Does the Economy." *Wall Street Journal,* January 20, p. 1.

Malinvaud, E. 1980. *Statistical Methods of Econometrics.* 3rd rev. ed. Amsterdam: North-Holland.

Malthus, T. R. 1966. Reprint of 1798 edition. *First Essay on Population.* New York: St. Martin's Press.

Martin, J. J. 1967. *Bayesian Decision Problems and Markov Chains.* New York: John Wiley & Sons.

Martin, R., and M. Selowsky. 1984. "Energy Prices, Substitution and Optimal Borrowing in the Short Run: An Analysis of Adjustment in Oil Importing Developing Countries." *Journal of Development Economics,* 14:331-50.

Mayer, L. S., and C. E. Horowitz. 1979. "The Effect of Price on the Residential Demand for Electricity: A Statistical Study." *Energy,* 4:87-99.

Mayo, J. W. 1984. "The Technological Determinants of the U.S. Energy Industry." *Review of Economics and Statistics,* 66:51-58.

McColl, G. D., and D. R. Gallagher. 1982. "Prospective Trends in the Demand for Energy in New South Wales." Report prepared for the Energy Authority of New South Wales. Energy Information. EANSW ESP-8. Sydney: New South Wales government.

McFadden, D. 1963. "Constant Elasticity of Substitution Production Functions." *Review of Economic Studies,* 30:73-83.

McFadden, D., C. Puig, and D. Kirshner. 1977. "Determinants of the Long-Run Demand for Electricity." *Proceedings of the American Statistical Association,* 1:109-19.

McGillivray, R. G. 1976. "Gasoline Use by Automobiles." *Transportation Research Record,* 561:45-66.

McGuire, A. 1982. "Excess Capacity and the Demand for Electricity in Scotland." *Scottish Journal of Political Economy,* 29:45-58.

McLaren, K. R. 1982. "Estimation of Translog Demand Systems." *Australian Economic Papers,* 21:392-406.

McLeod, A. I., K. W. Hipel, and W. C. Lennox. 1977. "Advances in Box-Jenkins Modeling 2: Applications." *Water Resources Research,* 13:577-86.

McNown, R. F., and K. R. Hunter. 1980. "A Test for Autocorrelation in Models with Lagged Dependent Variables." *Review of Economics and Statistics,* 62:313-17.

McRae, R. N. 1979. "Primary Energy Demand in Canada." *Energy Economics,* 1:203-10.

Mead, W. J. 1974. "Wilson's 'Electricity Consumption: Supply Requirements, Demand Elasticity and Rate Design': Discussion." *American Journal of Agricultural Economics,* 56:433-35.

Meadows, D. H., D. L. Meadows, J. Randers, and W. W. Behrens. 1972. *The Limits to Growth.* New York: Universe Books.

Mehta, J. S., G. V. L. Narasimham, and P. A. V. B. Swamy. 1978. "Estimation of a Dynamic Demand Function for Gasoline with Different Schemes of Parameter Variation." *Journal of Econometrics,* 7:263-79.

Miernyk, W. H., F. Giarratani, and C. F. Socher. 1978. *Regional Impacts of Rising Energy Prices.* Cambridge, MA: Ballinger.

Miller, R. M., and A. S. Kelso. 1985. "Analyzing Government Policies: Economic Modeling with Lotus 1-2-3." *Byte,* 10:199-210.

Mitchell, B. M., W. G. Manning, and J. P. Action. 1978. *Peak-Load Pricing.* Cambridge, MA: Ballinger.

Mittelstadt, A., and V. B. Hall. 1981. "Price and Income Elasticities of Final Energy Demand in OECD Countries." Paper presented at the Third Annual IAEE Conference on International Energy Issues, University of Toronto, Toronto, June 21-24. Mimeo.

Mizon, G., and S. Nickell. 1983. "Vintage Production Models of U.K. Manufacturing Industry." *Scandinavian Journal of Economics,* 85:295-310.

Mohabbat, K. A., A. Dalal, and M. Williams. 1984. "Import Demand for India: A Translog Cost Function Approach." *Economic Development and Cultural Change,* 32:593-605.

Moore, T. G. 1970. "The Effectiveness of Regulation of Electric Utility Prices." *Southern Economic Journal,* 36:365-75.

Morgenstern, O. 1963. *On the Accuracy of Economic Observations.* 2nd ed. Princeton: Princeton University Press.

Mork, K. A., ed. 1979. *Energy Prices, Inflation and Economic Activity.* Cambridge, MA: Ballinger.

Morlan, T. H. 1984. "Northwest Power Planning Council Issue Brief: The Utility Death Spiral." Portland: Northwest Power Planning Council. Mimeo.

——. 1983. "A New Approach to Regional Power Planning: The Northwest Power Planning Council." Paper presented at the International Association of Energy Economists–North American Conference, Washington, D.C., June 9-10. Mimeo.

Moroney, J. R., and J. M. Trapani. 1981. "Alternative Models of Substitution and Technical Change in Natural Resource Intensive Industries." In *Modeling and Measuring Natural Resource Substitution,* edited by E. R. Berndt and B. C. Field, pp. 48-69. Cambridge, MA: MIT Press.

——. 1981. "Factor Demand and Substitution in Mineral Intensive Industries." *Bell Journal of Economics,* 12:272-84.

Morris, G. E. 1978. "An Optimization Model of Energy Related Economic Development in the Upper Colorado River Basin Under Conditions of Water and Energy Resources Scarcity." Los Alamos: Los Alamos Scientific Laboratory.

Moses, L. E. 1981. "Keynote Address: One Statistician's Observation Concerning Energy Modeling." In *Energy Policy Planning,* edited by B. A. Bayraktar, E. A. Cherniavsky, M. A. Laughton, and L. E. Ruff, pp. 17-33. NATO Advanced Research Institute on the Application of Systems Science to Energy Policy Planning, New York, 1979. New York: Plenum Press.

Mount, R. I., and H. R. Williams. 1981. "Energy Conservation, Motor Gasoline Demand, and OECD Countries." *Review of Business and Economic Research,* 16:48-57.

Mount, T. D. 1974. "Wilson's 'Electricity Consumption: Supply Requirements, Demand Elasticity and Rate Design': Discussion." *American Journal of Agricultural Economics,* 56:427-29.

Mount, T. D., and L. D. Chapman. 1979. "Electricity Demand, Sulfur Emissions and Health: An Econometric Analysis of Power Generation in the United States." In *International Studies of the Demand for Energy,* edited by W. D. Nordhaus, pp. 95-114. Amsterdam: North-Holland.

Mount, T. D., L. D. Chapman, and T. J. Tyrrell. 1973. "Electricity Demand in the United States: An Econometric Analysis." ORNL-NSF-49. Oak Ridge, TN: Oak Ridge National Laboratory.

Mountain, D. C. 1981. "The Spatial Distribution of Electricity Demand: Its Impact upon Input Usage." *Land Economics,* 57:48-62.

Munasinghe, M., and J. J. Warford. 1982. *Electricity Pricing: Theory and Case Studies.* Published for the World Bank. Baltimore: Johns Hopkins University Press.

Murphy, F. H., and A. L. Soyster. 1982. "The Averch-Johnson Model with Leontief Production Functions." *Energy Economics,* 4:169-79.

Murray, M. P., R. Spann, L. Pulley, and E. Beuavais. 1978. "The Demand for Electricity in Virginia." *Review of Economics and Statistics,* 60:585-600.

Musgrove, A. R. 1980. "An Australian Motor Spirit Demand Model." Lucas Heights, NSW: Australian Atomic Energy Commission. Mimeo.

Musgrove, A. R., K. J. Stocks, P. Essam, D. Le, and J. V. Hoetzl. 1983. "Exploring Some Australian Energy Alternatives Using MARKAL." Division of Energy Technology Technical Report No. TR 2. Lucas Heights, NSW: Commonwealth Scientific and Industrial Research Organization.

Nadiri, M. I. 1970. "Some Approaches to the Theory of Measurement of Total Factor Productivity: A Survey." *Journal of Economic Literature*, 8:1137-77.

Nelson, C. R. 1973. *Applied Time Series Analysis for Managerial Forecasting.* San Francisco: Holden-Day.

Nelson, D. C. 1965. "A Study of Elasticity of Demand for Electricity by Residential Consumers in Nebraska." *Land Economics*, 41:92-95.

Nelson, J. P. 1975. "The Demand for Space Heating Energy." *Review of Economics and Statistics*, 57:508-12.

Nelson, R. A. 1984. "Regulation, Capital Vintage and Technical Change in the Electric Utility Industry." *Review of Economics and Statistics*, 66:59-69.

Nerlove, M. 1968. "Returns to Scale in Electricity Supply." in *Readings in Economic Statistics and Econometrics*, edited by A. Zellner, pp. 409-39. Boston: Little, Brown.

———. "Recent Empirical Studies of the CES and Related Production Functions." In *The Theory and Empirical Analysis of Production, Vol. 31*, edited by M. Brown, pp. 55-136. New York: National Bureau of Economic Research. Distributed by Columbia University Press.

Neufeld, J. L., and J. M. Watts. 1981. "Inverted Block on Lifeline Rates and Micro-efficiency in the Consumption of Electricity." *Energy Economics*, 3:113-21.

Nguyen, D. T. 1976. "The Problems of Peak Loads and Inventories." *Bell Journal of Economics*, 7:242-48.

Niall, J., R. Smith, and P. Wilson. 1982. "The Economic Impact of Electricity Shortages: The Case of Victoria." *Australian Economic Review*, 54:62-76.

Nordhaus, W. D. 1977. "The Demand for Energy: An International Perspective." In *International Studies of the Demand for Energy*, edited by W. D. Nordhaus, pp. 239-85. Amsterdam: North-Holland.

———. 1976. "The Demand for Energy: An International Perspective." In *Proceedings of the Workshop on Energy Demand: May 22-23, 1975*, edited by W. D. Nordhaus, pp. 511-87. Laxenberg, Austria: International Institute for Applied Systems Analysis.

———. 1973. "World Dynamics: Measurement Without Data." *Economic Journal*, 83:1156-83.

———. 1978. *International Studies of the Demand for Energy.* Amsterdam: North-Holland.

Nordin, J. A. 1976. "A Proposed Modification of Taylor's Demand Analysis: Comment." *Bell Journal of Economics*, 7:719-21.

Norsworthy, R. R., and D. H. Malmquist. 1983. "Input Measurement and Productivity Growth in Japanese and U.S. Manufacturing." *American Economic Review*, 73:947-67.

Ohsfeldt, R. L. 1983. "Specification of Block Rate Price Variables in Demand Models: Comment." *Land Economics,* 59:365-69.

Oi, W. Y. 1952-53. "A Disneyland Dilemma: Two-Part Tariffs for a Mickey Mouse Monopoly." *Quarterly Journal of Economics,* 20:76-96.

Ostro, B. D., and J. L. Naroff. 1980. "Decentralization and the Demand for Gasoline." *Land Economics,* 56:169-80.

Ozatalay, S., S. Grubaugh, and T. V. Long. 1979. "Energy Substitution and National Energy Policy." *American Economic Review,* 369-71.

Pachauri, R. K. 1976. "A Dynamic Regional Analysis of Factors Affecting the Electrical Energy Sector in the U.S." In *Energy, Regional Science and Public Policy,* edited by M. Chatterji and P. Van Rompuy, pp. 90-104. Amsterdam: North-Holland.

Pagan, A. R. 1975. "A Note on the Extraction of Components from Time Series." *Econometrica,* 43:163-68.

——. 1974. "A Generalized Approach to the Treatment of Autocorrelation." *Australian Economic Papers,* 13:267-80.

Pagan, A. R., and A. D. Hall. 1983. "Diagnostic Tests, as Residual Analysis (with Comments)." *Econometric Reviews,* 2:158-254.

Pagan, A. R., and D. F. Nicholls. 1984. "Estimating, Predictions, Prediction Errors and Their Standard Deviations Using Constructed Variables." *Journal of Econometrics,* 24:293-310.

Pagoulatos, A., and J. F. Timmons. 1979. "Estimation and Projections of Demand for Crude Petroleum and Refined Petroleum Products." *Energy Economics,* 1:72-75.

Pakravan, K. 1981. "Exhaustible Resource Models and Predictions of Crude Oil Prices: Some Preliminary Results." *Energy Economics,* 3:169-77.

Panzar, J. C. 1976. "A Neoclassical Approach to Peak Load Pricing." *Bell Journal of Economics,* 7:521-30.

Panzar, J. C., and R. D. Willig. 1979. "Theoretical Determinants of the Industrial Demand for Electricity by Time of Day." *Journal of Econometrics,* 9:193-207.

——. 1977. "Economies of Scale in Mutli-output Production." *Quarterly Journal of Economics,* 111:481-93.

Parhizgari, A. M., and P. S. Davis. 1976. "The Residential Demand for Electricity: A Variant Parameters Approach." *Applied Economics,* 10:331-39.

Parzen, E., and M. Pagano. 1979. "An Approach to Modeling Seasonally Stationary Time Series." *Journal of Econometrics,* 9:137-53.

Pelaez, R. F. 1981. "Note: The Price Elasticity for Gasoline Revisited." *Energy Journal,* 2:85-89.

Peles, Y. C. 1981. "A Proposal for Peak Load Pricing of Public Utilities." *Energy Economics,* 3:187-90.

Petty, Sir William. 1964. Reprint of 1687 edition. *Five Essays in Political Arithmetick.* In *The Economic Writings of Sir William Petty,* edited by C. H. Hull. New York: August M. Kelley.

Phillips, A. 1955. "The *Tableau Economique* as a Simple Leontief Model." *Quarterly Journal of Economics,* 69:137-44.

Phlips, L. 1972. "A Dynamic Version of the Linear Expenditure Model." *Review of Economics and Statistics,* 54:450-88.

Pindyck, R. S. 1979. "Interfuel Substitution and the Industrial Demand for Energy: An International Comparison." *Review of Economics and Statistics,* 61:169-79.

——. 1979. *The Structure of World Energy Demand.* Cambridge, MA: MIT Press.

——. 1977. "Interfuel Substitution and the Industrial Demand for Energy: An International Comparison." Energy Laboratory Working Paper No. MIT EL 77-026WP. Cambridge, MA: Massachusetts Institute of Technology.

——. 1976. "International Comparisons of the Residential Demand for Energy: A Preliminary Analysis." MIT Energy Laboratory Working Paper No. MIT EL 176-023 WP. Cambridge, MA: Massachusetts Institute of Technology.

——. ed. 1979. *Advances in the Economics of Energy and Resources: The Structure of Energy Markets, Vol. 1.* Greenwich, CT: JAI Press.

Pindyck, R. S. and J. J. Rotemberg. 1983. "Dynamic Factor Demands and the Effects of Energy Price Shocks." *American Economic Review,* 73:1066-79.

Pindyck, R. S., and D. L. Rubinfeld. 1981. *Econometric Models and Economic Forecasts.* 2nd ed. New York: McGraw-Hill.

Pitt, M. M. 1985. "Estimating Industrial Energy Demand with Firm-Level Data: The Case of Indonesia." *Energy Journal,* 6:25-39.

Pitts, R. E., J. F. Willenborg, and D. L. Sherrell. 1981. "Consumer Adaptation to Gasoline Price Increases." *Journal of Consumer Research,* 8:322-30.

Pollak, R. A., R. C. Sickles, and T. J. Wales. 1984. "The CES-Translog: Specification and Estimation of a New Cost Function." *Review of Economics and Statistics,* 66:602-7.

Powell, A. A., and F. H. G. Gruen. 1968. "The Constant Elasticity of Transformation Production Frontier with Linear Supply System." *International Economic Review,* 9:315-28.

Prasad, R. 1983. "An Evaluation of the Lifeline-Rate Structure Using Total Welfare as a Measure of Efficiency." *Resources and Energy,* 5:201-17.

Proops, J. L. R. 1984. "Modelling the Energy-Output Ratio." *Energy Economics,* 6:47-51.

Quandt, R. E. 1958. "The Estimation of the Parameters of a Linear Regression System Obeying Two Separate Regimes." *Journal of the American Statistical Association,* 53:873-80.

Ramsey, J., R. Rasche, and B. Allen. 1975. "An Analysis of the Private and Commercial Demand for Gasoline." *Review of Economics and Statistics,* 57:502-7.

Rao, P. 1974. "Specification Bias in Seemingly Unrelated Regressions." In *Econometrics and Economic Theory: Essays in Honour of Jan Tinbergen,* edited by W. Sellekaerts, pp. 101-13. London: Macmillan.

Renshaw, E. F. 1979. "The Pricing of Off-Peak Power." *Energy Economics,* 1:144-47.

Reza, A. M., and M. H. Spiro. 1979. "The Demand for Passenger Car Transport Services and for Gasoline." *Journal of Transport and Economic Policy,* 13:304-19.

Richards, W. 1980. "Petroleum Product Pricing in Australia: The South Australian Prices Commission in a National Role." Petroleum Industry Research Project Working Paper No. 1. Department of Government. Sydney: University of Sydney.

Richter, M. K. 1966. "Invariance Axioms and Economic Indexes." *Econometrica,* 34:739-55.

Rodekohr, M. E. 1979. "Demand for Transportation Fuels in the OECD: A Temporal Cross-Section Specification." *Applied Energy,* 5:223-31.

Roth, T. P. 1983. "Electricity Demand Estimation Using Proxy Variables: Some Restrictions." *Applied Economics,* 15:703-4.

———. 1981. "Average and Marginal Price Changes and the Demand for Electricity: An Econometric Study." *Applied Economics,* 13:377-88.

Rowland, C. 1982. "A Direction for Energy Policy." *Scottish Journal of Political Economy,* 29:118-25.

Rowse, J. 1980. " 'World Energy Models: Survey and Critique': A Comment." *Energy Economics,* 2:249.

Rushdi, A. A. 1983. "Industrial Demand for Electricity in South Australia." Paper presented at the Twelfth Conference of Economists, University of Tasmania, Hobart. Mimeo.

Russell, C. S., and W. J. Vaughan. 1976. *Steel Production: Processes, Products, and Residuals.* Published for Resources for the Future. Baltimore: Johns Hopkins University Press.

———. 1974. "A Linear Programming Model of Residuals Management for Integrated Iron and Steel Production." *Journal of Environmental Economics and Management,* 1:17-42.

Saddler, H. D. W., J. Bennett, I. Reynolds, and B. Smith. 1980. *Public Choice in Tasmania: Aspects of the Lower Gordon River Hydro-electric Development Proposal.* Centre for Resource and Environmental Studies Monograph 2. Canberra: Australian National University.

Saddler, H. D. W., and W. A. Donnelly. 1983. "Electricity in Tasmania: Pricing, Demand, Compensation." Centre for Resource and Environmental Studies Working Paper No. 1983/24. Canberra: Australian National University.

———. 1982. "The Demand for Energy in Tasmania, with Particular Reference to Electricity." Senate Select Committee on South-West Tasmania, Canberra: Official Hansard Report, pp. 150-212.

Salant, S. W. 1982. *Imperfect Competition in the World Oil Market: A Computerized Nash-Cournot Model.* Lexington, MA: Lexington Books.

Salkever, D. S. 1976. "The Use of Dummy Variables to Compute Predictions, Prediction Errors and Confidence Limits." *Journal of Econometrics,* 4:393-97.

Samuelson, P. A. 1968. "Two Generalizations of the Elasticity of Substitution." In *Value, Capital, and Growth: Papers in Honour of Sir John Hicks,* edited by J. N. Wolfe, pp. 567-80. Chicago: Aldine.

Sargan, J. K. 1971. "Production Functions." In *Qualified Manpower and Economic Performance,* edited by P. R. G. Lagard, J. D. Sargan, M. E. Ager, and D. J. Jones, pp. 143-204. London: Penguin Press.

Sato, K. 1975. *Production Functions and Aggregation.* Amsterdam: North-Holland.

———. 1972. "Additive Utility Functions with Double-Log Consumer Demand Functions." *Journal of Political Economy,* 80:102-24.

Saunders, H. D. 1984. "The Macrodynamics of Energy Shocks, Short- and Long-Run." *Energy Economics,* 6:21-34.

Savin, N. E., and K. J. White. 1978. "Estimation and Testing for Functional Form and Autocorrelation." *Journal of Econometrics,* 8:1-12.

Scherer, C. R. 1977. *Estimating Electric Power Systems Marginal Costs.* Amsterdam: North-Holland.

———. 1976. "Estimating Peak and Off-Peak Marginal Costs for an Electric Power System: An Ex Ante Approach." *Bell Journal of Economics,* 7:575-601.

Schneider, A. M. 1977. "Elasticity of Demand for Gasoline Since the 1973 Oil Embargo." *Energy,* 2:45-52.

———. 1975. "Elasticity of Demand for Gasoline." *Energy Systems and Policy,* 1:277-85.

Schorsch, L. 1984. "The Abdication of Big Steel." *Challenge,* 27:34-40.

Schou, K., and L. W. Johnson. 1979. "The Short-Run Price Elasticity of Demand for Petrol in Australia." *International Journal of Transport Economics,* 6:357-64.

Schultz, H. 1938. *The Theory and Management of Demand.* Chicago: University of Chicago Press.

Seaks, T. G., and S. K. Layson. 1983. "Box-Cox Estimation with Standard Econometric Problems." *Review of Economics and Statistics,* 65:160-64.

Searl, M. F., ed. 1973. *Energy Modeling: Art, Science, Practice.* Published for Resources for the Future. Baltimore: Johns Hopkins University Press.

Sellekaerts, W., ed. 1974. *Econometrics and Economic Theory: Essays in Honour of Jan Tinbergen.* London: Macmillan.

Shalaby, A. S., and R. R. Waghmare. 1980. "A Model for Forecasting Passenger Car Gasoline Demand." In *Energy Policy Modeling: United States and Canadian Experiences: Specialized Energy Policy Models, Vol. 1,* edited by W. T. Ziemba, S. L. Schwartz, and E. Koenigsberg, pp. 117-27. Boston: Martinus Nijhoff.

Shephard, R. W. 1981. *Cost and Production Functions.* Reprint of the First Edition, Berlin: Springer-Verlag.

Simmons, P., and D. Weiserbs. 1979. "Translog Flexible Functional Forms and Associated Demand System." *American Economic Review,* 69:892-901.

Simon, J. L. 1981. "Global Confusion, 1980: A Hard Look at *The Global 2000*

Report." The Public Interest, Number 62, pp. 3-20.

Slater, M. D. E., and G. K. Yarrow. 1983. "Distortions in Electricity Pricing in the U.K." *Oxford Bulletin of Economics and Statistics,* 45:317-38.

Slesser, M. 1976. "Dynamic Energy Analysis as a Method for Predicting Energy Requirements." In *Proceedings of the Workshop on Energy Demand,* edited by W. D. Nordhaus, pp. 68-85. Laxenburg, Austria: International Institute for Applied Analysis.

Smith, A. 1982. Reprint of 1795 edition. *Essays on Philosophical Subjects.* Indianapolis: Liberty Press/Liberty Classics.

Smith, V. K. 1980. "Estimating the Price Elasticity of U.S. Electricity Demand." *Energy Economics,* 2:81-85.

Smith, V. K., and L. J. Hill. 1985. "Validating Allocation Functions in Energy Models: An Experimental Methodology." *Energy Journal,* 6:29-47.

Solow, J. L. 1984. "The Composition of Output and Aggregate Capital-Energy Complementarity." Department of Economics, University of Iowa, Working Paper Series No. 84-1. Paper presented in the Economics Department Seminar Series. Canberra: Australian National University. Mimeo.

____. 1979. "A General Equilibrium Approach to Aggregate Capital-Energy Complementarity." *Economic Letters,* 2:91-94.

Song, I. 1982. "The Revised Tax Rate on Gasoline and Its Impact on Tax Revenues from the Sale of Gasoline in the Commonwealth of Kentucky." *Regional Science Perspectives,* 12:69-76.

____. 1981-82. "The Impact of the Revised Fuel Tax Rate on the Tax Revenues from Gasoline Sales in the Commonwealth of Kentucky." *Kentucky Journal of Economics and Business,* 3:7-12.

Sorenson, J. R., J. T. Tschirhart, and A. B. Whinston. 1976. "A Game Theoretic Approach to Peak Load Pricing." *Bell Journal of Economics,* 7:497-520.

Sosin, K., and L. Fairchild. 1984. "Nonhomotheticity and Technical Bias in Production." *Review of Economics and Statistics,* 66:44-50.

Spann, R. M., and E. C. Beauvais. 1979. "Econometric Estimation of Peak Electricity Demands." *Journal of Econometrics,* 9:119-36.

Springer, R. K. 1978. "A Structural Model for Automobile Demand and Gasoline Usage." *Proceedings of the American Statistical Association,* pp. 698-703.

Springer, R. K., and R. W. Resek. 1981. "An Econometric Model of Gasoline Consumption, Vehicle Miles Travelled, and New Car Purchases." *Energy Systems and Policy,* 5:73-87.

Spulber, N., and I. Horowitz. 1976. *Quantitative Economic Policy and Planning: Theory and Models of Economic Control.* New York: Norton.

Stapleton, D. C. 1981. "Inferring Long-Term Substitution Possibilities from Cross-Section and Time-Series Data." In *Modelling and Measuring Natural Resource Substitution,* edited by E. R. Berndt and B. C. Field, pp. 93-118. Cambridge, MA: MIT Press.

Steiner, P. O. 1957. "Peak Loads and Efficient Pricing." *Quarterly Journal of*

Economics, 71:585-610.

Stewart, C. T., and J. T. Bennett. 1975. "Urban Size and Structure and Private Expenditures for Gasoline in Large Cities." *Land Economics,* 4:365-73.

Stine, R. A., and R. D. Small. 1981. "Some Factors Influencing Estimators in Motor Gasoline Demand Models." In *Proceedings of the 1980 Department of Energy Statistical Symposium,* edited by T. Truett et al. Oak Ridge, TN: Oak Ridge National Laboratories.

Stobaugh, R., and D. Yergin, eds. 1979. *Energy Future: Report of the Energy Project at the Harvard Business School.* New York: Random House.

Stocks, K. J. 1983. "The MARKAL Energy Model: A Linear Programming Approach to Energy Systems Analysis." Paper presented at the Australian Graduate School of Management, July 6-8. Kensington: University of New South Wales. Mimeo.

Stromback, C. T. 1983. "A Structural Dynamic Model of Electricity Demand in Western Australia." Paper presented at the First Annual Conference on Econometrics, Australian National University, Canberra. Mimeo.

Strout, A. M. 1961. "Weather and the Demand for Space Heat." *Review of Economics and Statistics,* 43:185-92.

Suits, D. B. 1984. "Dummy Variables: Mechanics v. Interpretation." *Review of Economics and Statistics,* 66:177-80.

Sutherland, R. J. 1983. "Instability of Electricity Demand Functions in the Post-Oil-Embargo Period." *Energy Economics,* 5:267-72.

Swan, P. L. 1983. "The Marginal Cost of Base-Load Power: An Application to Alcoa's Portland Smelter." *Economic Record,* 59:332-44.

——. 1981. "Pricing of Electricity to Alcoa at Portland, Victoria." Department of Economics, The Faculties. Canberra: Australian National University. Mimeo.

Sweeney, J. L. 1981. "Energy and Economic Growth: A Conceptual Framework." In *Mathematical Modeling of Energy Systems,* edited by I. Kavrakoglu, pp. 53-84. The Netherlands: Sijthoff Noordhoff.

——. 1979. "Effects of Federal Policies on Gasoline Consumption." *Resources and Energy,* 2:3-26.

——. 1979. "Passenger Car Gasoline Demand Model." Department of Engineering-Economics. Stanford: Stanford University. Mimeo.

——. 1978. "The Demand for Gasoline in the United States: A Vintage Capital Model." In *Workshops on Energy Supply and Demand,* pp. 240-77. International Energy Agency. Paris: Organization for Economic Co-operation and Development.

——. 1975. "A Vintage Capital Stock Model of Gasoline Demand." Washington, D.C.: U.S. Federal Energy Administration. Mimeo.

Tasmanian Directorate of Energy. 1980. "Report to the Co-ordination Committee on Future Power Development." Hobart: Tasmanian government.

Tasmanian Hydro-electric Commission. 1983. "Load Forecast." Hobart.

——. 1982. "Report for the Year 1981-82." Hobart: Parliament of Tasmania.

Taylor, L. D. 1979. "Decreasing Block Pricing and the Residential Demand for Electricity." In *International Studies of the Demand for Energy,* edited by W. D. Nordhaus, pp. 65-79. Amsterdam: North-Holland.

____. 1979. "On Modeling the Residential Demand for Electricity by Time-of-Day." *Journal of Econometrics,* 9:97-115.

____. 1977. "The Demand for Energy: A Survey of Price and Income Elasticities." In *International Studies of the Demand for Energy,* edited by W. D. Nordhaus, pp. 3-43. Amsterdam: North-Holland.

____. 1976. "Decreasing Block Pricing and the Residential Demand for Electricity." In *Proceedings of the Workshop on Energy Demand,* edited by W. D. Nordhaus, pp. 43-64. Laxenburg, Austria: International Institute for Applied Systems Analysis.

____. 1975. "The Demand for Electricity: A Survey." *Bell Journal of Economics and Management Science,* 6:74-110.

Taylor, L. D., and D. Weiserbs. 1972. "On the Estimation of Dynamic Demand Functions." *Review of Economics and Statistics,* 54:459-65.

Telsar, L. G. 1964. "Iterative Simultaneous Estimation of a Set of Linear Regressions." *Journal of the American Statistical Association,* 59:845-62.

Theil, H. 1980. *The System-wide Approach to Micro-economics.* Chicago: University of Chicago Press.

____. 1978. *Introduction to Econometrics.* Englewood Cliffs, NJ: Prentice-Hall.

____. 1975. *Theory and Measurement of Consumer Demand, Vol. 1.* Amsterdam: North-Holland.

Thursby, J. G., and C. A. K. Lovell. 1978. "An Investigation of the Kmenta Approximation to the CES Function." *International Economic Review,* 19:363-77.

Tinter, G., E. Deutsch, R. Rieder, and P. Posner. 1977. "A Production Function for Austria Emphasizing Energy." *De Economist,* 125:75-94.

Tishler, A. 1983. "The Demand for Cars and Gasoline." *European Economic Review,* 10:271-87.

____. 1982. "The Demand for Cars and the Price of Gasoline: The User Cost Approach." *Review of Economics and Statistics,* 64:184-90.

Train, K. 1986. *Qualitative Choice Analysis: Theory, Econometrics, and an Application to Automobile Demand.* Cambridge, MA: MIT Press.

Tsao, C. S., and R. H. Day. 1971. "A Process Analysis Model of the United States Steel Industry." *Management Science,* 17:588-608.

Tsvetanov, P. S., and W. D. Nordhaus. 1976. "Problems of Energy Demand Analysis." In *Proceedings of the Workshop on Energy Demand,* edited by W. D. Nordhaus, pp. 89-102. Laxenburg, Austria: International Institute for Applied Systems Analysis.

Turnovsky, M. H. L., and W. A. Donnelly. 1984. "Energy Substitution, Separability and Technical Progress in the Australian Iron and Steel Industry." *Journal of Business and Economic Statistics,* 2:54-63.

____. 1983. "Energy Substitution, Separability and Technical Progress in the

Australian Iron and Steel Industry." Centre for Resource and Environmental Studies Paper No. 9. Canberra: Australian National University.

——. 1982. "Energy Substitution, Separability and Technical Progress in the Australian Iron and Steel Industry." Centre for Resource and Environmental Studies Paper No. 5. Canberra: Australian National University.

——. 1982. "Energy Substitution and Separability in the Australian Iron and Steel Industry." Centre for Resource and Environmental Studies Paper No. 2. Canberra: Australian National University.

Turnovsky, M. H. L., M. Folie, and A. Ulph. 1982. "Factor Substitutability in Australian Manufacturing with Emphasis on Energy Inputs." *Economic Record,* 58:61-72.

——. 1978. "Factor Substitutability in Australian Manufacturing Emphasising Energy Inputs." Centre for Resource and Environmental Studies Working Paper No. R/WP31. Canberra: Australian National University.

Turnovsky, S. J. 1966. "The New Zealand Automobile Market, 1948-63: An Econometric Case-Study of Disequilibrium." *Economic Record,* 42: 256-73.

Turvey, R., and A. R. Nobay. 1965. "On Measuring Energy Consumption." *Economic Journal,* 75:787-93.

Ulph, A. M. 1980. "World Energy Models: A Survey and Critique—Reply." *Energy Economics,* 2:250-51.

——. 1980. "World Energy Models: A Survey and Critique." *Energy Economics,* 2:46-59.

United Nations. 1985. *1983 Energy Statistics Yearbook.* New York: UN.

U.S. Department of Energy. 1979. *State Energy Fuel Prices by Major Economic Sectors from 1960-1977.* DOE/EIA-0190. Washington, D.C.: U.S. Department of Energy.

U.S. Department of Energy, Energy Information Administration. 1981. *Residential Energy Consumption Survey: 1979-1980 Consumption and Expenditures, Part II: Regional Data.* DOE/EIA-0262/2. Washington, D.C.: U.S. Department of Energy.

——. 1981. *Price Elasticities of Demand for Motor Gasoline and Other Petroleum Products: Analysis.* DOE/EIA-0291. Washington, D.C.: U.S. Department of Energy.

——. 1981. *Residential Energy Consumption Survey: 1979-1980 Consumption and Expenditures, Part I: National Data (Including Conservation).* DOE/EIA-0262/1. Washington, D.C.: U.S. Department of Energy.

——. 1980. *Residential Energy Consumption Survey: Consumption and Expenditures, April 1978 Through March 1979.* DOE/EIA-0207/5. Washington, D.C.: U.S. Department of Energy.

——. 1980. *Residential Energy Consumption Survey: Conservation.* DOE/EIA-0207/3. Washington, D.C.: U.S. Department of Energy.

——. 1980. *Residential Energy Consumption Survey: Characteristics of the Housing Stocks and Households.* DOE/EIA-0207/2. Washington, D.C.:

U.S. Department of Energy.

———. 1979. *Single-Family Households: Fuel Inventories and Expenditures: National Interim Energy Consumption Survey.* DOE/EIA-0207/1. Washington, D.C.: U.S. Department of Energy.

U.S. Department of Energy, Energy Information Administration, Office of Energy Markets and End Use. 1981. "Technical Documentation for the Residential Energy Consumption Survey: National Interim Energy Consumption Survey 1978–1979, Household Monthly Energy Consumption and Expenditures–Public Use Data Tapes–User's Guide." Washington, D.C.: U.S. Department of Energy.

U.S. Federal Energy Administration. 1976. *National Energy Outlook.* FEA-N75/713. Washington, D.C.: U.S. Government Printing Office.

———. 1976. *National Petroleum Product Supply and Demand 1976–1978.* National Energy Information Center. FEA/B-76/281. Washington, D.C.: National Technical Information Service.

Uri, N. D. 1983. *Econometric Studies of Energy Demand and Supply.* Greenwich, CT: JAI Press.

———. 1982. *The Demand for Energy and Conservation in the United States.* Greenwich, CT: JAI Press.

———. 1982. "The Demand for Energy in the United Kingdom." *Bulletin of Economic Research,* 34:43-56.

———. 1982. "Testing for the Stability of the Investment Function." *Review of Economics and Statistics,* 64:117–25.

———. 1981. "Regional Forecasting of the Demand for Fossil Fuels by Electric Utilities in the United States." *Regional Science and Urban Economics,* 11:87–100.

———. 1980. "Energy as a Determinant of Investment Behaviour." *Energy Economics,* 2:179–83.

———. 1980. "The Stability of the Demand for Refined Petroleum Products in the United States." *Energy Systems and Policy,* 4:197–216.

———. 1979. "Energy Substitution in the U.K., 1948-64." *Energy Economics,* 1:241-44.

———. 1979. "A Mixed Time-Series/Econometric Approach to Forecasting Peak System Load." *Journal of Econometrics,* 9:155–71.

———. 1978. "The Demand for Electric Energy in the United States." *Energy Systems and Policy,* 2:233–43.

———. 1978. "Regional Interfuel Substitution by Electric Companies: The Short-Term Prospects." *Annals of Regional Science,* 12:4–15.

———. 1977. "An Integrated Box-Jenkins/Econometric Model for Forecasting Time Series." *Proceedings of the American Statistical Association,* 1: 404-7.

———. 1976. "An Application of Spatial Equilibrium Analysis to Electrical Energy Allocation." In *Econometric Dimensions of Energy Supply and Demand,* edited by A. B. Askin and J. Kraft, pp. 65-80. Lexington, MA:

Lexington Books.

——. 1976. "Intermediate-Term Forecasting of System Loads Using Box-Jenkins Time Series Analysis." In *Proceedings on Forecasting Methodology for Time-of-Day and Seasonal Electric Utility Loads,* edited by J. W. Boyd. Electric Power Research Institute. EPRI-SR-31. Palo Alto: EPRI.

——. 1976. "A New Approach to Load Forecasting in the Electrical Energy Industry." *Cycles,* 27:59-62.

——. 1976. "Short-Run Variations in the Demand for Electric Energy." *Proceedings of the American Statistical Association,* pp. 618-21.

Uzawa, H. 1964. "Duality Principles in the Theory of Cost and Production." *International Economic Review,* 5:216-20.

——. 1962. "Production Functions with Constant Elasticities of Substitution." *Review of Economic Studies,* 30:291-99.

Van Hoa, T., D. S. Ironmonger, and I. Manning. 1983. "Energy Consumption in Australia: Evidence from a Generalized Working Model." *Economics Letters,* 12:383-90.

Verleger, P. K., and D. P. Sheehan. 1976. "A Study of the Demand for Gasoline." In *Econometric Studies of U.S. Energy Policy,* edited by D. W. Jorgenson, pp. 179-241. Amsterdam: North-Holland.

Vermetten, J. B., and J. Plantinga. 1953. "The Elasticity of Substitution of Gas with Respect to Other Fuels in the United States." *Review of Economics and Statistics,* 35:140-43.

Vickrey, W. 1955. "Pricing in Transportation and Public Utilities: Some Implications of Marginal Cost Pricing for Public Utilities." *Papers and Proceedings, American Economic Association,* 45:605-20.

Victoria Ministry for Economic Development. 1982. "Submission to the Committee of Inquiry into the State Electricity Commission of Victoria."

Vinals, J. M. 1984. "Energy-Capital Substitution, Wage Flexibility and Aggregate Output Supply." *European Economic Review,* 26:229-45.

Vincent, D. P., P. B. Dixon, and A. A. Powell. 1980. "The Estimation of Supply Response in Australian Agriculture: The Cresh/Creth Production System" *International Economic Review,* 21:221-42.

Vinod, H. D. 1968. "Econometrics of Joint Production." *Econometrica,* 36: 322-36.

Wales, T. J. 1977. "On the Flexibility of Functional Forms–An Empirical Approach." *Journal of Econometrics,* 5:183-93.

Wallis, K. F. 1967. "Lagged Dependent Variables and Serially Correlated Errors: A Reappraisal of Three-Pass Least-Squares." *Review of Economics and Statistics,* 49:555-67.

Walters, A. A. 1963. "Production and Cost Functions: An Econometric Survey." *Econometrica,* 31:1-66.

Watkins, G. C., and E. R. Berndt. 1983. "Energy-Output Coefficients: Complex Realities Behind Simple Ratios." *Energy Journal,* 4:105-20.

Watts, P. E. 1955-56. " 'Block Tariffs': A Comment." *Review of Economic*

Studies, 23:42-45.

Webb, M. G., and M. J. Ricketts. 1980. *The Economics of Energy.* New York: John Wiley & Sons.

Weinberg, A. M. 1979. "Limits to Energy Modeling." Institute for Energy Analysis Occasional Papers. ORAU/IEA-79-16(0). Oak Ridge, TN: Oak Ridge Associated Universities.

Wenders, J. T. 1984. "Two-Part Tariffs and the 'Spiral of Impossibility' in the Market for Electricity." *Energy Journal,* 5:177-79.

____. 1976. "Peak Load Pricing in the Electricity Utility Industry." *Bell Journal of Economics,* 7:232-48.

Wenders, J. T., and L. D. Taylor. 1976. "Experiments in Seasonal-Time-of-Day Pricing of Electricity to Residential Users." *Bell Journal of Economics,* 7:531-52.

Wheaton, W. C. 1982. "The Long-Run Structure of Transportation and Gasoline Demand." *Bell Journal of Economics,* 13:439-54.

White, H. 1980. "A Heteroskedasticity-Consistent Covariance Matrix Estimator and a Direct Test for Heteroskedasticity." *Econometrica,* 40:817-38.

____. 1980. "Using Least Squares to Approximate Unknown Regression Functions." *International Economic Review,* 21:149-70.

White, K. J. 1978. "A Generalized Computer Program for Econometric Methods–SHAZAM." *Econometrica,* 46:239-40.

Wibe, S. 1984. "Engineering Production Functions: A Survey." *Economica,* 51: 401-12.

____. 1984. "Engineering Information for Production Studies." Umea Economic Studies No. 138. Umea, Sweden: University of Umea.

____. 1983. "Engineering Production Functions–a Survey." Umea Economic Studies No. 127. Umea, Sweden: University of Umea.

Wicksell, K. 1916. "Den 'kritiska pukten' i lagen fur jordbrukets aftagande produktivitet." *Ekonomisk Tidskrift,* 10:285-92.

Wicksteed, P. H. 1894. *An Essay on the Co-ordination of the Laws of Distribution.* London: Macmillan.

Wilder, R. P., and J. F. Willenborg. 1975. "Residential Demand for Electricity: A Consumer Panel Approach." *Southern Economic Journal,* 42:212-17.

Williams, J. W. 1945. "Professor Douglas' Production Function." *Economic Record,* 21:55-64.

Williams, M., and P. Laumas. 1981. "The Relation Between Energy and Nonenergy Inputs in India's Manufacturing Industries." *Journal of Industrial Economics,* 30:113-22.

Wills, J. 1981. "Residential Demand for Electricity." *Energy Economics,* 3:249-55.

____. 1979. "Technical Change in the U.S. Primary Metals Industry." *Journal of Econometrics,* 10:85-98.

Wilson, J. W. 1974. "Electricity Consumption: Supply Requirements, Demand Elasticity and Rate Design." *American Journal of Agricultural Economics,*

56:419-27.

——. 1971. "Residential Demand for Electricity." *Quarterly Review of Economics and Business,* 11:7-19.

Woodland, A. D. 1978. "On Testing Weak Separability." *Journal of Econometrics,* 8:383-98.

Wright, D. J. 1977. "The Energy Cost of Goods and Services: An Input-Output Analysis for the U.S.A., 1963." In *Energy Analysis,* edited by J. A. G. Thomas, pp. 62-70. Surrey: IPC Science and Technology Press.

Yoke, G. W. 1984. "Interpreting the International Workshop Survey Results—Uncertainty and the Need for Consistent Modeling." *Energy Journal,* 5:73-77.

Zellner, A. 1971. *An Introduction to Bayesian Inference in Econometrics.* New York: John Wiley & Sons.

——. 1963. "Estimators for Seemingly Unrelated Regression Equations: Some Exact Finite Sample Results." *Journal of the American Statistical Association,* 58:977-92.

——. 1962. "An Efficient Method of Estimating Seemingly Unrelated Regressions and Tests for Aggregation Bias." *Journal of the American Statistical Association,* 57:348-68.

Zellner, A., and D. S. Haung. 1962. "Further Properties of Efficient Estimators for Seemingly Unrelated Regression Equations." *International Economic Review,* 3:300-13.

Ziemba, W. T., S. L. Schwartz, and E. Koenigsberg, eds. 1980. *Energy Policy Modeling: United States and Canadian Experiences. Specialized Energy Policy Models, Vol. 1.* Boston: Martinus Nijhoff.

——. eds. 1980. *Energy Policy Modeling: United States and Canadian Experiences. Integrative Energy Policy Models, Vol. 2.* Boston: Martinus Nijhoff.

Glossary

MNEMONICS

ABS	Australian Bureau of Statistics
ACT	Australian Capital Territory
AIP	Australian Institute of Petroleum
ASIC	Australian Standard Industrial Classification
AWE	male average weekly earnings
BESOM	Brookhaven Energy System Optimization Model
BHP	Broken Hill Proprietary Company, Ltd.
BOS	basic oxygen steel furnace
CD	Cobb-Douglas
CDD	cooling degree day
CEC	California Energy Commission
CES	constant elasticity of substitution
CPI	consumer price index
DESOM	Dynamic Energy System Optimization Model
DNDE	Australian Department of National Development and Energy
DOE	U.S. Department of Energy
DRE	Australian Department of Resources and Energy
DRI	Data Resources Inc.
DS	Donnelly-Saddler
EA	electric arc furnace
EC1	Energy Crisis 1, OPEC 1973 oil embargo
EC2	Energy Crisis 2, 1978 Iranian revolution
EC3	Energy Crisis 3, 1986 oil glut
ElCom	Electricity Commission of New South Wales
EPRI	Electric Power Research Institute
ERDM	Econometric Regional Demand Model
ESAA	Electricity Supply Association of Australia
FEA	U.S. Federal Energy Administration
GPI	gasoline price index
HDD	heating degree day
HEC	Hydro-electric Commission of Tasmania

INFORUM	Interindustry Forecasting Model of the University of Maryland
I-O	input-output
LLF	log-likelihood function
L-P	linear programming
MARKAL	Market Allocation Model
MEFS	Mid-range Energy Forecasting System
NIECS	National Interim Energy Conservation Survey
OH	open hearth furnace
OPEC	Organization of Petroleum Exporting Countries
ORNL	Oak Ridge National Laboratories
PIES	Project Independence Evaluation System
QEC	Queensland Electricity Commission
READ	Regional, Energy, Activity, and Demographic
REC	Residential Energy Consumption
REEPS	Residential End-Use Energy Planning System
RES	Reference Energy System
SD	Saddler-Donnelly
SECV	State Electricity Commission of Victoria
SEE	Standard Error of the Estimate
SJ	Schou-Johnson
TD	Turnovsky-Donnelly
TESOM	Time-Stepped Energy System Optimization Model
TFU	Turnovsky-Folie-Ulph
tonne	metric ton, 2,204.6 pounds
translog	transcendental logarithmic production function

MATHEMATICAL AND STATISTICAL NOTATION

α, β, θ	unknown parameters
δ	constant elasticity of substitution production function distributional parameter or unknown variable
γ	constant elasticity of substitution production function efficiency parameter or unknown variable
λ	Box-Cox parameter or Lagrangian multiplier
ρ	autoregressive parameter or constant elasticity of substitution production function substitution parameter
σ	elasticity of substitution

σ^2	variance
η	demand elasticity
∂	partial derivative
ϵ	zero mean, white noise
Ω	maintained hypothesis
χ^2	chi-square statistic
$\hat{}$	predicted value
!	factorial
$\tilde{}$	denoted as
$^{-1}$	matrix inverse operator
a	input-output coefficient or time-series root defined outside the unit circle
A	technical change variable, matrix of processes, a diagonal matrix of backward shift operators, or administrative workers
AES	Allen partial elasticity of substitution
AR	autoregressive
ARIMA	integrated autoregressive, moving average
ARMA	autoregressive, moving average
b	vector of parameters or time-series root defined outside the unit circle
B	diagonal matrix of backward shift operators
BC	Box-Cox
BCE	extended Box-Cox
BLUE	best linear unbiased estimate
c	vector of costs or a complementary good
C	total cost or average consumption per vehicle
d	integrated differencing parameter or total derivative
DW	Durbin-Watson
e	base of the natural logarithm or electricity
E	energy or exports
$E(\ )$	expected value
F	F statistic
FG	Farrar-Glauber
g	motor gasoline or natural gas
G	gasoline demand
h	Durbin's h statistic
H_0	null hypothesis
i, j, k	industry, region, or energy subscript
I	identity matrix

K	capital
KAPE	capital, administrative employees, production workers, and energy
KAPEM	capital, administrative employees, production workers, energy, and materials
KLE	capital, labor, and energy
KLEM	capital, labor, energy, and materials
KPEM	capital, production workers, energy, and materials
L	labor
LM	Hall-Pagan Lagrangian multiplier
ln	natural logarithm
LS	lump-sum payment
m	number of exogenous variables or outputs
M	materials, factor shares, or imports
MA	moving average
MARIMA	multivariate, integrated, autoregressive, moving average
MC	marginal cost
MLE	maximum likelihood estimator
MPPC or **MPPK**	marginal physical product of capital
MPPL	marginal physical product of labor
MRT	marginal rate of transformation
n	number of observations or inputs
N	new vehicle registrations
$\mathbf{N_G}$	occupied houses
$\mathbf{N_H}$	unoccupied houses
$\mathbf{NO_e}$	occupied houses connected to electricity
$\mathbf{NO_{eg}}$	occupied house connected to both electricity and natural gas
o	oil
OLS	ordinary least squares
p	autoregressive order of process parameter
P	price or production workers
q	moving-average parameter or degrees of freedom
Q	production or output
r	capital rental rate or growth rate
R	vehicles scrapped
R^2	coefficient of multiple determination

$\bar{R}^2$	coefficient of multiple determination adjusted for degrees of freedom
RMSE	root-mean-square error
s	substitute good, solid fuel, or trend term
S	factor share or stock of vehicles
SCE	scale elasticity
soe	solid fuel, oil, and electricity
soeg	solid fuel, oil, electricity, and natural gas
SURE	seemingly unrelated equations (Zellner, 1962)
t	time or Student's *t* statistic
T	matrix transpose operator
u	stochastic error term
U	ordinary least squares sum of squares of the error terms or Theil's inequality coefficient
v	rate of vehicle depreciation
V	value added
w	wage rate
W	weather variable
x	independent or exogenous variable or data
X	independent variable observation matrix or vector of final demand
y	dependent variable, dependent variable observation matrix, or intermediate demand
Y	income or vector of total or gross output
Z	objective function or Lagrangian function

Index

About the Author

William A. Donnelly holds joint appointments as Professor of Economics at the University of Wisconsin–La Crosse and a Senior Research Fellow at the Australian National University in Canberra. From 1974 through 1978 he was a supervisor economist serving as Division Chief of Modeling and Forecasting at the U.S. Federal Energy Administration and responsible for the creation of the Regional Energy, Activity and Demographic Model (READ) and direction of the Econometric Regional Demand Model (ERDM) and the Regional Demand Forecasting Model (RDFOR).

Dr. Donnelly has published widely in the area of policy modeling in *Applied Economics, Energy Economics, Energy Journal, Economic Record, Australian Economic Papers, Journal of Business & Economic Statistics,* and *International Journal of Transport Economics.*

He holds B.S., M.A. and Ph.D. degrees from the University of Maryland, College Park, Maryland.